Second Global Conference on Recent Developments in Computer and Communication Technologies (GC-RDCT 2022)

IOP Conference Series: Materials Science and Engineering
Volume 1272

Mangalore, India
27 – 29 July 2022

ISBN: 978-1-7138-6518-6
ISSN: 1757-8981

Printed from e-media with permission by:

Curran Associates, Inc.
57 Morehouse Lane
Red Hook, NY 12571

Some format issues inherent in the e-media version may also appear in this print version.

Institute of Physics
Dirac House, Temple Back
Bristol BS1 6BE UK

Phone: 44 1 17 929 7481
Fax: 44 1 17 920 0979

techtracking@iop.org

Additional copies of this publication are available from:

Curran Associates, Inc.
57 Morehouse Lane
Red Hook, NY 12571 USA
Phone: 845-758-0400
Fax: 845-758-2633
Email: curran@proceedings.com
Web: www.proceedings.com

TABLE OF CONTENTS

IOP Conf. Series: Materials Science and Engineering 1272 (2022) 011001 doi:10.1088/1757-899X/1272/1/011001

Preface

Computer and Communication Technology (CCT) are now involved in governmental, industrial, and business domains more than ever. Thus, it is important for CCT personnel to continue academic research to improve technology and its adoption to modern applications. The up-to-date research and technologies must be distributed to researchers and CCT community continuously to aid future development. The intent of the activity is to let the participants acquaint with transcendental growth, recent trends, innovations and security issues involved in the domain of communication technologies, high performance computing, big data, social media, hardware & software design, advanced software engineering, Internet of Things (IoT), e-governance etc. and their impact on societal applications through various brainstorming sessions. The conference will facilitate the young researchers, industries and research agencies especially, those, who are carrying out their research work in the aforesaid domain of Computer Science, Information Technology, Electronics and Communication Engineering with valuable discussions in order to make the outcomes more realistic. The technical committee of GC-RDCT 2022 (second edition) received 150+ papers, out of which 38 are shortlisted for the Conference for oral. Through this Conference, we have tried to bring in researchers from various universities and institutions to interact and deliberate within and get ideas for their future scientific endeavors. Due to the prevailing COVID-19 pandemic, the conference was organized in FULL VIRTUAL MODE. The conference was held in Google Meet online platform. All the speakers and the presenters have been connected through goggle meet platform and the inaugural function of the conference was held physically at AJIET and was live streamed through Google Meet and Youtube.

I am pleased to acknowledge the unstinting support from my colleagues at AJIET, the International Advisory, the Technical Program, and the Local Organization Committee members. I am incredibly grateful to our Management, Principal & Administrative Staffs, for providing required support.

Dr. Sangeetha D M

Chair, GC-RDCT 2022

IOP Conf. Series: Materials Science and Engineering 1272 (2022) 011001 doi:10.1088/1757-899X/1272/1/011001

Committees of GC-RDCT 2022

Chief Patron

Dr. A. J.Shetty

President, LMET, Mangaluru.

Patron

Mr. Prashanth Shetty

Vice President, LMET, Mangaluru.

Conference Chair

Dr. Shantharama Rai C.

Principal, AJIET, Mangaluru.

Organizing Chair(s)

Dr. Kiran Kumar V.G., AJIET, Mangaluru.

Dr. Laxmi Gulappagol, AJIET, Mangaluru.

Mr. John Prakash Veigas., AJIET, Mangaluru.

Dr. Sangeetha D M, AJIET, Mangaluru

Guest Editor – Proceedings

Dr. Kiran Kumar V.G.

A. J. Institute of Engineering and Technology, Mangaluru.

Dr. Antony P. J.

A. J. Institute of Engineering and Technology, Mangaluru.

Dr. Sangeetha D M.

A. J. Institute of Engineering and Technology, Mangaluru.

Dr. Varadraj P. Gurupur

University of Central Florida, Orlando, United States.

Information Contact

Dr. Kiran Kumar V.G., AJIET, Mangaluru.

IOP Conf. Series: Materials Science and Engineering 1272 (2022) 011001 doi:10.1088/1757-899X/1272/1/011001

Dr. Laxmi Gulappagol, AJIET, Mangaluru.

Mr. John Prakash Veigas., AJIET, Mangaluru.

Publicity and Promotions

Dr. Basappa B. Kodada, AJIET, Mangaluru.

Dr. Suresh D., AJIET, Mangaluru.

Dr. Gnane Swarnadh Satapathi, AJIET, Mangaluru.

Mr. Chanchal Antony, AJIET, Mangaluru.

Dr. Sankappa Rai, AJIET, Mangaluru.

Local Arrangements Committee

Mr. Ganesh U G., AJIET, Mangaluru.

Mrs. Pratheksha Rai N, AJIET, Mangaluru.

Mr. Vinayak G Pai., AJIET, Mangaluru.

Mr. Siju V Soman, AJIET, Mangaluru.

Finance Committee

Mr. Manjukiran B., AJIET, Mangaluru.

Mrs. Snitha Shetty, AJIET, Mangaluru.

Mr. Rakesh M R, AJIET, Mangaluru.

Sponsorship Committee

Prof. Prakash K Shetty, AJIET, Mangaluru.

Mrs. Nikhila., AJIET, Mangaluru.

Mr. Vijaykumar Dudhanikar, AJIET, Mangaluru.

Editorial Committee

Dr. Kiran Kumar V.G., AJIET, Mangaluru.

Dr. Antony P. J., AJIET, Mangaluru.

Dr. Sangeetha D M, AJIET, Mangaluru.

Dr. Sankappa Rai, AJIET, Mangaluru.

Technical Program Committee – Chairs

IOP Conf. Series: Materials Science and Engineering 1272 (2022) 011001 doi:10.1088/1757-899X/1272/1/011001

Prof. K. Chandrasekaran, National Institute of Technology Surathkal, Karnataka

Dr. Varadraj P. Gurupur, University of Central Florida, Orlando, United States

Technical Program Committee

Dr. Das B B, National Institute of Technology, Surathkal, Karnataka

Dr. Neena S P Panandikar, Don Bosco College of Engineering, Goa

Dr. Amjad Anvari Moghaddam, Aalborg University, Denmark

Dr. Rashed Al Mamun, Sylhet Agricultural University, Bangladesh

Dr. Goutham Sarang, VIT University, Chennai

Dr. Anotony P J, AJIET, Mangalore

Dr. Murugappan, Kwait institute of Science and Technology, Kwait

Dr. Manjaiah, National Institute of Technology Warangal, Andhrapradesh

Dr. Shrikanth S Rao, National Institute of Technology, Surathkal, Karnataka

Dr. Adake C V, KLE Dr. M.S. Sheshgiri College of Engineering and Technology, Belgaum

Dr. Rajesh Rai, AJ Institute of Engineering and Technology, Mangalore

Dr. Faridin Naim Tajuddin, Universiti Malaysia Perlis, Malaysia

Dr. Thangaprakash Sengodan, Diligentec Solutions, Coimbatore, Tamil Nadu

Dr. Shrikanth Bontha, National Institute of Technology, Surathkal, Karnataka

Dr. Kumar G N, National Institute of Technology, Surathkal, Karnataka

Dr. Sreeramkumar R, King Abdulaziz University Jeddah, Saudi Arabia.

Dr. Patel R N, National Institute of Technology, Raipur

Dr. Joseph Xavier R, Akshaya College of Engineering and Technology, India

Dr. Yasser S Abdalla, Suez University, Suez, Egypt

Dr. Senthilkumar S, SVS College of Engineering, Coimbatore

Dr. Sanjeevikumar Padmanaban, Aalborg University, Denmark

Dr. Vinod T D'Souza, AJIET, Mangalore

Dr. Shobha M E, Manipal Institute of Technology, Manipal

Dr. Moorthi S, S R Engineering College, Warangal, India

Dr. Rajamani R, PSG College of Technology, Coimbatore, India

Dr. Venkataramani N, Indian Institute of Technology Bombay

Dr. Swaminathan K, National Institute of Technlogy Surathkal, Karanataka

Dr. Bishakh Bhattacharya, Indian Institute of Technology Kanpur

Dr. Kotteswaran, Veltech University, Chennai

IOP Conf. Series: Materials Science and Engineering 1272 (2022) 011001 doi:10.1088/1757-899X/1272/1/011001

Dr. Balasubramanian M, Indian Institute of Technology Madras

Dr. Deshannavar U B, KLE Dr MSSCET, Belagavi

Dr. Srinivasan P, Birla Institute of Technology and Science (BITS), Pilani

Dr. Mohd Fareq Abdul Malek, University of Wollonong, Dubai, UAE

Dr. Shanthi N, Kongu Engineering College, Tamil Nadu

Dr. Sanjay Misra, Covenant University, OTA, Nigeria

Dr. rer. nat. Ravi Kumar, Indian Institute of Technology-Madras

Dr. Baharudin Ismail, Universiti Malaysia Perlis, Malaysia

Dr. Mani V N, C-MET, Government of India, Hyderabad

Dr. Rahul Mitra, Indian Institute of Technology Kharagpur

Dr. Sampath V, Indian Institute of Technology Madras

Dr. Patil S F, KLE Dr. M.S. Sheshgiri College of Engineering and Technology, Belgaum

IOP Conf. Series: Materials Science and Engineering 1272 (2022) 011002 doi:10.1088/1757-899X/1272/1/011002

Peer Review Statement

All papers published in this volume have been reviewed through processes administered by the Editors. Reviews were conducted by expert referees to the professional and scientific standards expected of a proceedings journal published by IOP Publishing Publishing.

- **Type of peer review:** Double Anonymous
- **Conference submission management system:** Morressier
- **Number of submissions received:** 157
- **Number of submissions sent for review:** 78
- **Number of submissions accepted:** 38
- **Acceptance Rate (Submissions Accepted / Submissions Received × 100):** 24.2
- **Average number of reviews per paper:** 2.2
- **Total number of reviewers involved:** 30
- **Contact person for queries:**
  Name: Dr. Sangeetha D M
  Email: dr.sangeetha@ajiet.edu.in
  Affiliation: A J Institute of Engineering and Technology, Mangalore, Karnataka, India

LMS Adaptive FIR Filter-Predictive Controller Based On D-Q Control Theory for Grid-Connected Solar PV System

Nalini Karchi[1,2] , Deepak Kulkarni[3]

[1]Reasearch Scholar, Dept. of Electrical and Electronics Engg., Gogte Institute of Technology, Belagavi, 590008
[2] Faculty, Dept. Electrical and Electronics Engg., KLEDr MSS CET, Belagavi, 590008
[3] Professor, Dept. of Electrical and Electronics Engg., Gogte Institute of Technology, Belagavi, 590008

Abstract. A Least Mean Square (LMS) adaptive Finite Impulse Response (FIR) filter-predictive controller is presented for the control of a grid-connected solar PV energy system. The objective of the work is to reduce total harmonic distortion (THD) in the grid and load current. It follows the systematic development and analysis of the implementation of an adaptive LMS control based on the direct-quadrature control theory. The analysis is done for the proposed adaptive control technique for grid-connected solar photovoltaic systems under the variation of solar irradiation and load. In this proposed work, the reference current is estimated by developing the LMS control technique. The proposed system is modeled in d-q frame form and simulated on the MATLAB Simulink platform. Developing an LMS control algorithm using the p-q theory reduces computational hard work. The reduction in load current harmonics is improved. The simulation results are used to discuss the convergence rate, harmonics reduction, and advantages of the adaptive LMS algorithm d-q control theory.

Keywords: Adaptive LMS algorithm, Converter, Controller, Solar PV panel, D-Q theory

1. Introduction

Solar photovoltaic systems (SPV) have become popular to generate electrical energy. It is also an alternate solution for the modern power system to meet electrical energy demand. The major advantage of this SPV is that there is no rotating part, availability of sunlight energy and conversion of photon energy into electrical energy is possible using a semiconductor diode. The interconnection of the SPV system to the grid causes many power-quality issues like voltage-current waveform distortion, harmonics introduction, flickers, and many more with load conditions. The synchronization of a PV system to the grid is enabled through power electronics devices like a converter. These power electronics switches introduce harmonics in the grid. The control of grid-connected PV systems has become more important to face technical challenges. This has created motivation and interest for the researcher. Effective control is required to solve these power problems to ensure the system's smooth operation. Conventional control techniques like Proportional Integral (PI) controller are well-practiced in the control of a solar PV system. They have certain disadvantages of improper tuning of controller gains. Many soft computing control techniques are developed based on fuzzy logic, particle optimization techniques, etc to tune the parameters of the PI controller. These techniques need more computation time. Sometimes they fail due to the slow convergence. The adaptive algorithms are simple, easy and require less computational time. The adaptive algorithm for PV inverters can result in great accuracy and faster dynamic response. These algorithms can be implemented on hardware using a DSP processor.

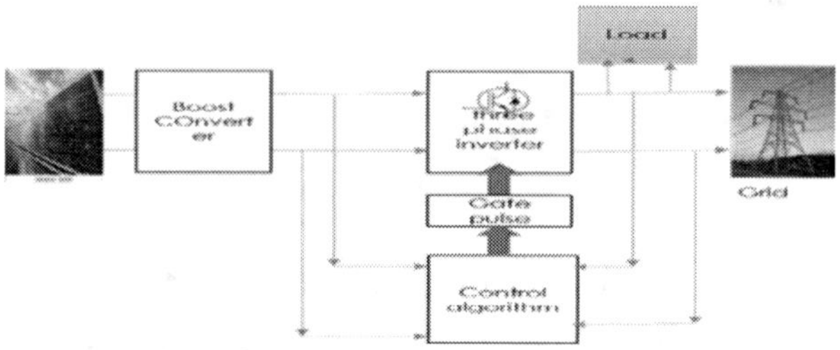

Figure 1. Block diagram of a Solar PV system

Control techniques based on adaptive algorithms are becoming more popular. Their performance is effective in the real-time operation of the system with dynamic load conditions. The proposed setup is assisting in reducing harmonics and maintaining the shape of AC voltage and current.

An adaptive finite impulse response filter is a self-learn filter, As the signal into the filter continues, the adaptive filter coefficients regulate themselves to achieve the desired result. An adaptive algorithm is an algorithm that changes its behavior at the time it is run, based on information available [1-2].

This paper discusses system formation, control technique, implementation of an adaptive algorithm for the grid-connected solar PV system, simulation results, discussion, and conclusion. The involvement of this paper is as given below

- The double-stage grid-connected solar panel energy system using the dc booster and voltage source inverter
- The reference dc current is obtained from reference dc voltage and dc bus voltage at maximum power extraction from the Solar PV system.
- An Adaptive LMS control is developed to extract the active weight component of load current.
- The d-q theory is considered to develop an LMS control scheme for this system.

2. System Formulation

The proposed system is a grid-connected solar PV system. It has different interconnected subsystems like solar PV array, DC-DC converter, three-phase DC-AC converter, inverter filter, grid, and load as shown in figure 1.

2.1. Formation of Solar PV energy system

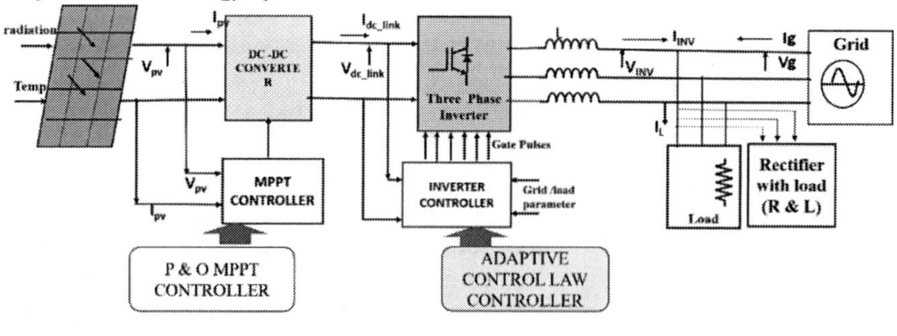

Figure 2. Block Diagram of LMS Adaptive FIR Filter-Predictive Controller Based On D-Q Control Theory for Grid-Connected Solar PV System

The expected electrical energy generation from the PV system is 51kW. The selection of parallel and series modules is decided as per the given equation (1).

The number of series modules in the panel (1) ns,

$$n_s = \frac{V_{dc}}{V_{mp}} = \frac{400}{29.5} = 14 \tag{1}$$

The number of parallel modules in the panel (2), n_p

$$n_p = \frac{P_{\max}/V_{dc}}{I_{mp}} = \frac{51kW/406.}{7.4} = 17 \quad (2)$$

Where V_{dc} is the total dc voltage of the PV panel, V_{mp} is the voltage at the maximum power point (29.5 V), P_{max} is the total output power from the PV panel, and I_{mp} is the current at the maximum PowerPoint (7.4 A). Therefore, the solar PV array with a peak power capacity of 51kW is modeled with 14 modules connected in series and 17 modules connected in parallel [8].

2.2 DC bus voltage selection

The selection of DC bus voltage is done by calculating the minimum value of dc bus voltage for the Voltage source inverter which has to be maintained at the common interconnection point or grid. The dc bus voltage is in (3).

$$V_{dc} = \frac{2\sqrt{2}V_{LL}}{\sqrt{3}m} = 620.537 \quad (3)$$

V_{LL} is grid line voltage (380V), and m is the modulation index (m=1). The selected value of DC bus voltage is 700 V.

2.3 Formulation of DC-link capacitor and inductor

The L_boost and C_boost are an inductor and a capacitor of the boost converter. They are estimated by given equations (4) and (5). There V_o is the DC converter output voltage (700 V) and Vin is the input DC voltage of the DC converter is 406 V. F_{sw} is the switching frequency of the IGBT switch used in the DC-DC converter. ΔI is the current ripple factor and ΔV is the voltage ripple factor in dc converter current and voltage respectively.

$$L\_boost = \frac{V_{in}(V_o - V_{in})}{F_{sw} * \Delta I * V_o} \quad (4)$$

$$C\_boost = \frac{I_o(V_o - V_{in})}{F_{sw} * \Delta V * V_o} \quad (5)$$

2.4 Inverter filter inductance formulation

The inverter output filter inductance is evaluated by (6)

$$L_f = \frac{0.1 * U^2}{2 * \pi * f * P/3} \quad (6)$$

$R_{Lf} = L_f * 25$

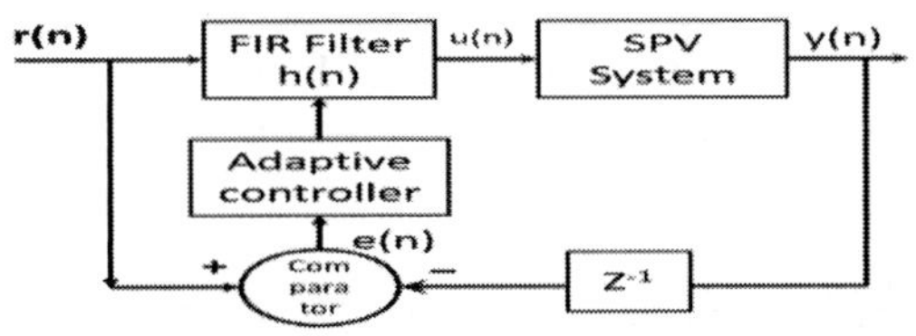

Figure 3. Adaptive control system using LMS algorithm

where P=Power capacity of inverter in kW, f= grid frequency in Hz, U= inverter output line voltage in Volt. The basic schematic of the three-phase grid-connected solar PV system in d-q form is shown in figure 2.

3. Control Theory

3.1 Adaptive Control theory

An Adaptive control system using the LMS algorithm is proposed in this paper based on the Finite Impulse Response (FIR) filter concept as shown in Figure.3. The output signal u is the control input to the inverter control circuit. The control input is given by the convolution of the reference input r(n) and filter function h(n). The step size u (8) is selected between the range 0 to 1, which is any real number. The advantage of the adaptive LMS algorithm is easy and simple to calculate. The output signal y(n) is delayed and used as a feedback signal, later it is compared to the reference signal r(n) to determine the error signal e(n). The weight function h(n+1) is updated by using the LMS algorithm in (8). u(n) is the output signal of the FIR filter which is the result of convolution between weight function h(n) and r(n). The following steps are used to implement the LMS Adaptive FIR Filter as a controller.

- Sense the output signal y(n)
- Calculate the error, [r(n)-y(n-1)]
- Determine the h(n+1) using eq. (8)
- Repeat the process for a given number of iterations

$$u(n) = \sum_{i=0}^{N} h_i r(n-i) \tag{7}$$

$$h(n+1) = h(n) + \mu e(n) r(n) \tag{8}$$

3.2 Overview of d-q theory (d-q-0)

The d-q theory is the product of Clarke and Park transforms. This can be used to rotate the reference frames of AC waveforms so that they get converted into dc signals. The three-phase quantities and transformed into DC frame quantities. Simplified calculations can be done on dc quantities before performing the inverse transform. It is mostly used for the analysis of the three-phase system. In this work, the sensed three-phase system parameters are transformed into d-q components by using the d-q frame control theory.

4. Proposed Control Scheme

The proposed algorithm's schematic is expressed in Figure 3. In a Solar PV System, the Perturb and observe MPPT controller is applied to track the maximum power point and the developed adaptive LMS control technique in sub-section 4.2.3 is implemented with a three-phase inverter. The MPPT algorithm is to extract maximum power from the solar panel array at all conditions. Due to simplicity and ease, the MPPT algorithm Perturb and observe is selected. On the other side, the control scheme of the voltage source converter (VSC) emphasizes generating gating signals for the appropriate switching of the Converter's IGBTs under different conditions [8].

4.1 MPPT Control

The maximum power point tracking (MPPT) method is used in the Solar PV DC-DC converter. It continuously adjusts the impedance of the converter circuit by looking into the solar array to track the PV system operating at the peak power point of the SPV panel under changing solar irradiance, temperature, and load. The P-O algorithm is based on finding the maximum power point by taking the derivative of power with respect to voltage [1-9].

4.2 Control method for switching of VSC

4.2.1. D-Q component estimation

The control structure is divided into sub-section as shown in figure.3. The grid voltage is sensed at the interconnection point of the grid, load and inverter. This voltage is converted into the α-β phase system.

IOP Conf. Series: Materials Science and Engineering 1272 (2022) 012001 doi:10.1088/1757-899X/1272/1/012001

The phase angle ϱ is found by (9) and (10). Vα and Vβ are alpha-beta components of grid voltage. The d-q components of load current and grid current with respect to phase angle ϱ are found by eq. (9).

$$\begin{bmatrix} V_\alpha \\ V_\beta \end{bmatrix} = \begin{bmatrix} V_a - \frac{V_b}{2} - \frac{V_c}{2} \\ \frac{\sqrt{3}V_b}{2} - \frac{\sqrt{3}V_c}{2} \end{bmatrix} \tag{9}$$

$$\rho = wt = \cos^{-1}\left(\frac{V_a - \frac{V_b}{2} - \frac{V_c}{2}}{\frac{\sqrt{3}V_b}{2} - \frac{\sqrt{3}V_c}{2}} \right) \tag{10}$$

The q component and q component are determined by applying Clarke's -park transformation theory (11). Determine I_{gd} and I_{gq} which are the d-q component of grid current by (11).

$$\begin{bmatrix} I_d \\ I_q \end{bmatrix} = \begin{bmatrix} \cos\rho & \sin\rho \\ -\sin\rho & \cos\rho \end{bmatrix} \begin{bmatrix} 1 & -1/2 & -1/2 \\ 0 & \frac{\sqrt{3}}{2} & \frac{-\sqrt{3}}{2} \end{bmatrix} \begin{bmatrix} I_a \\ I_b \\ I_c \end{bmatrix} \tag{11}$$

4.2.2 DC component estimation

I_d* set point dictates the current that is being pumped into the grid. It indicates the total space current vector that is fed into the grid as the grid voltage is fixed. The control point of view I_q* is made zero. I_d* is the I_{dc_ref} dc current which is obtained from dc bus voltage and reference voltage eq (12) or the output of the MPPT algorithm [13].

$$Id^* = V_{dce} \cdot {}^* K_p + K_I \int V_{dce} dt \tag{12}$$

Where V_{dce}=V_{dc_bus} - V_{dc_busref}

4.2.3 Extraction of the reference current

The instantaneous error at k[th] instant is determined from sensed grid current and weighted current for the d-q component by applying the FIR filter concept in section 3. The weighted component of the load current has been evaluated.

$$e_g(n) = I_d - \varphi(n) I_{dc\_ref}(n) \tag{13}$$

$$\varphi(n) = \varphi(n-1) + \eta e(n) I_{dc\_ref}(n) \tag{14}$$

The reference compensator current component is found by (15)

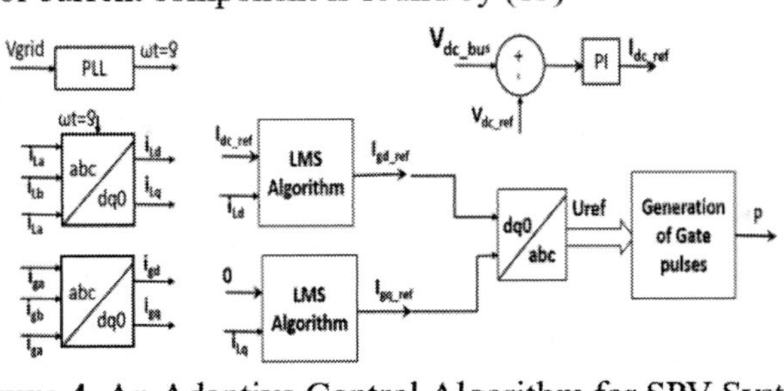

Figure 4. An Adaptive Control Algorithm for SPV System

$$I_d(n) = \varphi(n) I_d^*(n) \tag{15}$$

$$e_{q_g}(n) = I_q - \varphi(n) I_q^*(n) \tag{16}$$

$$\varphi(n) = \varphi(n-1) + \eta e(n) I_q^*(n) \quad (17)$$

$$I_q(n) = \varphi(n) I_q^*(n) \quad (18)$$

Where, $\varphi(n)$ is the weight value at the nth instant, η is increment size in the weight function. The reference quadrature component is made zero. The q-component (I_q^*) of the reference compensator component is calculated by (16), (17), and (18) The generated reference components are transformed into a-b-c three-phase form. There are used to generate gate pulses for the multifunctional inverter [11].

5. Simulation Results and Discussion

The proposed work is modeled, simulated, and verified on the MATLAB Simulink platform. The LMS Adaptive FIR Filter-Predictive Controller is tested under linear and nonlinear conditions and variations in solar irradiation. The system details are given in appendix A. The total harmonics distortion (THD) analysis is obtained by the FFT tool.

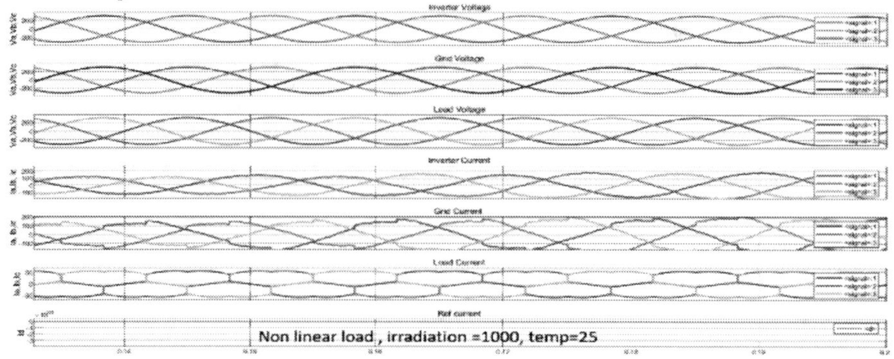

Figure 5. Performance of inverter voltage, grid voltage, load voltage, inverter current, grid current, load current under non-linear.

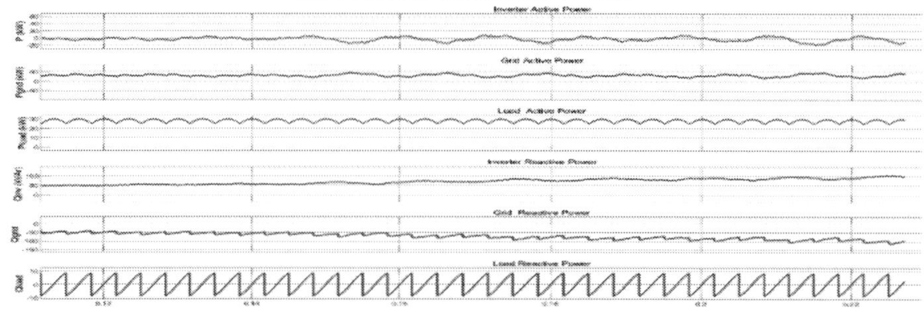

Figure 6. Performance of inverter active and reactive power, grid active and reactive power, load active and reactive power, under non-linear.

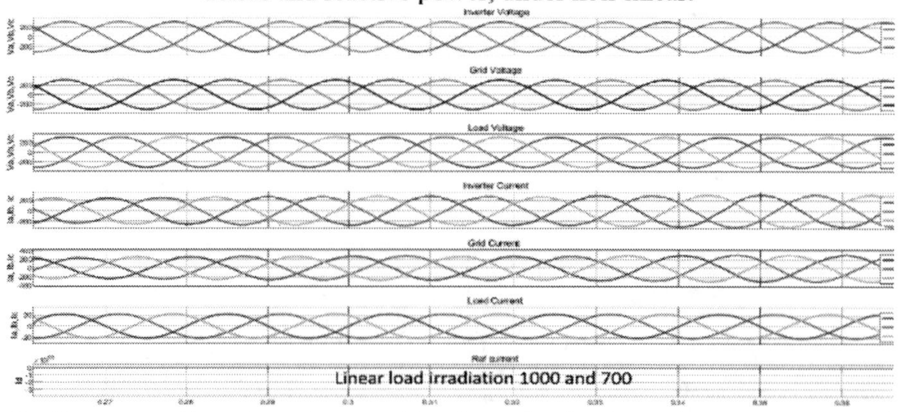

Figure 7. Performance of inverter voltage, grid voltage, load voltage, inverter current, grid current, load current under linear load

IOP Conf. Series: Materials Science and Engineering 1272 (2022) 012001 doi:10.1088/1757-899X/1272/1/012001

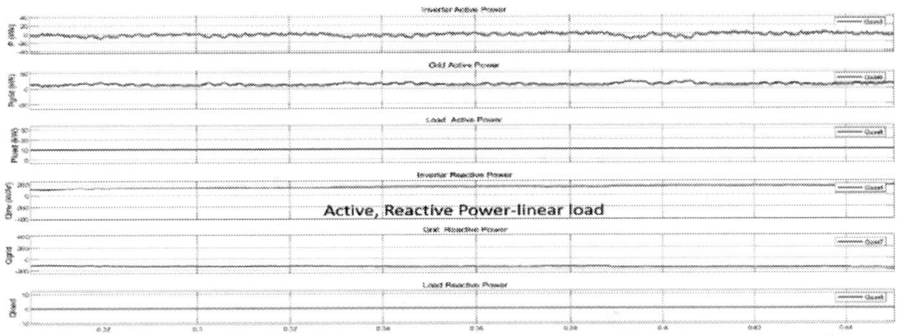

Figure 8. Performance of inverter active and reactive power, grid active and reactive power, load active and reactive power, under linear.

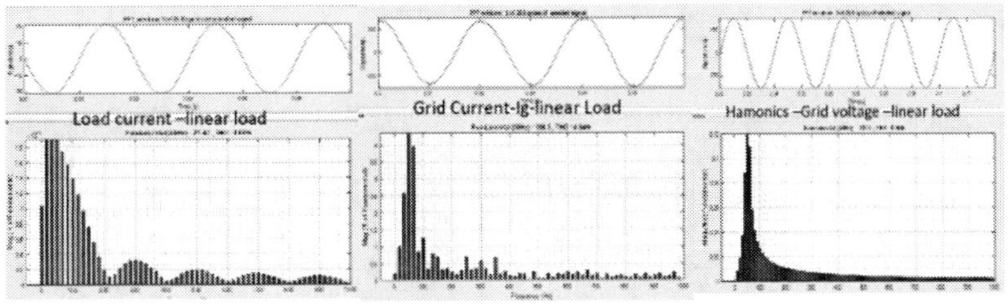

Figure 9. Harmonics analysis for load current, grid current and grid voltage under linear load.

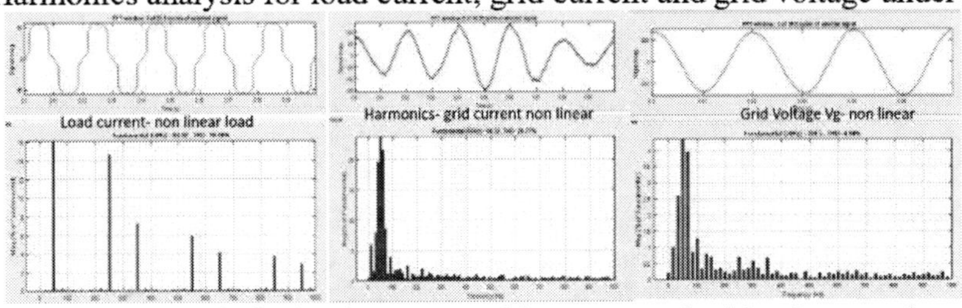

Figure 10. Harmonics analysis for load current, grid current and grid voltage under nonlinear load.

5.1 Performance under linear and nonlinear load

The performance of LMS Adaptive FIR Filter-Predictive Controller for a grid-connected PV system under linear and non-linear load is observed from results shown in figure 5 to figure 10. It is observed that the waveform of voltage and current are maintained in sinusoidal form with a change in linear to nonlinear load.

Table 1. Harmonics results obtained under linear and nonlinear load

Load Condition	Grid voltage (THD) in %	Grid current (THD) in %	Load current (THD) in %	Convergence rate	P Power	Q power
Linear load	310.5, 0.15%	107.19A, 4.9%	21.47, 0%	0.05 sec	P_g=10 kW P_{inv}=0 Pl=	Q_g=0 Q_{inv}=0
Non-linear load	310.5, 0.00	133.5A, 4.70%	60.52, 19.89%	0.1sec	P_g=20-40 kW P_{inv}=10 kW P_L=25-30 kW	Qg=50-80 Q_{inv}=50-80 Q_L=10 VAR

Under linear load, the grid supplies the active power to the load. Whereas the reactive power is supplied from the inverter to the grid. The THD measured in grid current, I_g is 4.9%, grid voltage (V_g) is 0.15%, and load current, I_L is 0% under linear load. Similarly, THD in grid voltage V_g is 0%, grid current THD I_g is 4.70%, and load current THD is 19.89% under nonlinear, as mentioned in table 1. Grid voltage harmonics have been reduced drastically. It is observed that the efficiency of the LMS adaptive control algorithm is better for the nonlinear load. The simulation is started with the non-linear load. So, this control algorithm has proved that it works systematically under a change in load. It has a fast convergence rate of less than 0.01 sec. Under the nonlinear condition, the grid supplies active power as well as reactive power. This reactive power is compensated by inverter reactive power. During a change in load conditions from linear to nonlinear, the convergence rate of the adaptive controller is quite fast.

5.2 Performance under solar irradiance variation

The entire system with LMS Adaptive FIR Filter-Predictive Controller is tested for various solar irradiation 600 to 1000 W/m². The DC bus is retained at a constant value Figure 11 (a). The same adaptive algorithm is implemented for single-stage grid-connected PV systems. The results are shown in Figures 12 and 13. The control algorithm works satisfactorily for both single and two-stage grid-connected solar PV systems.

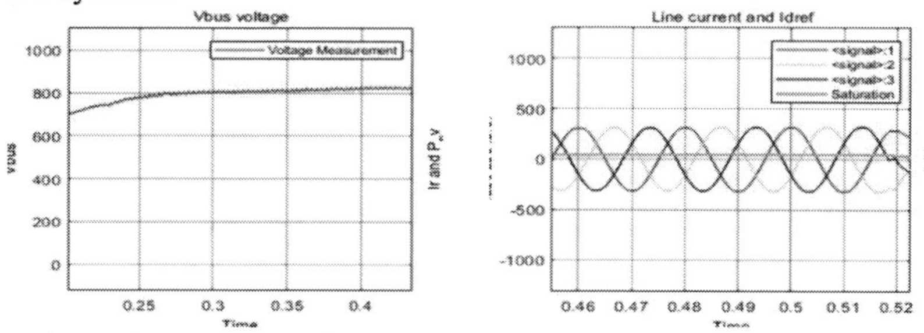

Figure 11. (a) DC bus voltage, (b) dc reference current and inverter line current –irradiation 1000w/m²

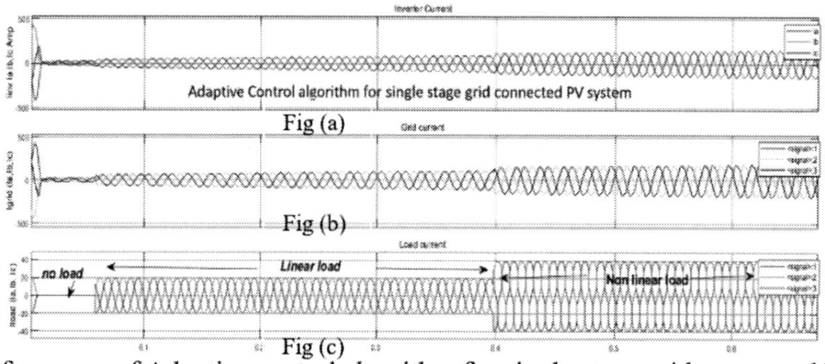

Figure 12. Performance of Adaptive control algorithm for single stage grid-connected PV system, (b) Inverter current, (c) grid current and (d) load current

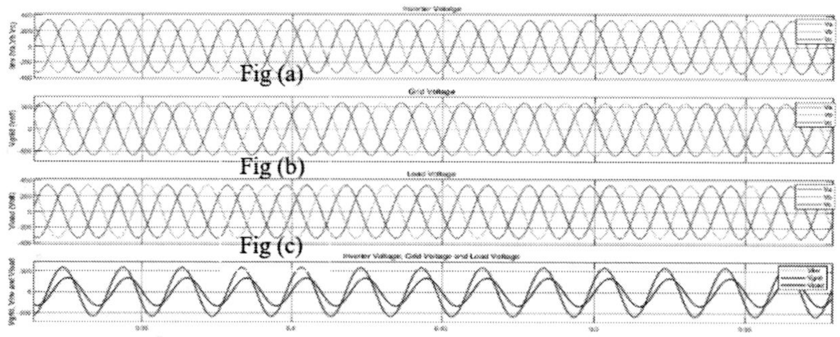

Figure 13. Performance of Adaptive control algorithm for Single Stage grid-connected PV system, (a) inverter voltage, (b) grid voltage (c) load voltage.

Table 2. Steady-state and dynamic comparison of other LMS family controllers and LMS d-q control

Control algorithms	Steady-state	Amplitude of oscillation	Convergence rate
LMS d-q	Less	Less	Fast
Sign Error- LMS [5]	Less	Less	Slower
Sign Data-LMS	Medium	More	Medium
Sign Sign-LMS	Medium	Less	Medium

5.3 Summary

The number of equations derived to build the control scheme are less in number. The convergence time is less. If the sampling time is more than 30 microseconds and the hardware processor is of high speed, convergence time reduces. LMS Adaptive FIR Filter-Predictive Controller is a better substitute for the PI controller. This controller does not require tuning the parameters. The selection of step size is a difficult task.

5.4 Future Scope

Implementation of LMS Adaptive FIR Filter-Predictive Control on DSP Processor can experiment. The effect on parameters of change in step size of LMS algorithm can be observed,

6. Conclusion

The objective of the work was to develop LMS Adaptive FIR Filter-Predictive Controller to control a Solar PV system. Performance was analyzed by implementing LMS Adaptive FIR Filter-Predictive Controller with a Solar PV system. The development of the LMS Adaptive FIR Filter-Predictive Controller was based on the d-q theory for a grid-connected PV system observed under linear, nonlinear load and variation in solar irradiation. It was found that the system operation was more efficient for nonlinear load and change in solar irradiation from 600 to 1000 W/m^2. This control algorithm was a good solution for the conventional controller. Building an LMS control algorithm based on p–q theory has reduced computational cost and the results obtained were satisfactory. The load current THD was 19.89 % which was less than 25% and grid current THD was 4.70 % also less than 5 % as per IEEE 519 standards. The waveform of grid current and inverter current was retained in sinusoidal.

Appendix A

PV system parameters: nominal power:51 kW, DC Output voltage:406 V, nominal current:150 A
Interfacing inductor:0.027 H, resistor:0.0676 ohm
DC- DC converter: Inductor (Henry):00.0014e-4, Capacitor (farad):8.73690x10^{-4}
Input voltage=406V, Output voltage=700V
DC-AC converter: 51kVA, 380 V, DC input voltage:700V
Linear load:10 kW, Nonlinear Load: rectifier with load (P=10 kW, Q=100 VAR)

7. References

[1] Sunaina Singh and Seema et.al. 2020 Adaptive based Leaky LMS Control Technique of Grid Connected SPV System *IEEE International Conference on Power Electronics, Drives and Energy Systems* 978-1-7281-5672-9

[2] Pallavi Verma, Priya Mahajan et.al 2020 Smooth LMS-based adaptive control of SV system tied to the grid for enhanced power quality *IET Power Electronics* September

[3] Avdhesh Kumar and Rachana Garg et.al 2021 Control of Grid Integrated Photovoltaic System using new Variable Step size Least Mean Square adaptive filter *Springer Nature* March

[4] Manoj Badoni, Alka, Ankit Kumar Singh et.al 2020 Grid Tied Solar PV system with Power Quality Enhancement Using Adaptive Generalized Maximum Versoria Criterion DOI:10:17775/CSEEEJPES.2020.04820

IOP Conf. Series: Materials Science and Engineering 1272 (2022) 012001 doi:10.1088/1757-899X/1272/1/012001

[5] Satish Choudhury and Byomakesh dash et.al 2020 Comparative Analysis of LMS Based Control Algorithms for Grid Integrated System Innovation *in Electrical Power Engineering, Communication and Computing Technology* Springer Nature Singapore Pte Ltd. 2020) Lecture Notes in Electrical Engineering 630, doi.org/10.1007/978-981-15-2305-2_46.
[6] Naki Guler and Erdal Irmak 2019 MPPT Based Model Predictive Control of Grid-Connected Inverter for PV Systems'. *International Conference Renewable Energy Research and Applications* ICERER Brasov Romana 978-1-7281-3587-8/19
[7] Eugene Walach and B. Widrow 1984 The least mean fourth LMF adaptive algorithm and its family', *IEEE Transactions on Informative Theory*
[8] M Kalai Arasi and Dr.S Shivananaitha 2018 Harmonic Reduction by LMF Algorithm in Grid-Connected SPV System *4th International Conference on Energy Efficient Technologies for Sustainability ICEETS18*
[9] Sachin Devassy and Bhim Singh Design and Performance Analysis of Three-Phase Solar PV Integrated UPQC' 978-1-5090-0128-6/16-IEEE
[10] Neha Beniwal and Ikhlaq et.al 2018 Implementation of DSTATCOM with i-PNLMS Based Control Algorithm under Abnormal Grid Conditions *IEEE Transactions on Industry Applications* DOI;10.1.1094/TIA.2018
[11] Gunjan Varshney and Madhukar Dave et.al 2019 Unit Template-based Control of PV DSTATCOM *Recent Advances in Electrical and Electronics Engineering* (Formerly Recent Patents on Electrical & Electronics Engineering) 12 1-7 DOI:10.2174/2352096511665181108112853
[12] Rahul Kumar Agarwal and Ikhlaq Hussain et.al 2017 Application of LMS-Based NN Structure for Power Quality Enhancement in a Distribution Network Under Abnormal Conditions *IEEE Transactions on Neural Networks and Learning Systems* DOI; 10.1109/TNNLS.2017.2677961
[13] Priyabrat Garanayak and Gayadhar Panda, Pravat Kumar Ray 2015 Harmonic estimation using RLS algorithm and elimination with improved current control technique based SAPF in a distribution network doi.org/10.1016/j.ijepes.2015.04.023

A Cascaded Multilevel Inverter design based Harmonic Optimization using TLBO Algorithm for Wind Energy System

Sekhar V[a], Bakthavastalam Deepa[b], Murali Muniraj[c]

[a,b] *Department of EEE, Kuppam Engineering College, Kuppam ,India*
[C] *Department of EEE, Sona College of Technology, Salem, India.*

Abstract. In order to remove undesirable lower order harmonics from a cascaded H-bridge multilevel inverter and solve a constrained in design optimization issue with parameters involved in a changeable objective function for the given system, the TLBO approach is provided. To test the stability of the system, the recommended system output was designed to be supplied to various load drive systems. The entire efficiency of wind grid inverters has now decreased due to the low wind speed, making it unable to utilise wind energy effectively. To solve this issue, it is necessary to improve both the control method and the inverter side's topological structure. A cascaded H-bridge multilevel inverter architecture is used to implement the entire system's functionality, with thefirst stage circuit being built in MATLAB-Simulink.

Keywords: TLBO, controller, wind, MATLAB, multilevel, inverter

1. Introduction

Due to the decreased voltage stress between the power switches and the low THD (Total Harmonic Distortion) output voltage waveform, multilayer inverters are now widely used. High output performance, low output harmonics, and minimal commutation losses are characteristics of multi-level converters. Its intricacy is their biggest flaw. This calls for a significant number of power devices, passive parts, and very complex control circuitry. This work offers a novel MLI with additive and subtractive topologies to decrease the number of switching devices. Additionally, this lowers the cost of production [1-3]. This project's primary objective is the development of a single-phase MLI with nine levels and fewer components.

Renewable energy is acknowledged as one of the most significant and significant energy resources in the world and is one of the cleanest kinds of energy. Major power networks are also growing more sophisticated as the world's energy usage rises. However, the usage of alternative energy sources has rapidly increased as a result of rising worries about global pollution and the finite supply of fossil fuels [4-6]. Due to its capacity to lower pollution, renewable energy has recently gained popularity and currently provides a sizeable amount of the world's power [7- 8]. One of the most promising energy sources for reducing greenhouse gas (GHG) emissions and global warming, and preserving ecological balance, is wind power. Finding the best output from each generation source over a 24- hour period is the aim of Dynamic Economic Load Dispatch (DLED), which is defined as a real-time optimization problem. This is done while adhering to operational constraints and taking into account the uncertainties associated with load demand, as shown in [9-11]. The TLBO control technique for multi-level cascaded H-bridge inverters proposed in wind turbine systems is studied in this research using a newly created efficient supervised-learning- based optimization method.

IOP Conf. Series: Materials Science and Engineering 1272 (2022) 012002 doi:10.1088/1757-899X/1272/1/012002

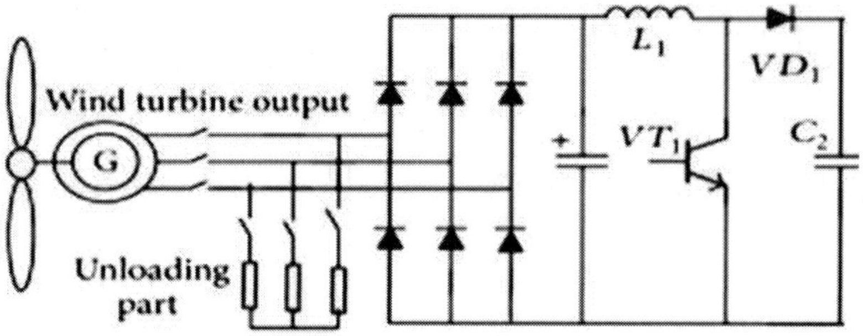

Figure 1. WTS System

2. Topology and Design

Systems for wind turbines typically consist of wind turbines, generators, and equipment for protecting against discharges, converters, and inverters. To achieve the highest output intensity possible, the wind turbine's 3-segment AC output is rectified to DC by an unregulated rectifier unit using a DC/DC circuit, as illustrated in Figure 1.

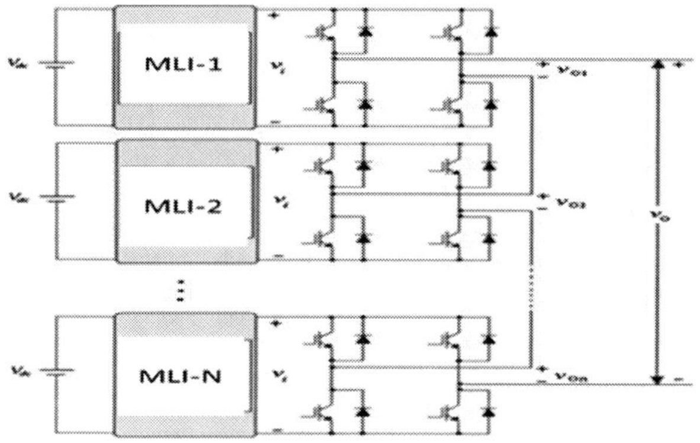

Figure 2. Multi-level inverter

Using distinct DC resources to produce L, a multi-level inverter (MLI) is a power electronics technique utilised for renewable energy buildings and power conversion, as seen in Figure 2. It is made up of H-bridge cells in sequence with output. S number of switching angles based on DC systems necessary for L-level inverters. Each mobile device in the MLI may produce both -VDC and +VDC. The 7-level inverter's staircase output voltage waveform is shown in Figure 3.

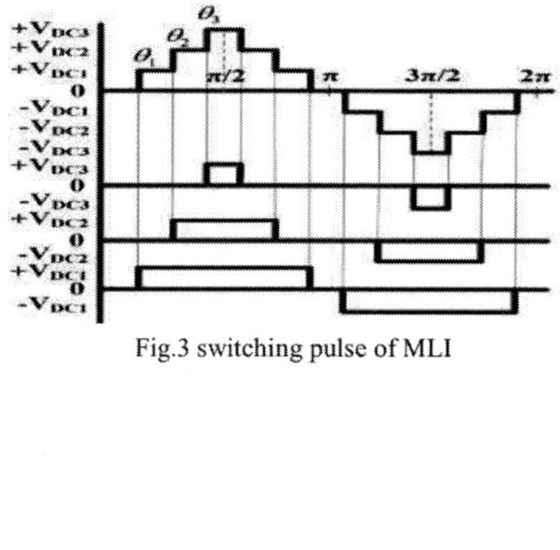

Fig.3 switching pulse of MLI

The result side of the conventional business matrix mapped inverter is primarily a frame mapped structure without limitations, the multi-level inverter adopts a hierarchical inverter circuit, the result voltage is 220V, and the transport voltage is 400V. Given the ensuing current coherence, the DC/DC circuit's commitment ratio can only be between 0.2 and 0.9, and the matrix-induced DC input voltage must be more than 40 V, often between 50 and 70 V. Although wind energy normally has a breakdown voltage of 30 to 50 volts, wind energy is not typically supportedby commercial matrix inverters, and the low wind speed signal is not accurate enough. The inverter will operate in grid-tied mode in accordance with the criteria of the plan as long as the DC input voltage is greater than 30V. The projected beginning voltage in this framework is 40 V, and several control techniques are used in conjunction with suggested fuzzy models, adaptive fuzzy models, and neural network models to maximize power transmission. Overall, based on the control parameters and instantaneous changes in the control signal, the input power voltage and current are sampled based on control structure in Fig.4.

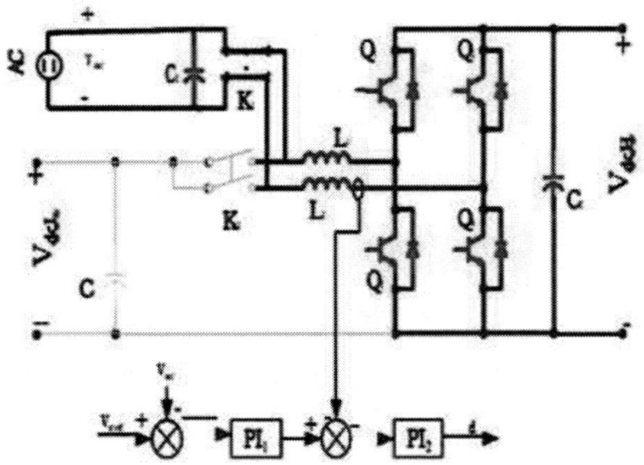

Fig.4 Inverter mode and current control circuit

2.1 Control Algorithm

The regulation of the current on the load side has also been suggested using a current feedback control loop. Here, his TLBO algorithm an algorithm that draws inspiration from the teaching-learning process is put out in an effort to influence teachers' perceptions of their students' performance as stated based on algorithm described in Fig.5. The design variables are really the variablesinvolved in the objective function of a specific optimization problem,

$$Total\ Difference\_\ Average\ Mean_{j,k,i} = r_i\ (T_{j,kbest,i} - L_F U_{j,i}) \quad (1)$$

The optimum solution is the ideal value of the objective function, and the best solution throughout the population is thought of as the teacher.

Where Xj,kbest,i represents the outcome for the top student in topic j. ri is a random number in the range[0, 1], and TF is the teaching factor that determines the value of the mean to be altered. TF's value can either be 1 or
2. TF's value is chosen at random with an equal chance are,

$$L_F = round\,[1+rand\,(0,1)\{2-1\}] \quad (2)$$

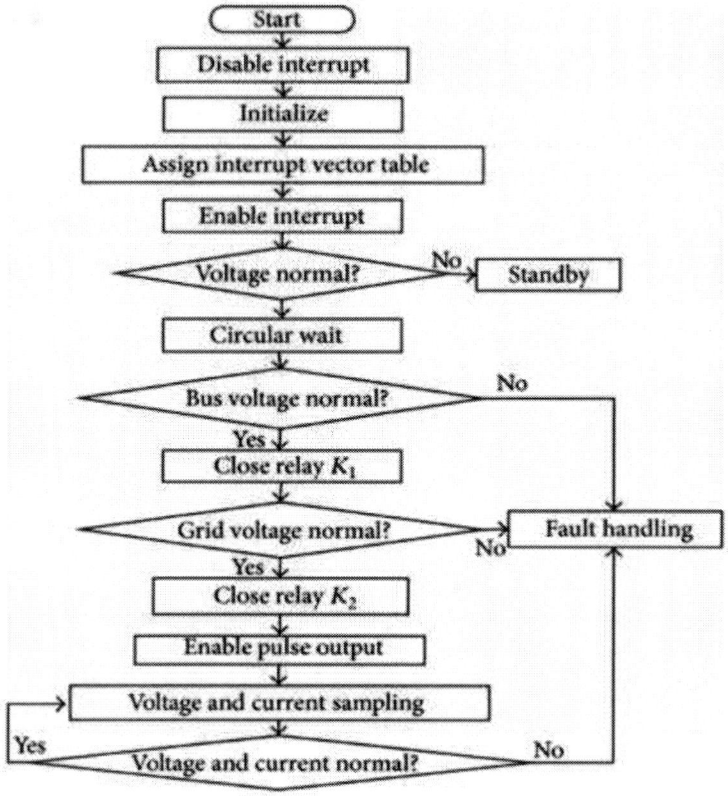

Fig.5 Flowchart algorithm process of TLBO algorithm

The algorithm does not receive the value of LF as input; instead, it chooses at random what that value should be.

$$R''_{j,P,i} = X'_{j,P,i} + r_i\,(X'_{j,Q,i} - X'_{j,P,i}),\ \text{If}\ X'_{total\text{-}Q,I} < X'_{total\text{-}P,i} \quad (3)$$

There are "m" topics and "n" learners at each iteration I and Mj,i is a specific subject ("j" = 1, 2,..., m). R" j,P,Iare acceptable if they provide superior function values based on Eq. (3), which is employed in minimization-related issues.

3. Simulation of Cascaded Multi Level Inverter

The MATLAB Simulink environment may be used to modulate multilayer inverter topologies in a multicarrier PWM scheme, as demonstrated in Figure 6. Switching devices employ multi-carrier PWM methods, where the carrier signal is contrasted with a reference signal, and the generated pulses are mapped to each voltage level.

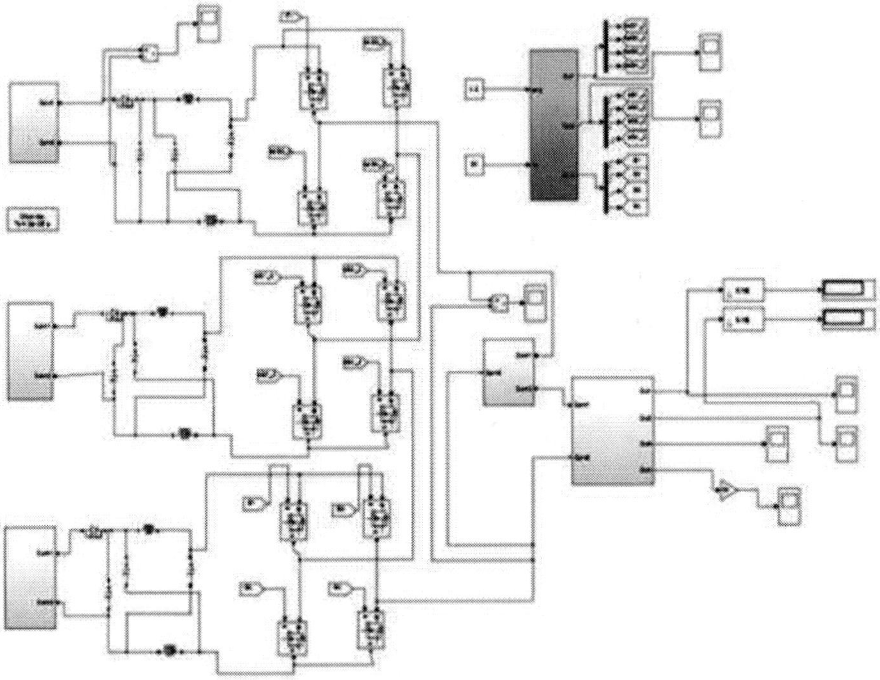

Fig 6 Simulation Circuit of Proposed System

In this modulation method, a reference sine wave is compared to each carrier waveform on the timeline to determine whether it is a "1" or a "0." The reference sine wave is compared to all carriers below the time axis, and the result is either 0 or 1.

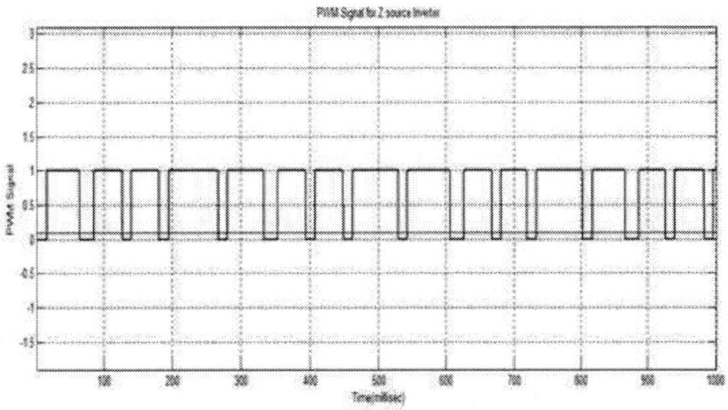

Figure. 7 PWM Pulse signals fed to MLI

The switching signal from the based driver circuit's sum signal, which was determined based on the control signal modified by TLBO. Combining the switching patterns and the saturation limit results in the switching signal based on PWM signals. Using the one-to-one correspondence between the levels present in the aggregate signals and the corresponding levels in the output waveform, extract the switching pulse from the aggregate signal.

IOP Conf. Series: Materials Science and Engineering 1272 (2022) 012002 doi:10.1088/1757-899X/1272/1/012002

As illustrated in fig.7, the output of the sum signal is delivered to a switch that corresponds to the necessarylevel after being compared to a constant level. The proposed algorithm was also confirmed using the simulation model with variable output current in Fig. 8 and 9 the designed algorithm model.

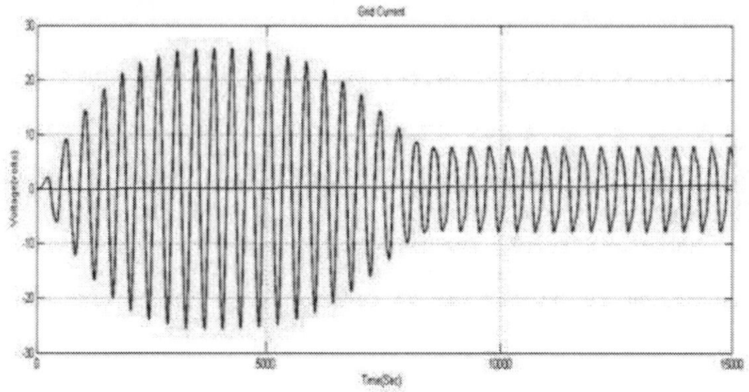

Figure.8 Output Current fed to load

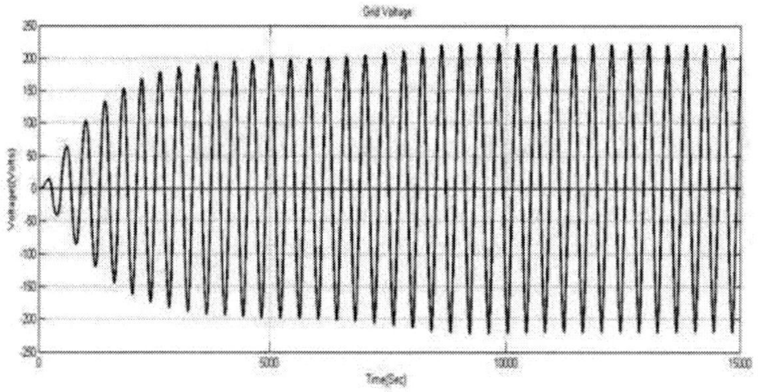

Figure.9 Output voltage fed to load

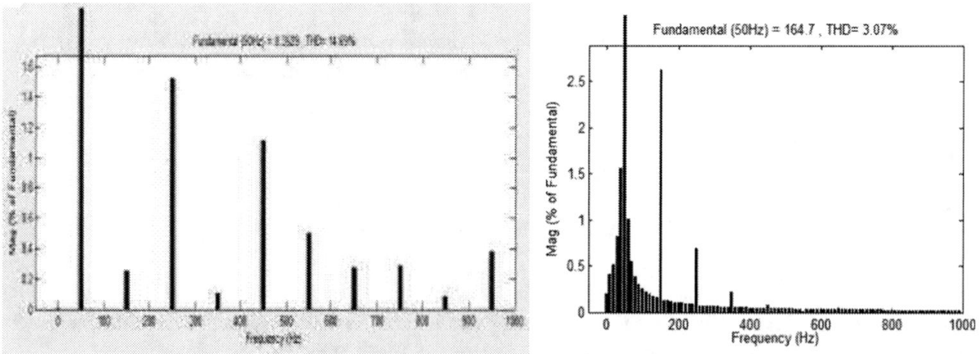

Figure.10 THD analysis of output current

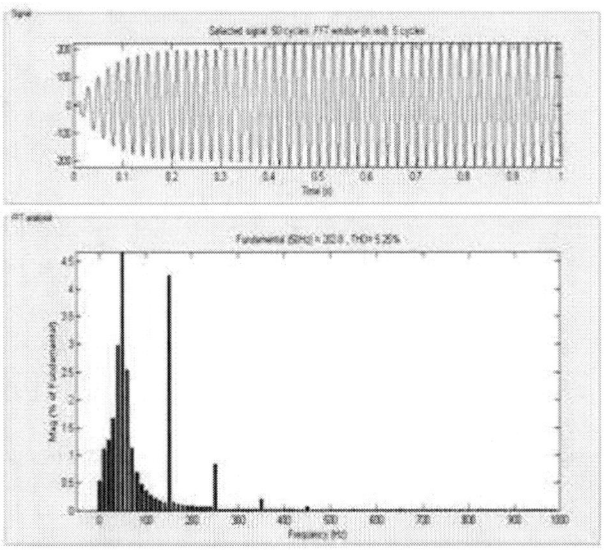

Figure.11 THD analysis of Output voltage

TABLE.1 Output Performance of Control Parameters

Parameters for Comparison	MLI with A-PSO Control Algorithm	MLI with TLBO Control Algorithm
Input Voltage	V1=V2=V3=50 V	V1=V2=V3=50 V
Output Voltage	182V	218V
THD for Output Voltage	14.69 %	5.25 %
THD for Output Current	12.67 %	3.07 %

With inputs that are multiples of three, a generated simulation of the suggested multilayer inverter utilising the TLBO algorithm generates an output voltage level of 218V for a given input of 50V. The harmonic spectra of the feed load for output current and voltage are displayed in Figures 10 and 11. In comparison to other traditional controllers, this prototype was created with fewer components and a reduced THD.

4. Conclusion

Simulated design and analysis of a multi-level inverter topology carried out for wind turbine operation using the TLBO algorithm. Multiple layers of inverter and converter circuits seen in various architectures are reported to reduce voltage and current ripple compared to traditional converters. According to the findings of the simulation, TLBO computes the right answer with a high probability and can accomplish similar outcomes to other controller modules as the number of switching signals rises. The suggested approach perfectly eliminates chosen low-order harmonics from the output voltage waveform for unstable wind energy sources under a variety of real- time operating scenarios. The fluctuations between the number of voltage levels, THD, and its RMS voltage are being investigated and extensively explained by the simulation carried out using MATLAB/Simulink. The comparison of the suggested topology versus the previously given topologies are also undertaken to reinforce its originality.

References

[1] T. S, M. Murali and R. Arulmozhiyal, "Design of Multi-Mode Power Converter and Control Using TLBO Algorithm for Renewable Energy based EV-Charging Systems," 2022 Second International Conference on Advances in Electrical, Computing, Communication and Sustainable Technologies (ICAECT), 2022, pp. 1-5, doi: 10.1109/ICAECT54875.2022.9808033.

[2] L. K. Haw, N. A. Jefry and W. K. Ing, "The New Hybrid Multilevel Inverter with Reduced Number of Switches," 2021 IEEE 11th International Conference on System Engineering and Technology (ICSET), 2021, pp. 337-341, doi: 10.1109/ICSET53708.2021.9612532.

[3] T. Qanbari and B. Tousi, "Single-Source Three-Phase Multilevel Inverter Assembled by Three-Phase Two- Level Inverter and Two Single-Phase Cascaded H-Bridge Inverters," in IEEE Transactions on Power Electronics, vol. 36, no. 5, pp. 5204-5212, May 2021, doi: 10.1109/TPEL.2020.3029870.

[4] R Arulmozhiyal, P Thirumalini, M Murali, Implementation of Multilevel Z-Source Inverter using Solar Photovoltaic Energy system for Hybrid Electric Vehicles, - Computation and Communication Technologies, 2016.

[5] H. Khoun-Jahan et al., "Switched capacitor based cascaded half-bridge multilevel inverter with voltage boosting feature," in CPSS Transactions on Power Electronics and Applications, vol. 6, no. 1, pp. 63-73, March 2021, doi: 10.24295/CPSSTPEA.2021.00006.

[6] L. Wan, K. Gong, G. Zhang, C. Li, Z. Wang and X. Deng, "Ensemble Pruning of RF via Multi-Objective TLBO Algorithm and Its Parallelization on Spark," in IEEE Access, vol. 9, pp. 158297-158312, 2021, doi: 10.1109/ACCESS.2021.3130905.

[7] P. Samuel, R. Gupta and D. Chandra, "Grid Interface of Wind Power With Large Split-Winding Alternator Using Cascaded Multilevel Inverter," in IEEE Transactions on Energy Conversion, vol. 26, no. 1, pp. 299- 309, March 2011, doi: 10.1109/TEC.2010.2096538.

[8] G. Singh and V. K. Garg, "THD analysis of cascaded H-bridge multi-level inverter," 2017 4th International Conference on Signal Processing, Computing and Control (ISPCC), 2017, pp. 229-234, doi: 10.1109/ISPCC.2017.8269680.

[9] K. Ma, M. Liserre and F. Blaabjerg, "Reactive Power Influence on the Thermal Cycling of Multi-MWWind Power Inverter," in IEEE Transactions on Industry Applications, vol. 49, no. 2, pp. 922-930, March- April 2013, doi: 10.1109/TIA.2013.2240644.

[10] E. Gatavi, A. Hellany, M. Nagrial and J. Rizk, "An integrated reactive power control strategy for improvinglow voltage ride-through capability," in Chinese Journal of Electrical Engineering, vol. 5, no. 4, pp. 1-14, Dec. 2019, doi: 10.23919/CJEE.2019.000022.

[11] M. A. Memon, M. D. Siddique, S. Mekhilef and M. Mubin, "Asynchronous Particle Swarm Optimization- Genetic Algorithm (APSO-GA) Based Selective Harmonic Elimination in a Cascaded H-Bridge Multilevel Inverter," in IEEE Transactions on Industrial Electronics, vol. 69, no. 2, pp. 1477-1487, Feb. 2022, doi: 10.1109/TIE.2021.3060645.

Hybrid Electric Vehicle Design and Simulation Performance Using Isolated DC/DC Converter Based MFO Algorithm

Sekhar V[a,], Nithyamala K[b], K Renuka[c], Murali M[d]

[a,b,c] Department of EEE, Kuppam Engineering College, Kuppam and 517425,India
[d] Department of EEE, Sona College of Technology, Salem, India

Abstract: The accuracy and effectiveness of the vehicle's motor, DC-DC converter, controller, and electrical system are crucial for Hybrid Electric Vehicles (HEV). The study is given with a DC-DC converter architecture and motor based on the MFO control algorithm for smooth driving and quick charging of hybrid electric vehicles. The front-end AC-DC converter's output is translated by the suggested DC-DC converter architecture, which then provides a controlled supply to the EV propulsion battery. The suggested system consists of a drive train model created in MATLAB using the SIMULINK tool, a motor drive, a DC-DC converter, an inverter, and a drive train. A Simulink model with motor drive model was constructed and successfully simulated in order to operate and evaluate the performance of hybrid electric vehicles utilizing an MFO control method. The simulation findings are examined its performance of EV system and results clearly stated the hybrid car's fuel efficiency advantages over the conventional vehicle model.

Keywords: HEV, DC-DC, SIMULINK, DC-DC Converter, MFO

1. Introduction:

The use of electric vehicles is rising in emerging nations as a result of awareness of the significance of renewable energy and its application to automotive technology. Due to the depletion of renewable energy sources, the global market for electric vehicles is predicted to increase to 30% by 2030 [1]. Further infrastructure development for charging will be needed as the number of electric vehicles rises. The main components of the present charging infrastructure are direct current rapid charging stations, car chargers, and electric vehicle supply units (EVSEs) as shown in Fig.1. For the majority of automobiles, car chargers are the most typical source of power [2]. They are made up of AC-DC and DC-DC power conversion circuits for single stage application in two stage topologies [3] through [4]. The advantages of the include the option to choose when and where to charge, practically ubiquitous access to AC grid power, longer battery life [5], but the power levels are fairly low as a result of the converter's volume and weight limitations. As a result, the majority of EVs are utilized for overnight charging and aren't the ideal replacement for petrol stations or a way to allay EV users' range concern. AC-DC and DC- DC converters are used in DC rapid charging, also known as OFF board charging, and they are all mounted outside the car. Thus, high power levels may be built into the DC charging stations. Without extra hardware, the converter generates a voltage waveform with three levels. Without extra hardware, the converter generates a voltage waveform with three levels.

IOP Conf. Series: Materials Science and Engineering 1272 (2022) 012003 doi:10.1088/1757-899X/1272/1/012003

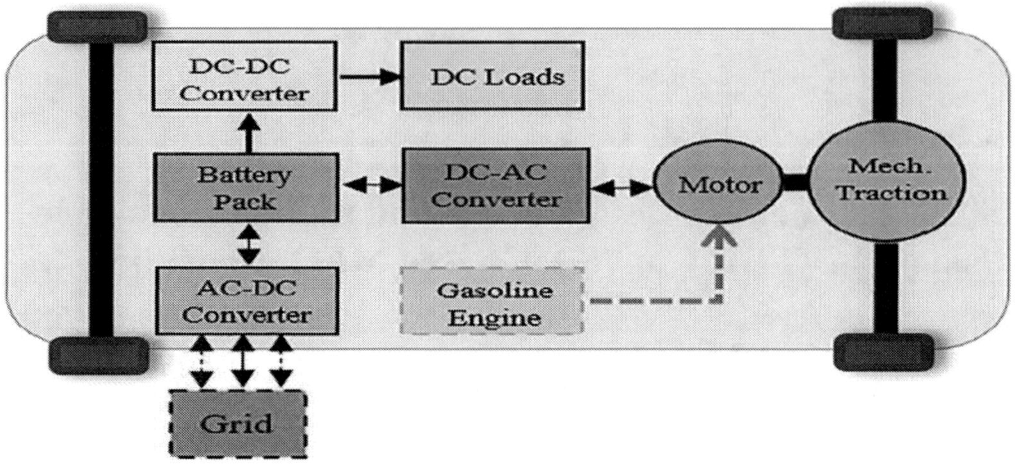

Figure 1. Electric Vehicle overview model

EVs has advantages such as choice of time and place for charging, almost universal availability of AC grid power supply, better battery life [6], the power levels are quite low due to restrictions on volume and weight of the converter. Thus, most EVs are used for overnight charging and aren`t exactly the best solution for replacing gas stations or alleviating range anxiety amongst EV consumers. DC fast charging or OFF board charging consists of ACDC and DCDC converters, all placed outside of the vehicle. Hence the DC charging stations can be designed for high power levels. The main architectural differences in DCFC are based on structures based on AC distribution grids or DC distribution grids [7]. AC distribution networks have the advantages of mature AC distribution and protection technologies and easily accessible components for such implementations. Despite these advantages, the DC grid architecture has several advantages over the AC grid and integration of renewable energy sources and energy storage, low cost, miniaturization, high efficiency. A typical DC distribution grid architecture consists of a line frequency ACDC converter and a DCDC converter for a regulated power supply to an EV battery. The proposed converter achieves high efficiency, but has a limited voltage conversion range and suffers from switching losses that limit high frequency operation. The two 3-level power converter units working together in parallel provide a multi-level, parallel structure that lessens the burden on the hardware. The suggested architecture has smooth switching operation and lessens current and voltage strains across the devices. Recent research on swarm intelligence approaches was driven by the desire to obtain more intelligent parameter estimates. The final value was likewise attained by MFO accelerated particle swarm optimizationmore quickly than by the other techniques [8-9]. Moths employ a constant angle with regard to the moon to fly during moderate moonlight conditions, which is a highly successful techniquefor moving in a straight line over vast distances. This is the primary source of inspiration for MFO.

2. Design of Converter Electric Vehicle:

Three different types of power sources' intricate mathematical models are introduced. The structure of the proposed DC-DC converter, shown in Fig. 2, includes any number of input and output levels. The actuator in this instance consists of two phases: modeling and controlling. The essential elements of this framework are a DC-DC converter, an inverter, and an MFO controller, together with an actuator linked to a battery and operating voltages. The detailed mathematical models of three kinds of power sources are introduced. The topology of the proposed DC-DC converter with arbitrary number of inputs and outputs levels and actuator comprises of two phases: modeling and controlling phases worked based on PWM signals as shown in Fig.3. This framework constitutes a DC-DC converter, inverter and MFO controller as main components with the modeling phase, the actuator associated with battery and operating voltages.

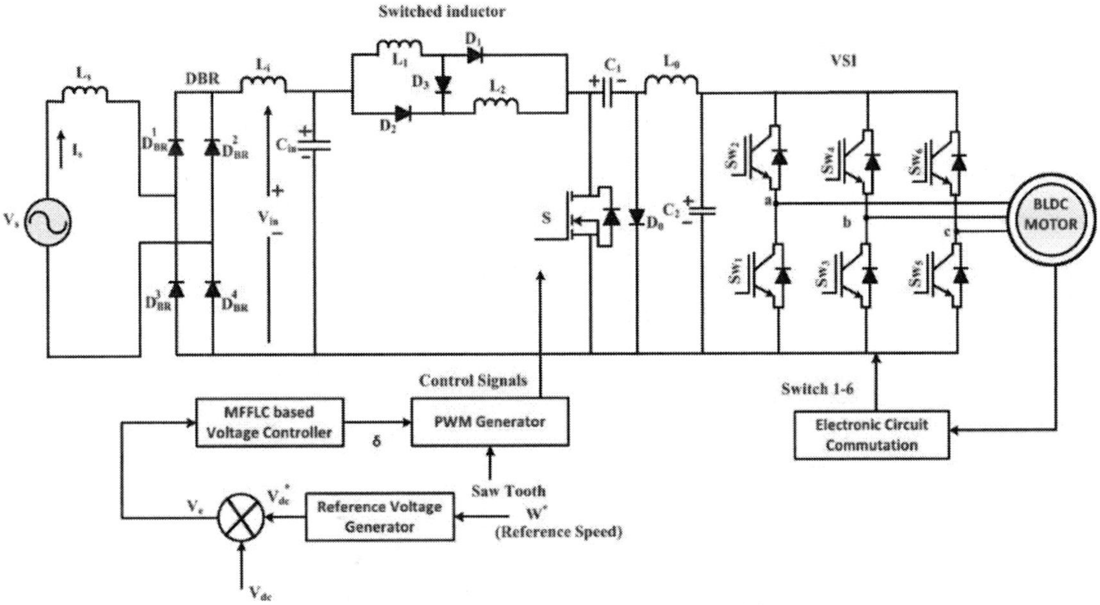

Figure 2. Schematic structure based control for motor

The speed estimator is used to estimate the feedback control loop mechanism in the actuator drive control system of electric vehicles. Equation is used to determine the speed error from actual and reference values, which is then transmitted to the voltage controller for speed control and error correction.

$$I= Kp*e(t)+Kt*e(t) \tag{1}$$

Based on gain parameter values stored from the output of the suggested controller algorithm, MFO strategy is implemented and updated. As a result, the best possible power value is regulated, and the controlled signal is sent to the PWM generator to regulate the performance of the actuator and inverter switching.

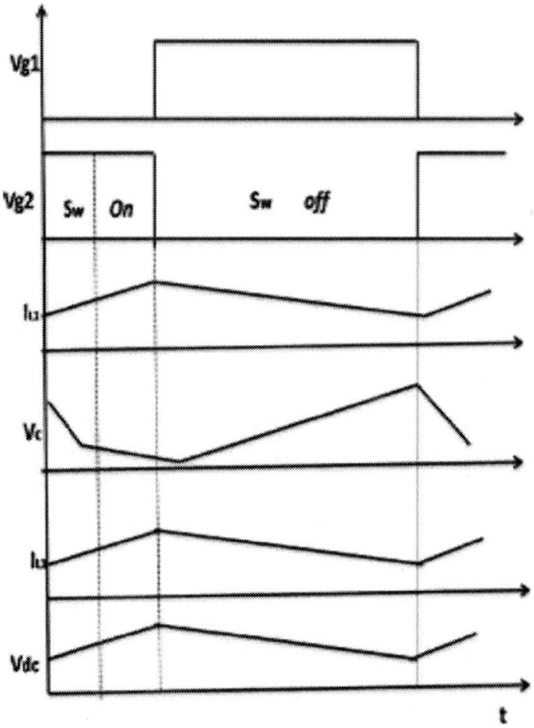

Figure 3. switching pulse of DC-DC converter

3. Proposed MFO strategy algorithm

Moths use a unique kind of nighttime navigation known as transverse orientation. They fly using a set angle in relation to the moon in this process, which is particularly useful for covering large distances in a straight path. Moths circle about a light source when it is nearby before ultimately converging toward it after a few adjustments. Each moth circles a flame and updates it if a better option is discovered. The following equation is used to update each moth's location with relation to a flame in order to quantitatively simulate this behavior:

$$Ls= S(Lm,Bk) \quad (2)$$

S- Spiral function m indicates moth and kth the flame optimization.

Considering the superior ability of the MFO algorithm in terms of local optima avoidance and convergence, this paper adopts it to solve multiple peaks and to track the global MPP of EV system and proposed with an algorithm and methodology to analysis is shown in Fig. 4.

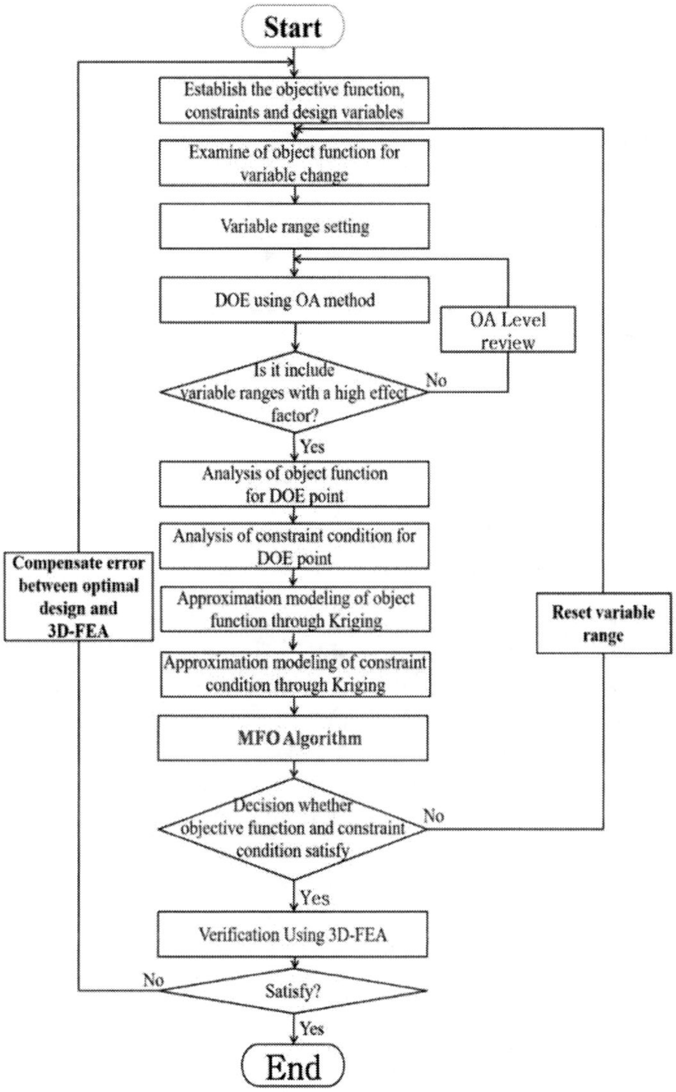

Figure. 4 Algorithms reorientation of MFO

4. Simulation result Analysis

MATLAB/Simulink software is used to implement a number of simulations. The simulation model of a PV MPPT system consists of five parts, including a boost convertor, a MPPT control module, and a linear load as shown in Fig. 5.

IOP Conf. Series: Materials Science and Engineering 1272 (2022) 012003 doi:10.1088/1757-899X/1272/1/012003

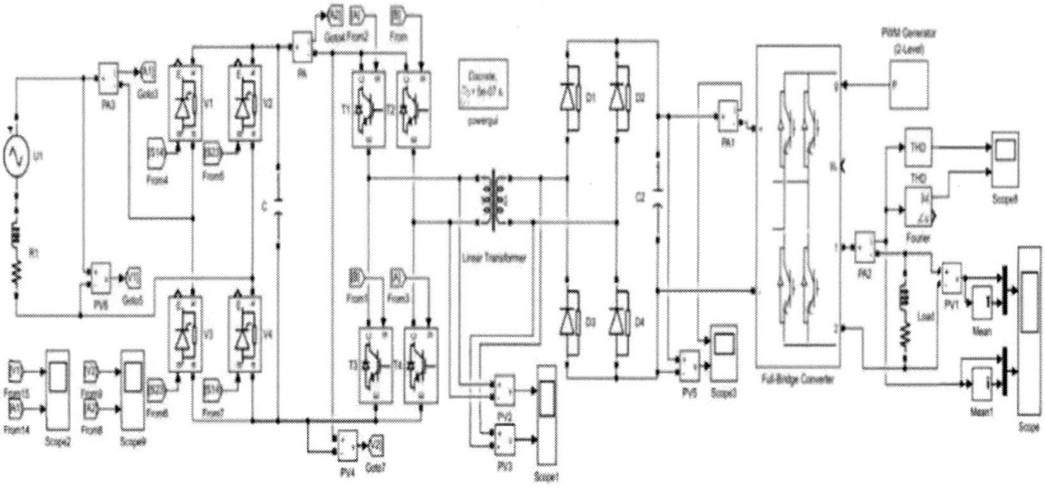

Fig.5 Simulation model in MATLAB

This work uses the MFO algorithm to solve numerous peaks and track the global MPP of an EV system due to its superior capability in terms of local optima avoidance and convergence. The selected MOSFET Frequency f = 50 kHz, C1 = 100 F, L = 0.5 mH, C2 = 100 F, and Rload =40 components are for the intended MPPT system. In order to run the simulation, an analogous model takes the place of the PV module. The tracking performances of the MFO are compared tovalidate the performance of the suggested method. Figures 6 and 7 demonstrate the effectiveness of MFO-based input battery voltage and PWM settings. The PSO and MFO control algorithms' output values are reported in the table. 1. Without any spikes, the peak value of the leakage inductor current is decreased within the specified limits. The current naturally decreases to zero, and the negative current indicates that an anti-parallel diode is conducting across the switch, turning off the ZCS. At lower power levels as well as throughout the cycle, soft-switching is successfully maintained.

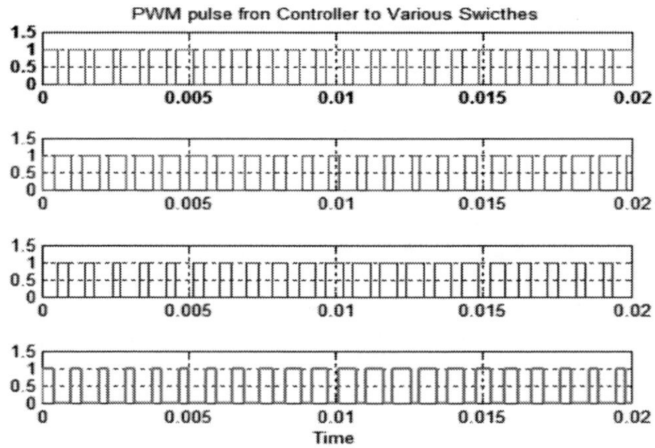

Fig.6 PWM pulse for converter for MFO Control algorithm

IOP Conf. Series: Materials Science and Engineering 1272 (2022) 012003 doi:10.1088/1757-899X/1272/1/012003

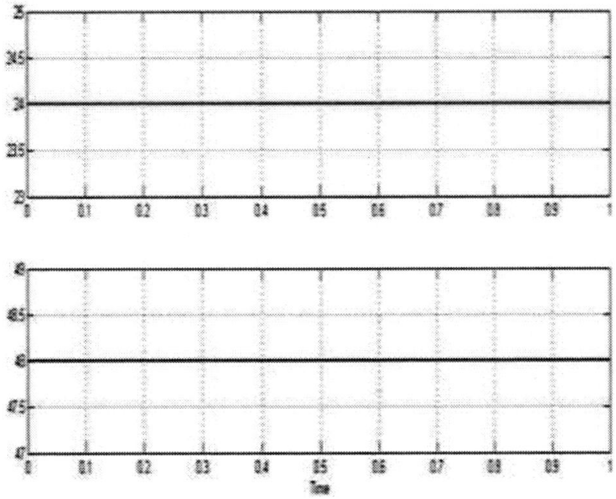

Fig.7 Output of DC/DC converter for MFO Control algorithm

Table 1. Performance of MFO and PSO Control in EV System

Controller Performance	Output Power watts	Speed RPM	Torque NM	Dc Link Voltage volts	Dc Link Current Amps	Battery Current Amps	Efficienc y %
MFO	120	522	2.75	58	2.82	7.25	92.42
PSO	108	514	1.25	41.5	3.12	7.5	81.9

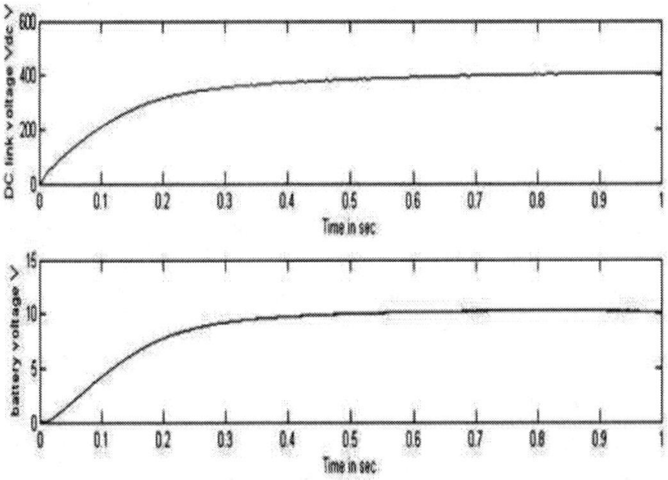

Fig.8 Output Analysis of Inverter to motor load

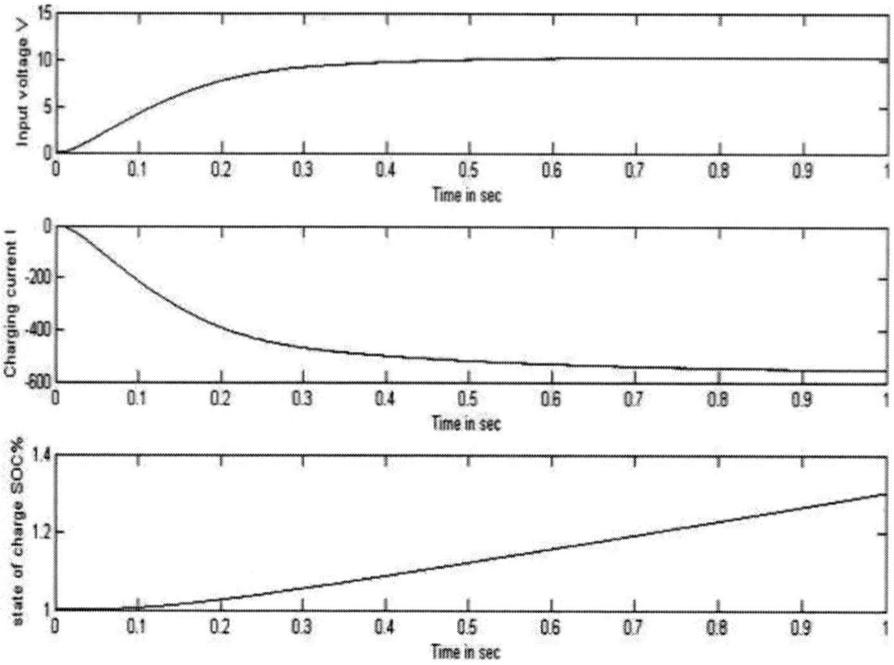

Fig. 9 Battery performance of MFO fed to EV System

The DC Link Voltage Characteristics are shown in for motoring and braking state of motor in battery performance during charging and discharging is shown in Fig.8 and 9. From the results it is inferred that the battery charging state at the voltage is boosting the voltage and maintains level 12 volts of battery capacity. The charging current get reduced once further the battery is charging its desired level. During the motor drive runs the E-vehicle at desired speed the battery get release state of charge level and SOC is maintained.

5. Conclusion

In this study, the dynamic behavior of the actuator-based EV was examined using the MFO control technique. The suggested approach was utilized to model and regulate the process, and the proposed strategy-based actuator was put into practice in the MATLAB/Simulink stage. During the modeling stage, the actuator was fed through the bidirectional converter and inverter devices. The voltage and current in the regulating part are managed using the suggested control approach. Three separate working scenarios—linear, non-linear, and speed and torque—were used to evaluate it. Utilizing already-in-use controllers like the PI controller and the MFO approach, the suggested controller was put into action. The generation of PWM signals also lessens torque ripples.

References

[1] M. Mao, L. Zhang, H. Huang, B. Chong, and L. J. J. o. C. P. Zhou, "Maximum power exploitation for grid-connected PV system under fast-varying solar irradiation levels with modified salp swarm algorithm," vol. 268, p. 122158, 2020.
[2] M. S. Bouakkaz, A. Boukadoum, O. Boudebbouz, N. Fergani, N. Boutasseta, I. Attoui, *et al.*, "Dynamic performance evaluation and improvement of PV energy generation

systems using Moth Flame Optimization with combined fractional order PID and sliding mode controller," vol. 199, pp. 411-424, 2020.

[3] A. Fathy, H. Rezk, and D. J. S. E. Yousri, "A robust global MPPT to mitigate partial shading of triple-junction solar cell-based system using manta ray foraging optimization algorithm," vol. 207, pp. 305-316, 2020.

[4] S. Arab Ansari, J. S. Moghani, M. J. I. J. o. C. T. Mohammadi, and Applications, "Analysis and implementation of a new zero current switching flyback inverter," vol. 47, pp. 103-132, 2019.

[5] Arulprakash.A, Arulmozhiyal.R and Murali.M, "FPGA controlled Single Phase Inverter in Solar Power Generation for a Domestic Load", International Journal of Applied Engineering-IJAER,ISSN 1087-1090,VOl.10,Issue 10,pp 9373-9377.

[6] W. M. Zhang, Y. X. J. I. J. o. S. Zhang, and P. Modelling, "A novel method of reactive voltage optimisation for photovoltaic systems," vol. 15, pp. 145-154, 2020.

[7] Murali, M&Arulmozhiyal, R, 'Investigation on Solar PVgeneration and design of switched reluctance motor for SmartAgriculture actuation system', Brazilian Archives of Biology andTechnology, Tecpar Publisher, ISSN : 1678-4324, vol.61, Oct. 2018,pp. 1-1.

[8] Murali, M & Arulmozhiyal, R, 'High Gain Dc-Dc ConverterWith Reduced Voltage Ripples Application For Smart Pv FarmAgriculture System', Ecology Environment and Conservation,EM International, ISSN: 0971-765X,vol.23,pp. 266-27,2017.

[9] J. Wang, S. Gao, Y. Sun, Z. Ji, L. Cheng, L. Li, *et al.*, "Single Phase Bidirectional H6 Rectifier/Inverter," *IEEE Transactions on Power Electronics,* vol. 34, pp. 10710-10719, 2019.

[10] Q. Wang, W. Qin, X. Han, P. Wang, L. Wang and Y. Zhang, "Robustness evaluation for harmonic suppression of LCL-type converter based on converter-side current feedback strategy under weak and distorted grid," in CPSS Transactions on Power Electronics and Applications, vol. 6, no. 2, pp. 166-177, June 2021, doi: 10.24295/CPSSTPEA.2021.00015.

[11] R Arulmozhiyal, P Thirumalini, M Murali, Implementation of Multilevel Z-Source Inverter using Solar Photovoltaic Energy system for Hybrid Electric Vehicles, - Computation and Communication Technologies, 2016

Design of low cost IoT enabled greenhouse control system for precision agricultural research application

Khushi Singh[1,2,5] and Malaya Kumar Hota[1,3,4]
[1] Department of Communication Engineering, School of Electronics Engineering, Vellore Institute of Technology, Vellore 632014, Tamil Nadu, India.

[2] khushi35b@gmail.com
[3] malaykumar.h@vit.ac.in
[4] ORCID_ID: 0000-0002-3669-1611
[5] ORCID_ID: 0000-0002-8858-7330

Abstract. Greenhouse is climate controlled transparent chamber made of materials like polycarbonate, in which crops are grown under controlled climatic conditions. This research paper aims to present a design of low-cost IoT interfaced Greenhouse Control Unit, where microclimate is controlled using various sensors and actuators to maintain desired environmental conditions as per crop need. It is used in precision agricultural research to study the impact of environmental parameters like humidity, temperature, ambient light and soil moisture on crop yield and its nutritional components. Temperature, humidity, ambient light and soil moisture like parameters are fetched by deploying suitable sensors at suitable locations in the chamber, and based on predefined threshold values of these parameters switching of cooling, heating system, light, fogger, dehumidifiers, and irrigation system are controlled using embedded controller device and desired climatic condition is maintained inside the Greenhouse chamber for the required duration. Such control units are required for agricultural research to study the impact of environmental parameters on crop yield and on the nutritional contents of the crop. It is also used in producing medicinal and herbal plants in the region which climate is very harsh for these plants. Authors have developed a low-cost cloud interfaced standalone embedded control unit around ARM Cortex M3 CPU, various sensors for controlling greenhouse actuators based on pre-defined threshold values of environmental parameters.

Keywords: Greenhouse, Precision farming, Digital Agriculture, Climate control unit, Smart agriculture.

1. Introduction

The development of Greenhouse based control units is in huge demand due to the increase in agricultural research. Many times, agricultural framing needs controlled climate conditions for more crop yield [1]. IoT based greenhouse control units send sensors data periodically to the internet cloud. Sensors data like ambient light, water flow pressure, ambient temperature, humidity, soil temperature, and humidity are collected in the greenhouse chamber in real-time and the cloud server is updated for remote monitoring applications [2]. Collecting sensor data through a wired network is not feasible in some setups, wireless sensor network (WSN) based deployment helps in such large greenhouse chambers. Each WSN nodes carry sensors, which may act as an end device to fetch the sensor data and send it to the gateway unit. ZigBee based WSN motes are deployed in the chambers in suitable positions to record environmental parameters more precisely with more coverage. ZigBee technology allows wireless networking of up to

255 nodes in a single WSN network [3-8]. Researchers are also using commercially available embedded control units along with simulation software to study the behaviour of climate control appliances with varying field conditions. The main emphasis is on more accuracy in maintaining desired ambient conditions inside the chamber in addition to optimizing the uses of electrical energy and irrigation water [9]. IoT-based greenhouse provides ease of remote monitoring of the greenhouse chamber through mobile applications or web-based monitoring [10-14]. Due to the enhanced computing power of the latest microcontroller, embedded control units are a new focus area for the researchers to develop compact yet efficient and accurate greenhouse control units for agricultural research applications. [15].

Farming lands are shrinking day by day due to growing urbanization and traditional farming is being taken over by aquaponics, vertical farming and terrace soilless farming. These smart farming technologies are mostly enabled by Greenhouse chambers by providing desired controlled climate for specific high value crops, so there is huge demand of low-cost greenhouse control units which are affordable to thefarmers. Existing greenhouse solutions are very costly due to lack of research in this domain. This research paper offers design and implementation of a low-cost Greenhouse climate controller unit, whichis developed around SAM3X8E ARM Cortex-M3 CPU with Wi-Fi development board which has inbuiltESP8266 IoT chip. Various sensors like humidity and temperature sensor (DHT11), ambient light sensor and sensor for soil moisture are interfaced with the microcontroller. Electrical loads like cooling systems, light, heater, foggers, and dehumidifiers are connected to relays which are controlled by microcontroller in closed-loop combination with sensors and predefined threshold values of environmental parameters as per crop need.

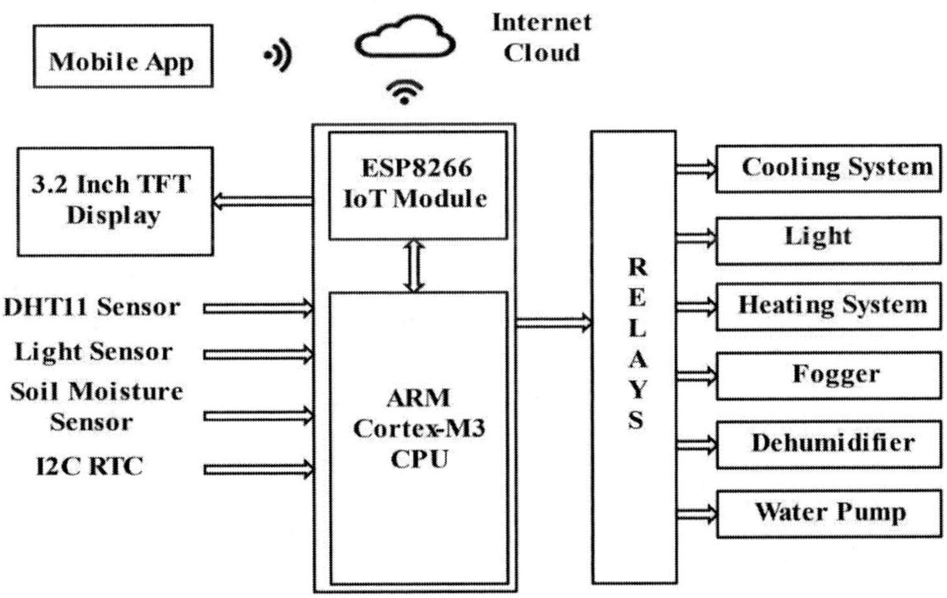

Figure 1. System Architecture

IOP Conf. Series: Materials Science and Engineering 1272 (2022) 012004 doi:10.1088/1757-899X/1272/1/012004

2. Literature Review

Siddiqui et al. (2017) conducted experiments by designing greenhouse control unit around Arduino mega 2560, ethernet shield, sensors and LAN network for remote monitoring of data using web-based application software. Tomato crop is used for experimenting with controlled climatic condition inside the greenhouse chamber. Azhar et al.(2019) conducted experiment inside transparent chamber of polyethylene sheet by using temperature and humidity sensor DHT22, BH-1750 ambient light sensor, SEN-13322 soil moisture sensor along with Arduino mega development board to control actuators for maintain desired environmental condition as per crop need. T. Bhuvaneswari and Yao (2014) implemented greenhouse automation using LM35 analog temperature sensor, rain sensor, LCD, servo motor, fan and microcontroller and Arduino IDE. DC loads are controlled for demonstration of proof of concept. Achour et al. (2018) and Huang et al.(2018) experimented with high speed computing microcontroller to fetch the sensor data and real-time actuator control is achieved in greenhouse without any lag time. Pahuja et al. (2013) found that application of wireless sensor network in greenhouse setting eliminates mesh of wired network for sensing the climatic condition. Deploying wired sensor network for fetching sensor data is a better option, if size of the chamber is bigger. Zigbee based motes are deployed for the purpose of sensing microclimate inside the enclosed chamber. Singh Khushi and Jindal SK (2022) found that closed loop automation yield efficient and optimized controlling of environmental parameters with use of cost-effective sensors and electrical actuators.

Greenhouse automation is very active field for researchers but still greenhouse control units are very costly and not comprehensive. This research work is focused to fill this research gap in greenhouse automation by offering low-cost greenhouse control unit with added advantages of IoT cloud interface to enable remote monitoring of the climatic condition of the greenhouse chamber.

3. Methodology

This research work is about design and implementation of hardware prototype of control unit of greenhouse chamber. Methodology used in this research work is based on closed loop control system, in which electrical loads are controlled based on real-time sensor`s inputs. Actual greenhouse chamber used in agricultural research application are of size starting from approximately 150 square feet and heavy air conditioners, foggers and lights are used to work in closed loop setup with sensors for management of micro climate inside the chamber. For proof of concept (POC), authors have used DC load like DC fan, LED light Panels instead of actual AC loads along with various sensors for sensing micro climate inside the chamber and controlling the switching of actuators connected to DC loads for maintaining desired climatic condition inside the chamber as per crop need.

4. System Architecture

The embedded system is developed using a Cortex-M3 microcontroller, ESP8266 Wi-Fi module, Sensors, Relays, I2C RTC module, and a 3.5inch TFT display screen. Sensors are used for fetching temperature, humidity, ambient light, and soil moisture data from the greenhouse chamber at a predefined sampling rate, and electrical loads are switched on/off to maintain the desired climatic condition inside chamber. It is closed-loop automated process to control electrical appliances as per crop need. sensor data is uploaded to internet Cloud periodically for remote monitoring using web-server or mobile application.

The authors used Atmel's sam3x8e arm cortex-m3 CPU; it is a 32-bit microcontroller from the arm core. It has 54 digital input-output pins with 512 K flash memory, suitable for having a greater number of interfaces which is a key requirement of the system design. A 3.5-inch TFT touch screen is used in design to display real-time parameters like current temperature, humidity, light and soil moisture of the greenhouse chamber. 3.5inch TFT LCD color screen display with 480 x 320 resolutions is used, which is interfaced with the microcontroller. The block diagram of the control unit is depicted in figure 1. The

IOP Conf. Series: Materials Science and Engineering 1272 (2022) 012004 doi:10.1088/1757-899X/1272/1/012004

microcontroller is interfaced with the ESP8266 module for providing IoT capabilities, which works well with AT commands. The control unit pushes real-time data to the internet cloud using a Wi-Fi module. Channel is created on the Thingspeak cloud server to store and display various greenhouse parameters. Various graphical interfaces are available in the Thingspeak cloud server for statistical analysis of the data and to visualize the real-time data with recorded timestamps. Every crop needs different and specific ranges of environmental conditions for healthy crop growth and optimized crop yield. These parameters vary with the growth stage of the crops. A lookup table is created in microcontroller for these parameters for the specific crop on which experiments are to be done in the greenhouse chamber. This lookup table is pre-programmed in the microcontroller. When the control unit receives environmental parameters values from deployed sensors, then it compares these values with the lookup table and controls heating, cooling, irrigation unit, or light units accordingly.

Figure 2. Picture of Greenhouse Controller Unit

5. Sensor and load interfacing

LM 393 optical photosensitive LDR is used as an analog light sensor, which is interfaced with the analog port of the microcontroller. This sensor is available in digital sensor format also, but it is not suitable in greenhouse application, as a continuous value of light in Lux is required to be monitored in the greenhouse chamber and to actuate additional light if required. LM393 Digital light sensor gives output 1 or 0 depending on the day/night mode only, which is not suitable for the application. DHT 11 is a digital sensor for reading temperature and humidity and it is interfaced with the microcontroller. It works on 3.5-5 V with a 20-95% range for humidity and 0-to-50-degree Celsius for detection of temperature, which is suitable for greenhouse application, as it falls within most of the crop need.

More crops per drop are the key in precision agriculture, so a soil moisture sensing-based drip irrigation system is implemented. The soil moisture sensor is interfaced at the analog port of the microcontroller, for sensing water contents in the soil. If soil is getting dry and if irrigation is required as per the crop growth model then the system will switch on the water pump. I2C RTC module DS3231 is interfaced with the I2C port of the microcontroller for timekeeping. Depending on the number of electrical loads, a six-port relay board is used in the design, to control the switching of electrical appliances. The number of relays can be increased if a greater number of electrical devices are required in the application. On-board optocouplers, isolate the microcontroller pins to save them from the back EMF of the electromagnetic relays. Relays with 250V/ 16Amp AC rating are suitable for driving heavy electrical loads of the greenhouse. Foggers are used to increase humidity levels if it is required for the crop growth model. Similarly, a dehumidifier is used to reduce the humidity of the greenhouse chamber. The exact specification of electrical loads is designed as per the size of the greenhouse chamber. An irrigation pump with drip irrigation facilities is installed in the chamber. Soil moisture sensor fetches soil moisture which water contents in the soil and switches ON/OFF irrigation pump as per crop need.

6. Experimental setup and Results

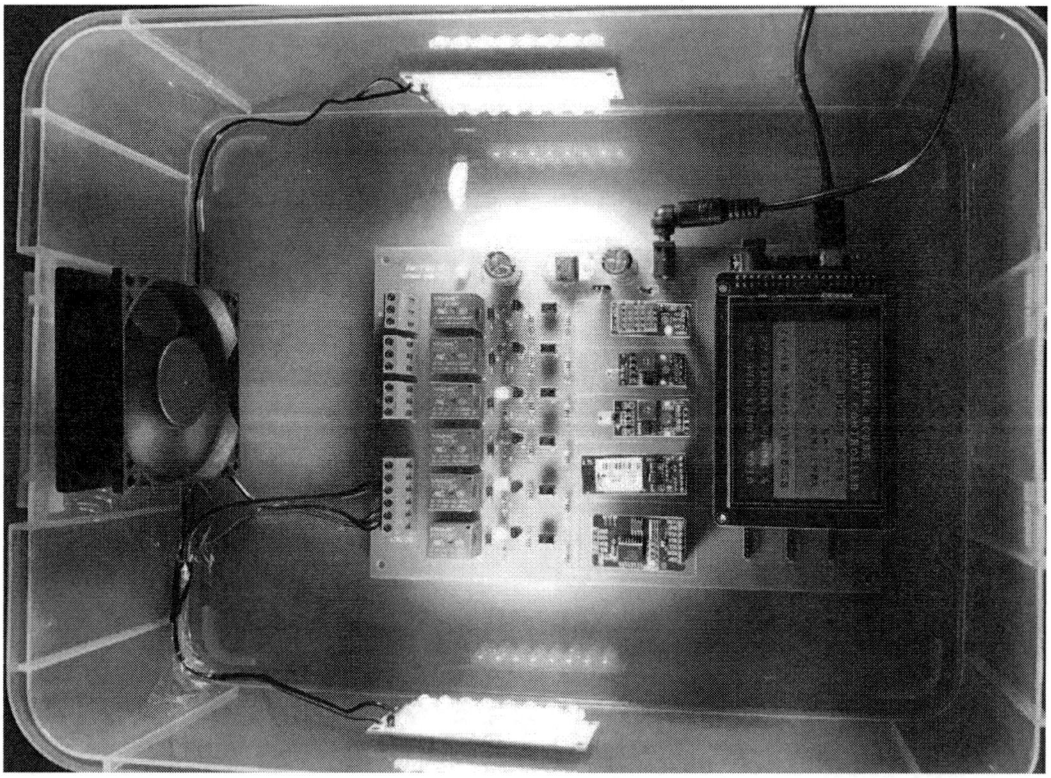

Figure 3. Experimental setup

A working prototype is developed and all required sensors and actuators are installed inside a transparent cabinet to do experiments related to closed-loop switching. For POC, DC pumps, DC LED array Lights, DC fans are used at the place of actual electrical loads. The prototype worked perfectly within design limits with desired performances. Picture of the developed prototype and experimental setup is shown in Figures 2 and 3. The greenhouse controller fetched data from various sensors periodically from the sensor module and based on predefined values of these parameters and crop growthmodel, the controller controls the switching of electrical appliances to maintain microclimate in the greenhouse as per the desired specification. For example, if DHT11 output is 25 degree Celsius, and weneed to maintain between 28-35 degree Celsius of temperature as per crop need; the heating appliance will be switched on by the controller unit and it will be switched off when the chamber reaches desiredclimate conditions.

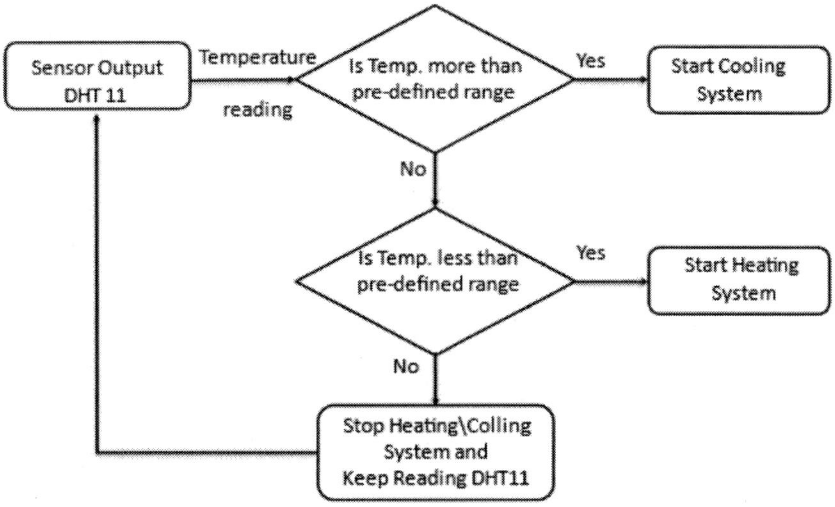

Figure 4. Temperature control logic flow diagram

Control units fetches real-time temperature data from DHT11 sensor, compares it with lookup table for pre-defined temperature for the specific crop and follows the logic flow depicted in figure 4 for switching control of cooling\heating units and desired range of temperature is maintained in the greenhouse chamber constantly. Similarly switching of dehumidifier, fogger and lighting are controlled in closed loop operation based on inputs of corresponding sensors and pre-deigned threshold values of these parameters as shown in figure 5-6-7. These threshold values are normally varying with crop growth stages and are pre- programmed in the control unit. Generally, such control units are customized for specific crop and cropcycles.

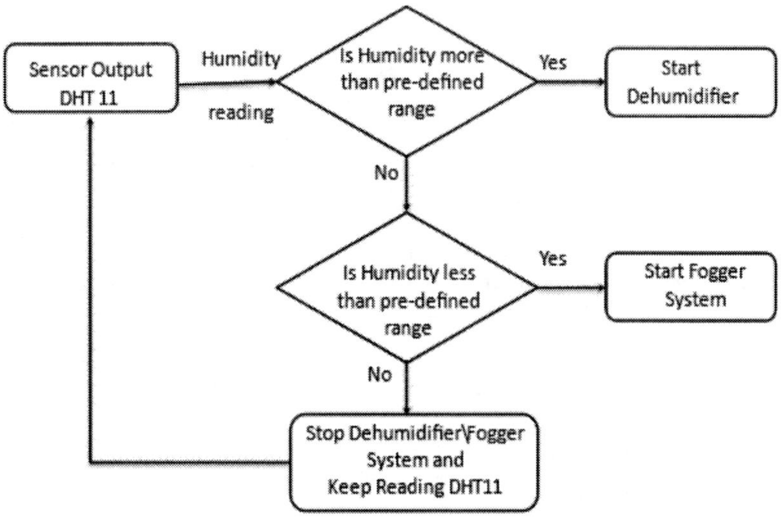

Figure 5. Humidity control logic flow diagram

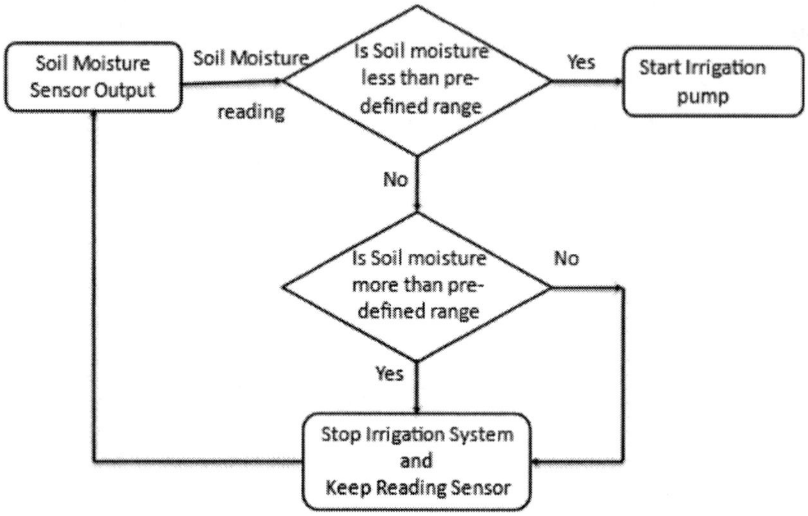

Figure 6. Irrigation control logic flow diagram

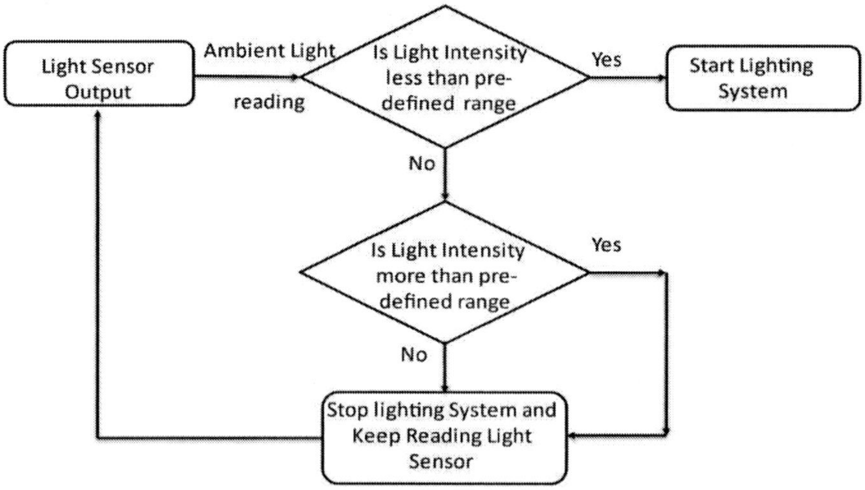

Figure 7. Lighting control logic flow diagram

More crops per drop is a key in greenhouse irrigation system to save water wastage and the solution is drip irrigation system as shown in figure 8. In drip irrigation system many drip nozzles are attached to water line and water comes out in form of drops which is absorbed from the soil for good health of the crop. Flooding of water is avoided using such irrigation system. Soil moisture is sensed by soil moisture sensor and if it drops below predefined threshold level then control unit start pumping water in drip irrigation as per logic flow in figure 6.

Figure 8. Picture of Drip Irrigation System

All the sensor data is uploaded to internet cloud periodically, so that Greenhouse operation can be monitored in real time even from distant locations using Mobile app or visiting internet Cloud server web pages. Picture of data representation at Thingspeak cloud is shown in figure 9.

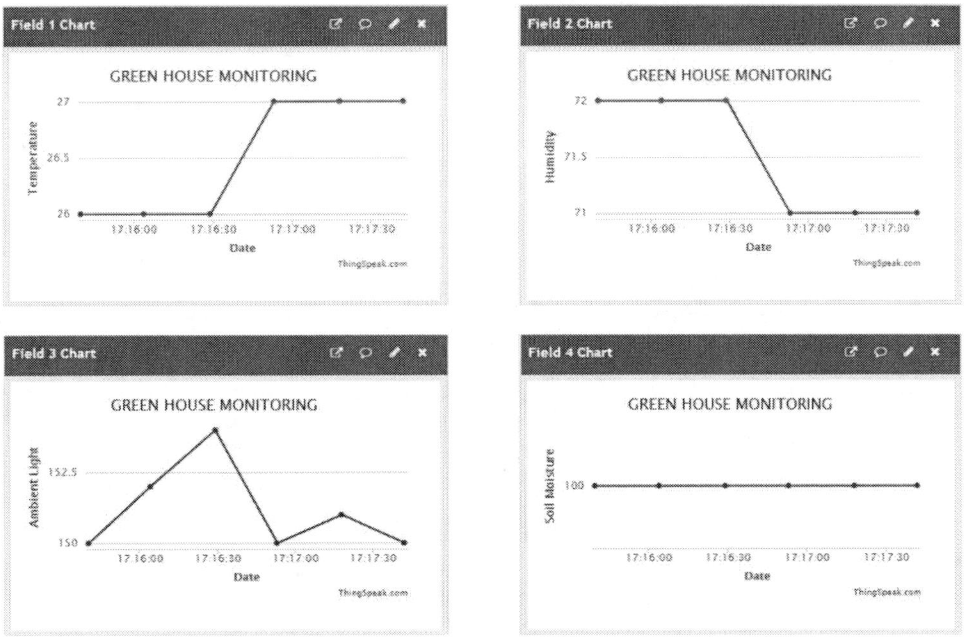

Figure 9. Greenhouse parameters data at Thingspeak cloud

7. Conclusion

Developed prototype does not include any costly component, so it is realized in very low-cost budget, without compromising precision of the switching control. Although full load switching of greenhouse involves heavy electrical loads like cooling and heating system including high power air conditioners, which is not tested on developed control units but control units is capable of driving such loads. Prototype can be further be upgraded to drive even more actuators as there are many unused pins left with microcontroller. IoT enabled cloud interfacing of the control units provides easy access for remote monitoring of greenhouse parameters using mobile application.

References

[1] Revathi S, Radhakrishnan T K and Sivakumaran N 2017 Climate control in greenhouse using intelligent control algorithms *American Control Conference (ACC)*, pp. 887-892, doi: 10.23919/ACC.2017.7963065.

[2] Zhao R, Ding Y, Ma S and Wang M 2019 Design of Intelligent Greenhouse Control System Based on Internet of Things *2nd International Conference on Information Systems and Computer Aided Education (ICISCAE)*, pp. 117-121, doi: 10.1109/ICISCAE48440.2019.221600.

[3] Lijun Liu and Yang Zhang 2017 Design of greenhouse environment monitoring system based on Wireless Sensor Network *3rd International Conference on Control, Automation and Robotics (ICCAR)*, pp. 463-466, doi: 10.1109/ICCAR.2017.7942739.

[4] Mekki M, Abdallah O, Amin M B M, Eltayeb M, Abdalfatah T and Babiker A 2015 Greenhouse monitoring and control system based on wireless Sensor Network *International Conference.on Computing, Control, Networking, Electronics and Embedded Systems Engineering (ICCNEEE)*, pp. 384-7, doi: 10.1109/ICCNEEE.2015.7381396.

[5] Kang B, Park D, Cho K, Shin C, Cho S and Park J 2008 A Study on the Greenhouse Auto Control System Based on Wireless Sensor Network *International Conference on Security Technology* pp. 41-4, doi: 10.1109/SecTech.2008.60.

[6] Pahuja R, Verma H K and Uddin M 2013 A Wireless Sensor Network for Greenhouse Climate Control in *IEEE Pervasive Computing*, vol. **12**, no. 2, pp. 49-58, doi: 10.1109/MPRV.2013.26.

[7] Li L, Yang S, Wang L and Gao X 2011 The greenhouse environment monitoring system based on wireless sensor network technology *IEEE International Conference on Cyber Technology in Automation, Control, and Intelligent Systems*, pp. 265-8, doi: 10.1109/CYBER.2011.6011806.

[8] Hou J and Gao Y 2010 Greenhouse Wireless Sensor Network Monitoring System Design Based on Solar Energy *International Conference on Challenges in Environmental Science and Computer Engineering*, pp. 475-9, doi: 10.1109/CESCE.2010.274.

[9] Alsharakhat A, Almoalem M, AlHawaj N, Papageorgiou C and Sadriwala A 2017 Temperature Control of a Greenhouse System Using Embedded Systems Technology *9th IEEE-GCC Conference and Exhibition (GCCCE)*, pp. 1-9, doi: 10.1109/IEEEGCC.2017.8447937.

[10] Vimal P V and Shivaprakasha K S 2017 IOT based greenhouse environment monitoring and controlling system using Arduino platform *International Conference on Intelligent Computing, Instrumentation and Control Technologies (ICICICT)*, pp. 1514-9, doi: 10.1109/ICICICT1.2017.8342795.

[11] Aafreen R, Neyaz S Y, Shamim R and Beg M S 2019 An IoT based system for telemetry and control of Greenhouse environment *International Conference on Electrical, Electronics and Computer Engineering (UPCON)*, pp. 1-6, doi: 10.1109/UPCON47278.2019.8980258.

[12] Nagesh Kumar D N 2015 ARM based remote monitoring and Control System for environmental parameters in greenhouse *IEEE International Conference on Electrical, Computer and Communication Technologies (ICECCT)*, pp. 1-6, doi: 10.1109/ICECCT.2015.7226117.

[13] Achour Y, Mernissi N El and Zejli D 2018 Design and Implementation of Greenhouse Remote Monitor System *IEEE 5th International Congress on Information Science and Technology (CiSt)*, pp. 1-6, doi: 10.1109/CIST.2018.8596660.

[14] Huang H, Bian H, Zhu S and Jin J 2011 A Greenhouse Remote Monitoring System Based on GSM *International Conference on Information Management, Innovation Management and Industrial Engineering*, pp. 357-60, doi: 10.1109/ICIII.2011.231.

[15] Singh S A J, Raviram P and ShanthoshKumar K 2014 Embedded based Green House Monitoring system using PIC microcontroller *International Conference on Green Computing Communication and Electrical Engineering (ICGCCEE)*, pp. 1-4, doi: 10.1109/ICGCCEE.2014.6922290.

[16] Siddiqui M F, Rehman A Ur Khan, Kanwal N, Mehdi H, Noor A and Khan M A 2017 Automation

and monitoring of greenhouse, *International Conference on Information and Communication Technologies (ICICT)*, pp. 197-201, doi: 10.1109/ICICT.2017.8320190
[17] Azhar M R A , Hamid M, Irfan M H, Awais M, Khan U S and Zeb A 2019 Automated greenhouse system, *2nd International Conference on Communication, Computing and Digital systems (C-CODE)*, pp. 215-219, doi: 10.1109/C-CODE.2019.8681013.
[18] Bhuvaneswari T and Yao JTH 2014 Automated greenhouse *IEEE International Symposium on Robotics and Manufacturing Automation (ROMA)*, pp. 194-199, doi: 10.1109/ROMA.2014.7295887.
[19] Singh Khushi and Jindal S K 2022 *International Conference on Multidisciplinary Aspects of Materials in Engineering (IC-MAME 2021) 8th-9th October 2021, UIET, Panjab University, Chandigarh* IOP Conf. Ser.: Mater. Sci. Eng. 1225 012058

IOP Conf. Series: Materials Science and Engineering 1272 (2022) 012005 doi:10.1088/1757-899X/1272/1/012005

Design of SIW based Frequency Reconfigurable Antenna for K/Ku band applications

Ankit Sharma[1], Sneha Prasad[1], Sourav Mishra[1], Rohan Pathak[1]

[1]Department of Electronics and Communication Engineering, Galgotias College of Engineering and Technology, Greater Noida, Uttar Pradesh – 201308, India.

Abstract. A frequency reconfigurable antenna based on substrate integrated waveguide (SIW) technology is proposed in this paper for K/Ku band applications. In most of the previous reconfigurable structures terminal SIW was used but here in this proposed structure, terminal as well as the blind vias is being implanted to achieve maximum reconfigurability and better efficiency of the miniature antenna. The antenna parameters are analysed thoroughly without diode as well as with diode. According to the measurements, the suggested antenna has an impedance bandwidth of 12 to 12.8 GHz in off condition and in addition to this from 22 to 22.8 GHz in ON condition with a peak gain of 7.5 dB. The radiation efficiency of proposed work is 90%.

Keywords. Frequency reconfigurable, SIW, microstrip feedline, satellite communication, terrestrial microwave communication.

1. Introduction

Since the introduction of new communication technologies in the 1990s, the demand for high-performance millimeter-wave systems has grown. These must be efficient, low-cost, portable, and high-frequency compatible. Unfortunately, because of the large insertion and radiation losses at these frequencies, the well-known microstrip and coplanar lines technologies cannot be employed above 10 GHz. The rectangular waveguide topology can solve these problems since it is highly resistant to radiation losses and has low insertion losses. Rectangular waveguides, on the other hand, are incompatible with the downsizing demanded by modern applications in their traditional form.

SIW technology emerges after a decade as mature technology for a variety of microwave and millimeter wave applications. In high frequency applications, microstrip devices are not efficient, microstrip device production demands extremely tight tolerances due to the short wavelengths at high frequencies. Waveguide devices are favoured at high frequencies, but their manufacture is difficult. As a result, a new concept called substrate integrated waveguide evolved. The transition between microstrip and dielectric-filled waveguide is known as SIW (DFW). Vias for the waveguide's side walls turn the dielectric filled waveguide to a substrate integrated waveguide (SIW).

There has always been a demand for smaller, lower weight antenna systems that can accomplish selectivity in frequency, bandwidth, polarisation, and gain in satellite and airborne communication. A reconfigurable antenna is a phenomenon with similar features that allows it to operate in many frequency bands.

To achieve reconfigurability in the SIW structure for various applications, several techniques have been proposed. One such structure is a cavity-backed modified dumbbell-shaped slot antenna based on substrate integrated waveguide (SIW) technology , it introduces several new resonant points and the overall gain is also improved[1]. The use of a slow-wave structure to construct a small SIW cavity-backed circular-polarized antenna is presented in PCB technology, allowing a 47 percent decrease in the antenna cavity compared to its conventional SIW antenna counterpart [2].

A small substrate integrated waveguide (SIW) antenna array with four SIW cavities constructed on a single sheet of substrate is presented for operation at 28 and 38 GHz. A rhombic slot and a triangular split-ring slot are implemented in each cavity. The mode bifurcation of the rhombic and triangular-split-ring slots produces resonance in the lower frequency range, but the higher resonant frequency may be adjusted only by adjusting the triangular-split-ring slot [3]. The paper [4] presents two-element multiple input multiple output (MIMO) reconfigurable antenna that can be switched among 600 MHz, 1.8, 2.4, 3.5, and 5.5 GHz bands. The reconfigurability of this antenna is not better than the other existing literature inspite of having a better ECC and DG values of the simulated antenna. Two single feed low-profile cavity-backed slot antennas and an antenna array for circular polarisation (CP) applications are introduced with additional refinements [5]. In the SIW cavity, a spoon-shaped slot structure was employed to stimulate CP radiation. However, there is no knowledge regarding how to apply and expand this structure to array applications for wireless or satellite communication systems.The paper [6] discusses techniques used to reconfigure SIW antennas with regard to resonant frequency including the latest improvements in the design.It was found that the cognitive radio is an emerging technology which needs further researches especially in frequency reconfigurable antenna for cognitive radio as stated in the studies.

In this paper, a cavity-backed triangular complementary split ring slot (TCSRS) antenna based on substrate integrated waveguide (SIW) [7] and implemented at 28 GHz and 45 GHz, although, the proposed antenna shows poor performance at 28 GHz. Similarly, in the paper [7]-[8],the frequency ranges of 28 GHz and 38 GHz, a dual band substrate integrated waveguide (SIW) is suggested. At 28 GHz, the main beam has moved somewhat in the H-plane, and the observed and simulated sidelobe levels at both frequencies do not match precisely. In [9], Frequency reconfigurable substrate integrated waveguide (SIW) cavity F-shaped slot antenna is ddesigned to switch between four separate narrowbands ranging from 3.172 GHz to 3.606 GHz. The highest efficiency recorded is about -3.5 db and need further improvements.

In the further research, a rectangular-ring slot and a lengthy slit engraved on the upper surface of this antenna's SIW resonant cavity structure. The patch is attached to the upper surface using many PIN diodes. A simpler bias arrangement controls the pin diodes. The S band lower frequencies mode has an 18-MHz bandwidth, while the C-band high frequency mode does have a 322-MHz bandwidth. When using multiple PIN diodes, additional bias structures are required, and bandwidth must be increased. Whereas, using several PIN diodes requires extra bias structures and bandwidth should be increased [10]. One short-circuited varactor diode per patched element is used to tune the frequency [11].The 45-degree linear polarization (LP) is achieved by providing the very same phase to all patches. Polarizability is only available in the frequency band of 3.6 GHz to 4 GHz.

The structures [11]-[12] are almost similar to each other. In [12], also four pair of varactor diodes is symmetrically placed along the slot. Resonant frequencies of the two modes can be simultaneously changed by varying the capacitance values of the varactor diodes with single DC voltage. The efficiency rises from 47% to about 61% only, as the operating frequency increases from 1.92 to 2.51GHz.The antenna is made up of four rectangular patch antenna elements and has a planar shape. The dual hexagonal shaped defective ground structure is placed into the ground plane to reduce antenna size. It is suitable for hand - held devices due to its small size and dual-band frequency tunable characteristics. The wide frequency reconfigurable adjustment range should be expanded from 1.3 to 2.6 GHz [13].

IOP Conf. Series: Materials Science and Engineering 1272 (2022) 012005 doi:10.1088/1757-899X/1272/1/012005

A frequency reconfigurable antenna based on SIW with and without PIN diodes is proposed in this paper. Square bracket shaped patches attached with the microstrip feed line helps in achieving S11 below -10dB. The PIN diodes help in achieving reconfigurability of the antenna. In addition, with the terminal SIW's, additional 5x5 grid of blind vias is also inserted in the patch at the back side of proposed structure. For reconfigurability, five PIN diodes are used on both sides of antenna. There can be further research work in this novel structure and the reconfigurability could be extended to other bands of operation as well by using appropriate lateral and blind SIWs as per requirement. The sections mentioned below give a detailed description of design parameters and results thus obtained.

2. Configuration and Design Analysis

2.1. Evolution of antenna

The proposed antenna is evolved in four stages. The figures given below depict the structural evolution of antenna proposed in this work.

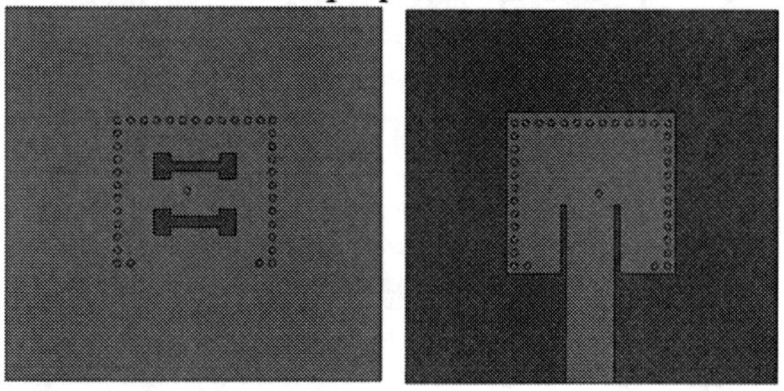

Figure 1. Evolution 1

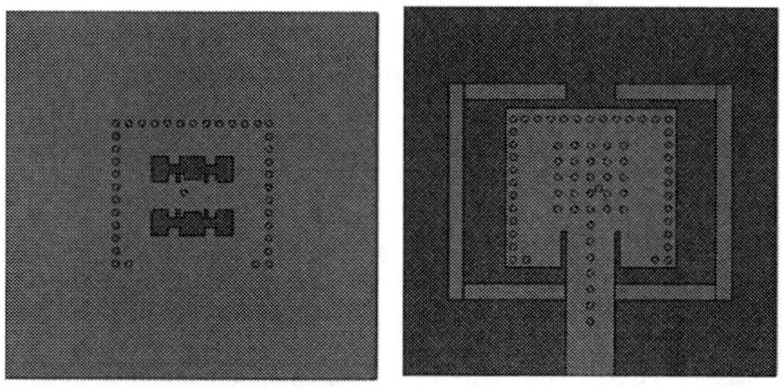

Figure 2. Evolution 2

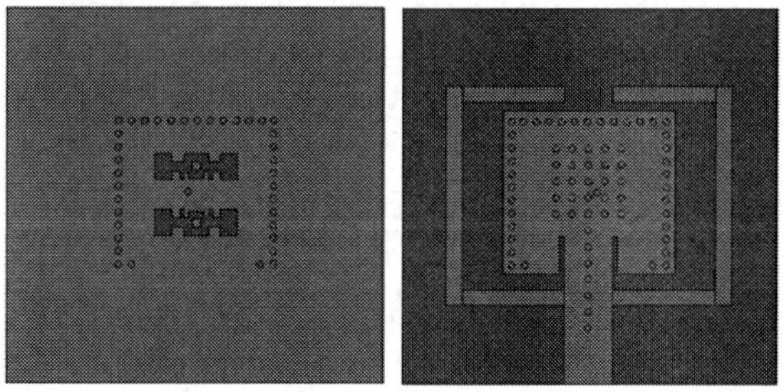

Figure 3. Evolution 3

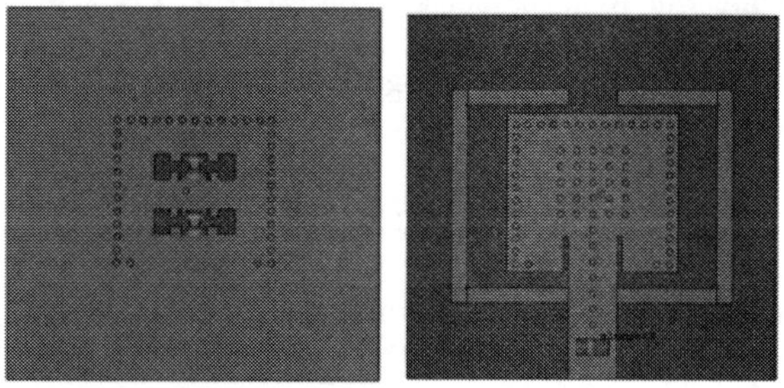

Figure 4. Evolution 4

2.2. Antenna in OFF condition

A single layer substrate is used to construct the suggested antenna and results are being simulated in OFF and ON condition respectively. The antenna in OFF condition is basically a sandwiched structure in which two Rogers RO4003 substrate layers (relative dielectric constant of 3.55, loss tangent of 0.0027) were used and stuck by using an adhesive layer Rogers 4450F (relative dielectric constant of 3.52, loss tangent of 0.004) as shown in figure 7. The top and bottom of the structure is covered with metallic layers having thickness of hm = 50 μm.

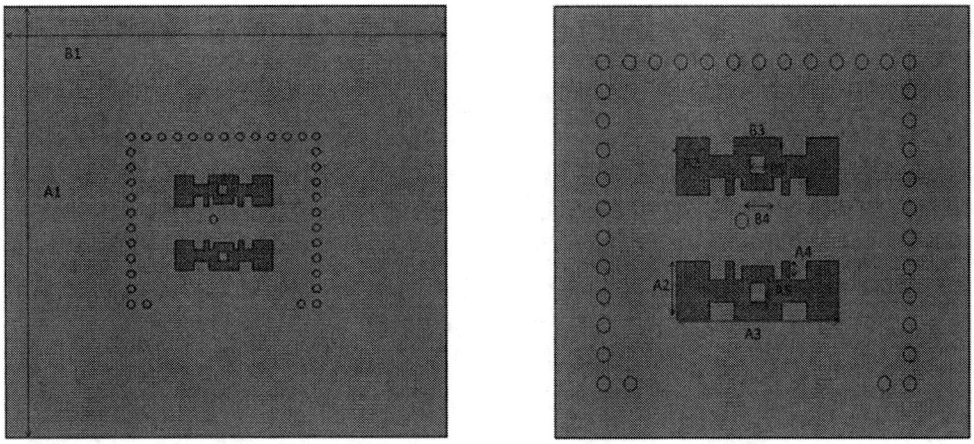

(a) (b)

Figure 5. Front side of proposed antenna

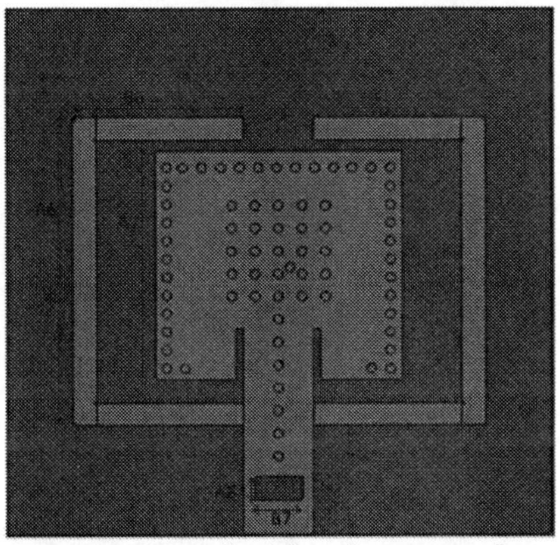

Figure 6. Back side of proposed antenna

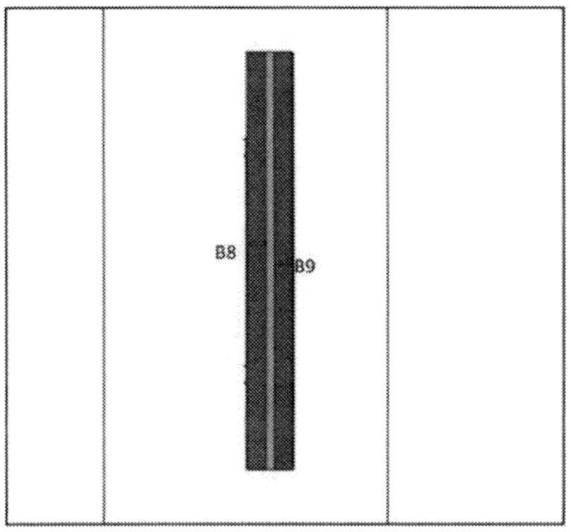

Figure 7. Side view of proposed antenna

The front side of the proposed antenna consists of two dumbbell shaped slots. In addition to a transverse slot, a pair of longitudinal slot is also added in the traditional dumbbell shaped slot. A square shaped slot is added in between the longitudinal slots as shown in figure 5. The back side consists of a square patch which is made up of PEC material. In addition, with the terminal SIW's, additional 5x5 grid of blind vias is also inserted in the patch as shown in figure 6. Apart from that, a rectangular slot of 1x2 mm is cut in the microstrip feed line. Two square bracket shaped patches are also attached with the microstrip feed line in order to achieve S11 below -10 dB.

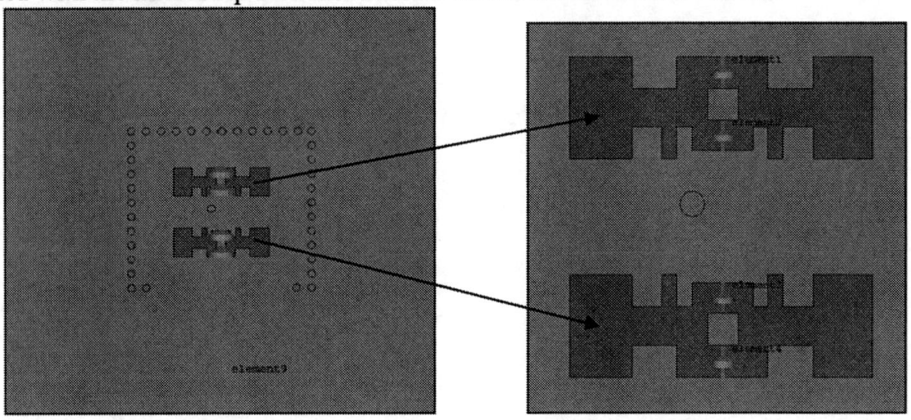

Figure 8. Position of diodes in front side of antenna

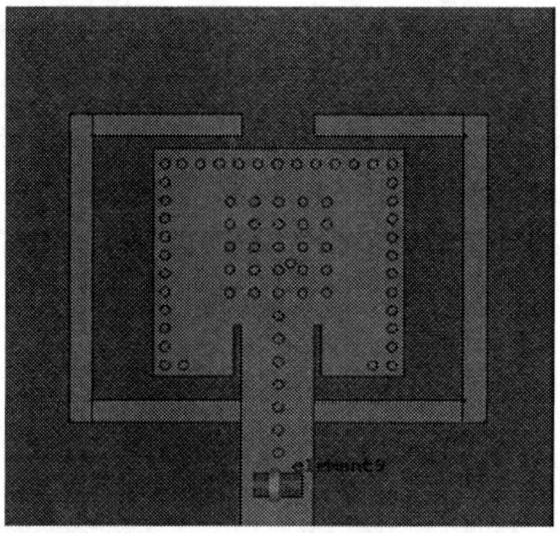

Figure 9. Diode positioned in microstrip feedline

2.3. Antenna in ON condition

After the successful implementation of antenna in OFF condition and simulated results, the same structure is used further for achieving reconfigurability by using PIN diodes. In this antenna, four diodes are added in the dumbbell shaped slot as shown in figure 8. In addition to this, a diode is also added in the rectangular slot positioned in the microstrip feed line as shown in figure 9. The antenna in ON condition is considered to operate in two frequency bands i.e., 12 GHz to 12.7 GHz and 22 GHz to 22.8 GHz. The size of substrate in the proposed design is 23mm x 23mm. The thickness of the substrate 1 and substrate 3 is 1.0mm and thickness of substrate 2 is 0.5mm given in table 1. The thickness of the metal layer is 50um and the height of the lateral SIW's is 1.05mm. The width of the applied microstrip feedline is 3mm and the size of the patch is 10.5mm.

The design parameters are given in table 1.

Table 1. Antenna design parameters

Symbol	Values (in mm)	Symbol	Values (in mm)
A1	23	B1	23
A2	1.595	B2	1.0
A3	5	B3	1.5
A4	0.5	B4	1.0
A5	0.5	B5	0.5
A6	13.5	B6	7.75
A7	10.5	B7	2
A8	1	B8	1
		B9	1

3. Results and discussions

The antenna is simulated in OFF and ON condition respectively. The proposed structure is simulated in a range of 8-24 GHz. In OFF condition, the impedance bandwidth comes out to be 12 – 12.7 GHz at centre frequency of 12.4 GHz. A good isolation of below -10 dB is achieved by imposing two square bracket shaped patches at the back side of simulated antenna as shown in figure 9.

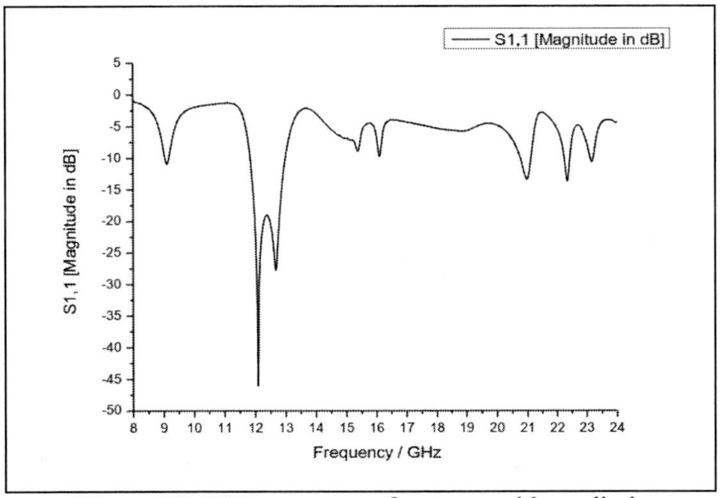

Figure 10. S11 parameter of antenna without diode

After implementing the same antenna in ON condition, reconfigurable frequency bandwidths are obtained. Two considerable bandwidths are obtained from 12 – 12.8 GHz and 22 – 22.8 GHz. The centre frequencies for both bandwidths are 12.5 GHz and 22.4 GHz respectively as shown in figure 10.

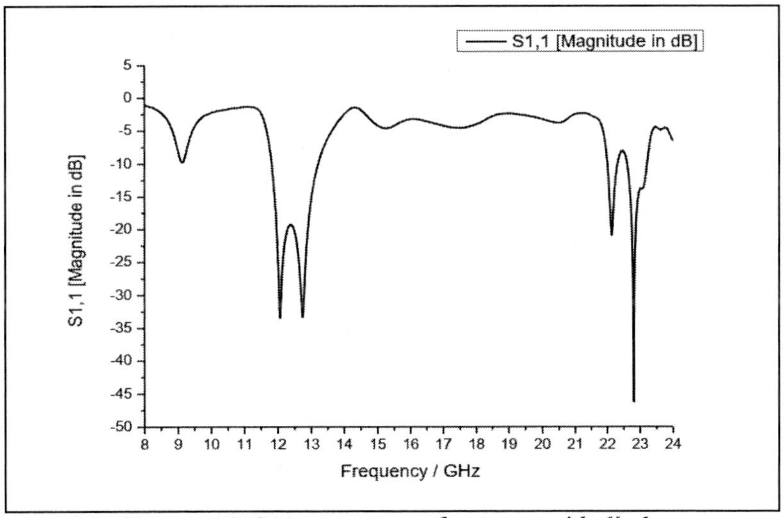

Figure 11. S11 parameter of antenna with diode

The farfield radiation pattern of antenna while OFF is shown in figure 12 and figure 13. At frequency of 12.088 GHz, the main lobe magnitude is 8.13 dBi while at frequency 12.705 GHz the main lobe has a magnitude of 7.23 dBi.

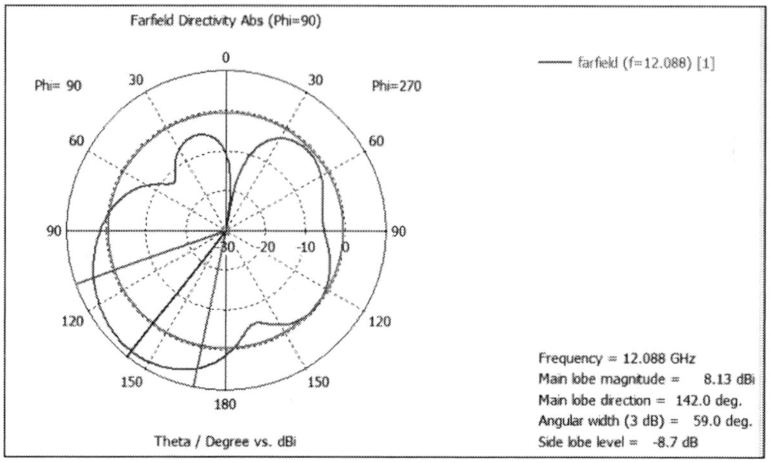

Figure 12. At frequency 12.088 GHz

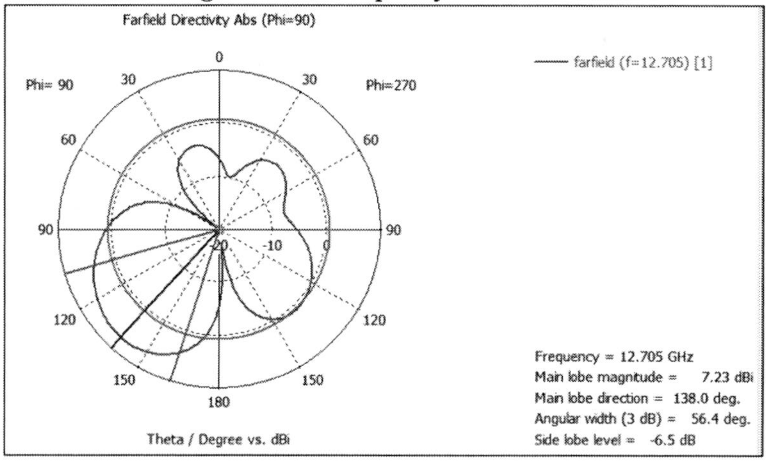

Figure 13. At frequency 12.705 GHz

For frequency reconfigurable antenna, far field radiation pattern is studied at four frequencies in the two obtained bandwidths respectively. The figures given below illustrate the main lobe and side magnitude at different frequencies.

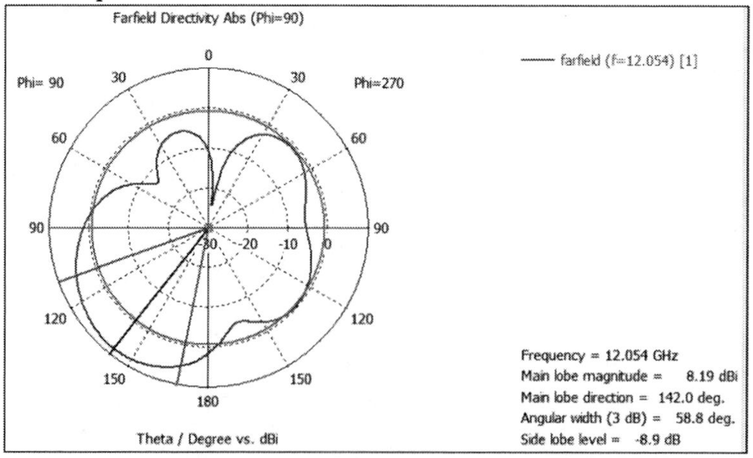

Figure 14. At frequency 12.054 GHz

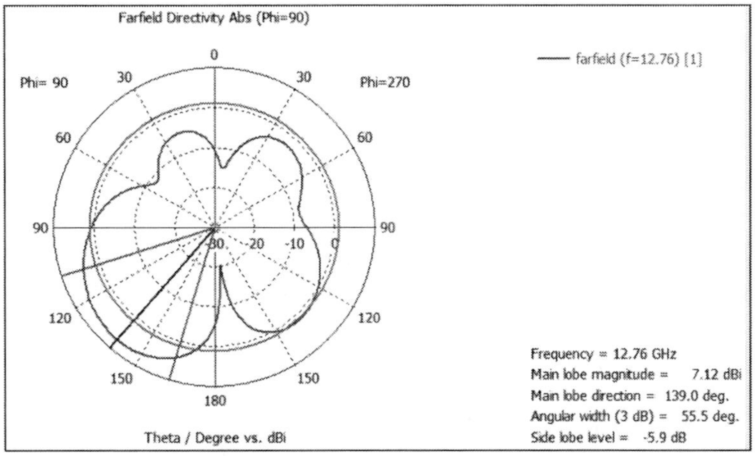

Figure 15. At frequency 12.76 GHz

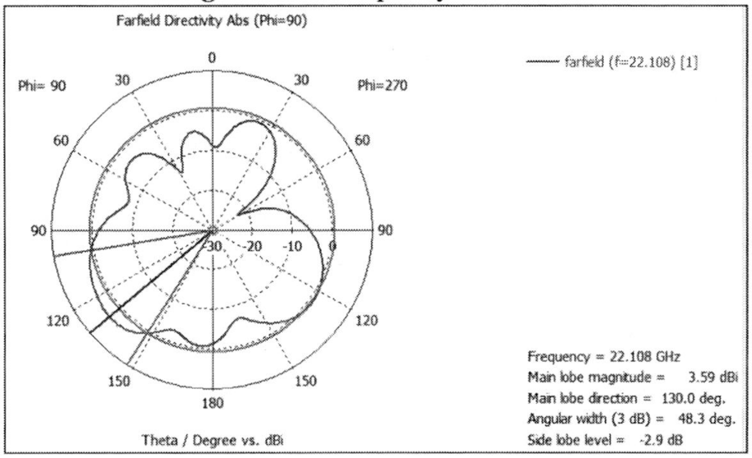

Figure 16. At frequency 22.108 GHz

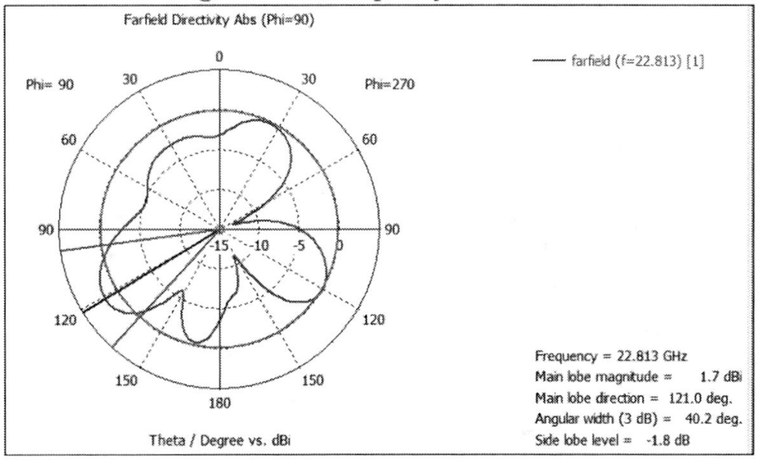

Figure 17. At frequency 22.813 GHz

The graph of gain of proposed antenna in both conditions is shown in figure 18 and figure 19. The numerical value of gain is 7.3 dB and 7.5 dB respectively.

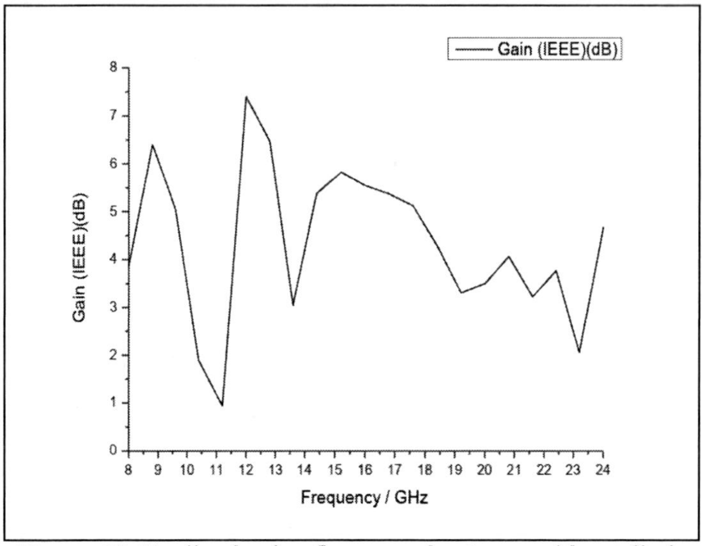

Figure 18. Realized gain of proposed antenna without diode

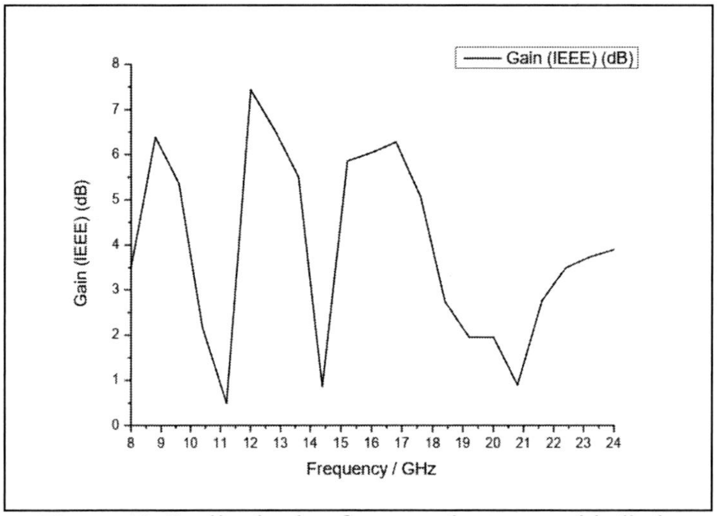

Figure 19. Realized gain of proposed antenna with diode

Radiation efficiency of antenna is 90% while it is ON or OFF. The graph of radiation efficiency of antenna in OFF condition is shown in figure 16 and that of antenna in ON condition is shown in figure 20.

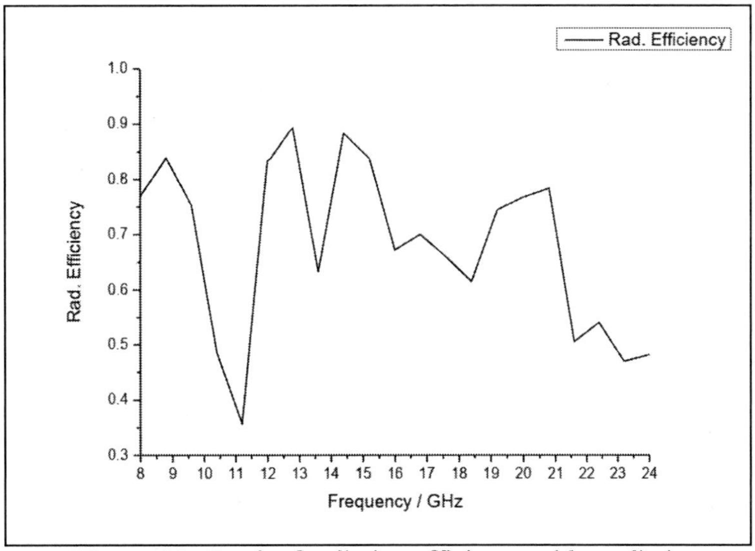

Figure 20. Graph of radiation efficiency without diode

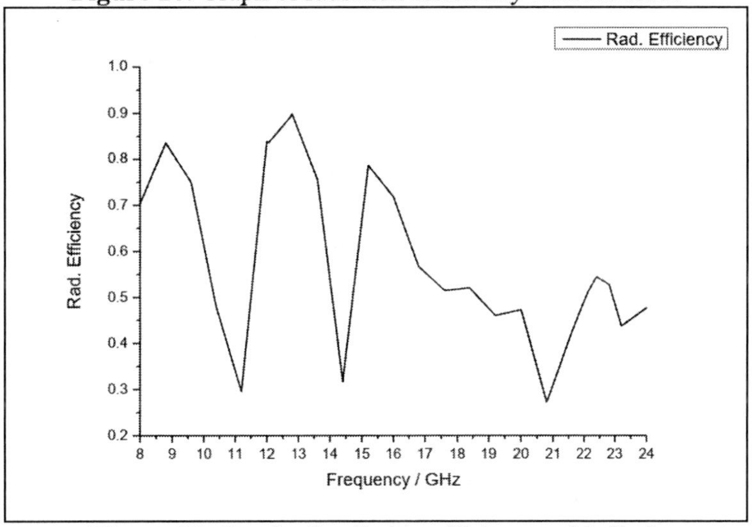

Figure 21. Graph of radiation efficiency with diode

A detailed parametric study is depicted in table 2 below -

Table 2. Comparison of antenna results

Parameters	Base paper [19]	Without diode	With diode
Impedance bandwidth	10-13 GHz	12-12.7 GHz	12-12.8 GHz 22-22.8 GHz
Gain	5.7 dB	7.3 dB	7.5 dB
Directivity	---	12.088 GHz – 8.13 dBi 12.705 GHz – 7.23 dBi	12.054 GHz – 8.19 dBi 12.76 GHz – 7.12 dBi 22.108 GHz – 3.59 dBi 22.813 GHz – 1.7 dBi
Radiation Efficiency	47%	90%	90%

4. Conclusion

A frequency reconfigurable antenna based on Substrate Integrated Waveguide with and without PIN diodes is proposed in this paper. Square bracket shaped patches attached with the microstrip feed line helps in achieving S11 below -10dB. The PIN diodes help in achieving reconfigurability of the antenna and can operate in two frequency bands also known as K band and Ku band. The radiation efficiency is considerably improved up to 90% when PIN diodes are applied. According to the measurements, the proposed antenna has an impedance bandwidth of 12 to 12.8 GHz in OFF condition and in addition to this 22 to 22.8 GHz in ON condition with a peak gain of 7.5 dB. The radiation efficiency of this work is 90%. Moreover, the major improvements are due to the use of blind vias in addition with terminal SIW which became one of the major reasons for reconfigurability in the antenna. Because of the advantages indicated, the suggested antenna is better suited for applications in satellite communication, direct-broadcast satellite television, terrestrial microwave communications, and radar, especially police traffic-speed detectors.

5. References

[1] Najib Mohammed Ahmed AL-Fadhali, Huda Abdul Majidd, Rosli Omar, Mohd Hezri Mokhtar, Najmaddin Abo Mosali," Frequency reconfigurable substrate integrated waveguide (SIW) cavity F-shaped slot antenna," Indonesian Journal of Electrical Engineering and Informatics, vol. 7, no .4, pp. 2832–2839,2019.

[2] AL-Fadhali, N., Majid, H. A., Omar, R., Abdul Sukor, J., Rahim, M. K. A., Zainal Abidin, Z., Abo Mosali, N. (2021). Review on frequency reconfigurable antenna using substrate-integrated waveguide for cognitive radio application. Journal of Electromagnetic Waves and Applications, 35(7), 958–990.

[3] Mudda, Gayathri, Frequency Reconfigurable Ultra Wide Band MIMO Antenna for 4G/5G Portable Devices Applications, International Journal on Emerging Technologies 11(3): 486-490(2020).

[4] Hassan, Muhammad Mateen; Zahid, Zeeshan; Khan, Adnan Ahmed; Rashid, Imran; Rauf, Abdul; Maqsood, Moazam; Bhatti, Farooq Ahmed (2020). Two element MIMO antenna with frequency reconfigurable characteristics utilizing RF MEMS for 5G applications. Journal of Electromagnetic Waves and Applications, (8), 1–15.

[5] Han, Liping; Ping, Yiwei; Liu, Yufeng; Han, Guorui; Zhang, Wenmei (2020). A Low-Profile Pattern Reconfigurable MIMO Antenna. IEEE Access, 8(34), 34500–34506.

[6] Q. Wu, H. Wang, C. Yu, and W. Hong, "Low-profile circularly polarized cavity-backed antennas using SIW techniques,"IEEE Transactions on Antennas and Propagation, vol. 64,no. 7, pp. 2832–2839, 2016.

[7] P. N. Choubey, W. Hong, Z.-C. Hao, P. Chen, T.-V. Duong,and J. Mei, "A wideband dual-mode SIW cavity-backed triangular-complimentary-split-ring-slot (TCSRS) antenna,"IEEE Transactions on Antennas and Propagation, vol. 64,no. 6, pp. 2541–2545, 2016.

[8] T. Deckmyn, M. Cauwe, D. VandeGinste, H. Rogier, and S. Agneessens, "Dual-band (28, 38) GHz coupled quartermode substrate-integrated waveguide antenna array for next generation wireless systems," IEEE Transactions on Antennas and Propagation, vol. 67, no. 4, pp. 2405–2412, 2019.

[9] Lai, F.-P., Chang, L.-W., & Chen, Y.-S. (2020). Miniature Dual-Band Substrate Integrated Waveguide Slotted Antenna Array for Millimeter-Wave 5G Applications. International Journal of Antennas and Propagation, 2020, 1–10.

[10] Rajmohan, R., Vishvaksenan, K. S., &Kalidoss, R. (2021). Compact four port slot based MIMO antenna with isolation enhancement. International Journal of Electronics, 1–14. doi:10.1080/00207217.2021.1908625

[11] Serhsouh, Imane; Himdi, Mohamed; Lebbar, Hassan; Vettikalladi, Hamsakutty (2020). Reconfigurable SIW Antenna for Fixed Frequency Beam Scanning and 5G Applications. IEEE Access, 8(44), 60084–60089.

[12] Qin, J., Fu, X., Sun, M., Ren, Q., & Chen, A. (2021). Frequency Reconfigurable Antenna Based on Substrate Integrated Waveguide for S-Band and C-Band Applications. IEEE Access, 9, 2839–2845.

[13] Li, Y., Wang, M., Wu, W., Zhou, X., &Qiu, C. (2021). SIW-Based Leaky Wave Antenna with High Radiation Efficiency. Journal of Physics: Conference Series, 1885(4), 042058.

[14] Ikram, Muhammad; Nguyen-Trong, Nghia; Abbosh, Amin (2019). A Simple Single-Layered Continuous Frequency and Polarization-Reconfigurable Patch Antenna Array. IEEE Transactions on Antennas and Propagation, (), 1–1.

[15] Gu, Hui; Wang, Jianpeng; Ge, Lei (2015). Circularly Polarized Patch Antenna with Frequency Reconfiguration. IEEE Antennas and Wireless Propagation Letters, (3), 1–1.

[16] Zhao, Xiongwen; Riaz, Sharjeel (2018). A Dual-Band Frequency Reconfigurable MIMO Patch-Slot Antenna Based on Reconfigurable Microstrip Feedline. IEEE Access, (3), 1–1.

[17] Zhang, Jin; Zhang, Shuai; Lin, Xianqi; Fan, Yong; Pedersen, Gert (2018). 3D Radiation Pattern Reconfigurable Phased Array for Transmission Angle Sensing in 5G Mobile Communication. Sensors, 18(12), 4204–4210.

[18] Ghaffar, A., Li, X. J., Awan, W. A., Nazeri, A. H., Hussain, N., &Seet, B. (2021). Compact multiband multimode frequency reconfigurable antenna for heterogeneous wireless applications. International Journal of RF and Microwave Computer-Aided Engineering, 31(7).

[19] Ho, A. T., Pistono, E., Corrao, N., & Ferrari, P. (2020). Circular polarized square slot antenna based on slow-wave substrate integrated waveguide. IEEE Transactions on Antennas and Propagation, 69(3), 1273-1282.

IOP Conf. Series: Materials Science and Engineering 1272 (2022) 012006 doi:10.1088/1757-899X/1272/1/012006

A Spear-head shaped UWB MIMO Antenna with triple notch-based characteristics for Wireless applications

Ranjana Kumari, Pragati Porwal, Prashant Kumar, Piyush Singh Katheria, Pragya Singh,

Department of Electronics and Communication Engineering, Galgotias College of Engineering and technology, Greater Noida, India

Abstract. In this article, a spearhead shaped dual element ultra-wide band (UWB) multiple input multiple output (MIMO) antenna with three notched bands is designed. The proposed antenna has the dimensions (28*18*1.6) mm^3. To achieve the three notch a L-shaped and spoon-shaped slots are etched out from the rhombic shape radiating patch for application ARN, INSATat (4.17-4.9) GHz and satellite communication at (7.1-7.75) GHz respectively and other inverted U- shaped slot is etched out from the feed line which achieves X-band application (8.5-9.6)GHz. In addition to this, the proposed antenna impedance bandwidth ranges from 3-11.3GHz with voltage standing wave ratio (VSWR>2) only for the selected notch band. The antenna achieves nil distortion with suitable omnidirectional radiation pattern which is good for ultra-wide band application. To show the diversity performance of the proposed antenna, the proposed antenna offers good isolation (s12/s21<-19db) with envelope correlation coefficient (ECC<0.005) and diversity gain of nearly 10db. Therefore, the proposed antenna with rejection capability is suitable for UWB application.

1. Introduction

In day today life, ultra-wideband (UWB) technology has recently gotten a lot of attention from researchers and the wireless communication industry. It was originally designed for radar imaging, but because of technical recent developments, this now allows users to access high-speed data transmission services. Ultra-wideband (UWB) technique is increasingly used because of many advantages which includes powerful anti-performance, consume less power, high positioning precision, good concealment, large communication capacity, and low cost [1-3]. Researchers have been paying close attention to design MIMO antennas which is made up of two structurally symmetrical antennas and is used to cover an 3–9 GHz impedance bandwidth with a VSWR < 2 [1]. To utilize ultra-wide-band spectrum authors [1-5] have designed only single element antennas but these antennas are facing the problem of multi-path fading so, this problem can be resolved by using more than one element of antenna on the same substrate. In recent, researchers have designed MIMO antennas for ultra- wideband applications in [2]-[5]. In [2], tree shape radiating element is utilized to enhance the bandwidth for UWB application. In [3], MIMO antenna is made up of two simple eye shape elements which has a 17.2 GHz impedance bandwidth 2.8-20 GHz which is considered as a good impedance matching and low mutual coupling is achieved by using an extruded T-shaped structure.A compact MIMO antenna system with a super-wide impedance bandwidth has been proposed with excellent gain [4-5]. Above mentioned MIMO antenna are not capable to remove interference of the antenna. So, to overcome this problem we use more than one element MIMO antenna. Therefore, in from [6]-[12] are designed for to reject the interference from the existing UWB. In [6] UWB MIMO antenna with single notch characteristics has been suggested with broad impedance range and has a bandwidth of 143 percent (3–11 GHz) notch band from 5 to 10 dB. In [7-13] authors present MIMO antennas with two notch bands. In [7], the given UWB MIMO antenna with two sharply rejected notchbands which are given with ultra-wide-band (UWB) applications. In [8], the MIMO antenna can be recommended as a potential candidate for WiMAX and military/radar applications from UWB sheet. The first notch is made rejecting for Wi-Max application while the second notch is made rejecting for bluetooth application. In [9], two notches for WLAN and WiMax applications are generated using reversed U-shaped slots at a frequency of 5.4 GHz. In [10],the work totally suggests a compact dual band- notched UWB-MIMO antenna with a second-order notched band of 5.15–5.825 GHz and effective isolation for the whole UWB spectrum. Above mention MIMO antenna are designed for ultra-wide-bandbut many applications like Wi-Max, X-band, WLAN provide interference. Therefore, to remove the interference single, double notched based MIMO antenna [6-10] are used. In [11], a two element MIMO antenna for dual notch applications for WLAN and X Band is presented. In [12], one more dual notch UWB antenna is designed with the size. In [14-19] authors present papers with three notch band rejections. Furthermore, to reject the triple band notch, researchers have focused to design these triple notched antennas. In [13], by adding two split ring resonators (SRR) of different sizes next to the microstrip feed lines, dual notches can be achieved to filter the interfere of WLAN and X-band communication satellites. In [14], a CPW feeding is used for Super-wideband MIMO antenna including three band rejection band is presented which ranges form 3.3-3.6GHz for WiMAX application, 7.1-7.9 GHz for X band application is achieved by inserting plus shape EBG structure and 5-6GHz for WLAN application achieved using mushroom EBG structure. In [15] presents triple band- rejection based UWB antenna for microwave access WiMAX band from 3.3 to 3.6 GHz, wireless local area network WLAN band from 5 to 6 GHz and X-Band satellite downlink communication band from 7.1 to 7.9 GHz. In [16], a minimized UWB-MIMO antenna attributes has been proposed effectively with triple band-rejection and radiation for rhombic slot to 4 microstrip- feed, and the triple band- rejection are achieved by c-slot, L-slot on the patch and also embedding H- slots on EBGS structure. [17] presents, a MIMO antenna for SWB application is presented with triple band notch characteristics. A step like slit removes the patch to reduce the interference at 3.5 GHz and ahexagonal shape CSRR on MESCA is used to eliminate the interference at 5.5 GHz and 8.5GHz. In [18], a minimized ultra-wideband with tri- band rejection is presented which rejects three notch bands for (UWB) ultra-wideband applications. [19] describes a two-element planar UWB MIMO antenna with triple rejection bands with an array that is highly isolated. Although, above mentioned [14-19] antenna are based on triple notch band but size was not compact and impedance bandwidth is not wider.

Therefore, to achieve triple band two slot (spoon-shaped and L- shaped) are etching out from radiating element and on more slot(U-shaped) is etching out from feed. Thus, In this article, a triple band notched UWB-MIMO antenna is presented. The average area of the structure is(28*18*1.6)mm^3. The proposedantenna can achieve good isolation which is less than -15db and also the performance is lower than
0.005 for envelope correlation coefficient (ECC) and diversity gain (DG) is nearly 10db which is exhibit the good performance according to the IEEE standards. The workflow of the antenna is divided into below sections where each aspect is clearly demonstrated and explained. The designing strategy of the proposed antenna is explained in Antenna design section where as its obtained results are demonstrated in results and discussion section.

2. Antenna Design

The antenna element is divided into two parts. one is single-element UWB MIMO antenna and the otheris dual element of UWB MIMO antenna.

2.1 Single element UWB antenna

In figure 1(a), depict that the square shape-head with the microstrip feed is put over the FR4-

IOP Conf. Series: Materials Science and Engineering 1272 (2022) 012006 doi:10.1088/1757-899X/1272/1/012006

Epoxy substrate after that the same square head procced with the microstrip feed converted into tapered feed which is seen in fig-b and then rotate the square head 45degree in the clockwise direction which is shownin figure 1.(c) and after that adjust both rhombus and tapered and merged them which gives the result as the impedance bandwidth (3.1–11.3 GHz) which comes under the UWB range and provide the enhancement of the bandwidth and improved isolation. As described in the figure 1. (d) the two-slot cutinside the patch of the antenna one rectangular strip and another is the spoon-shaped slot which providethe three notches which is seen in the s-parameter figure 2.

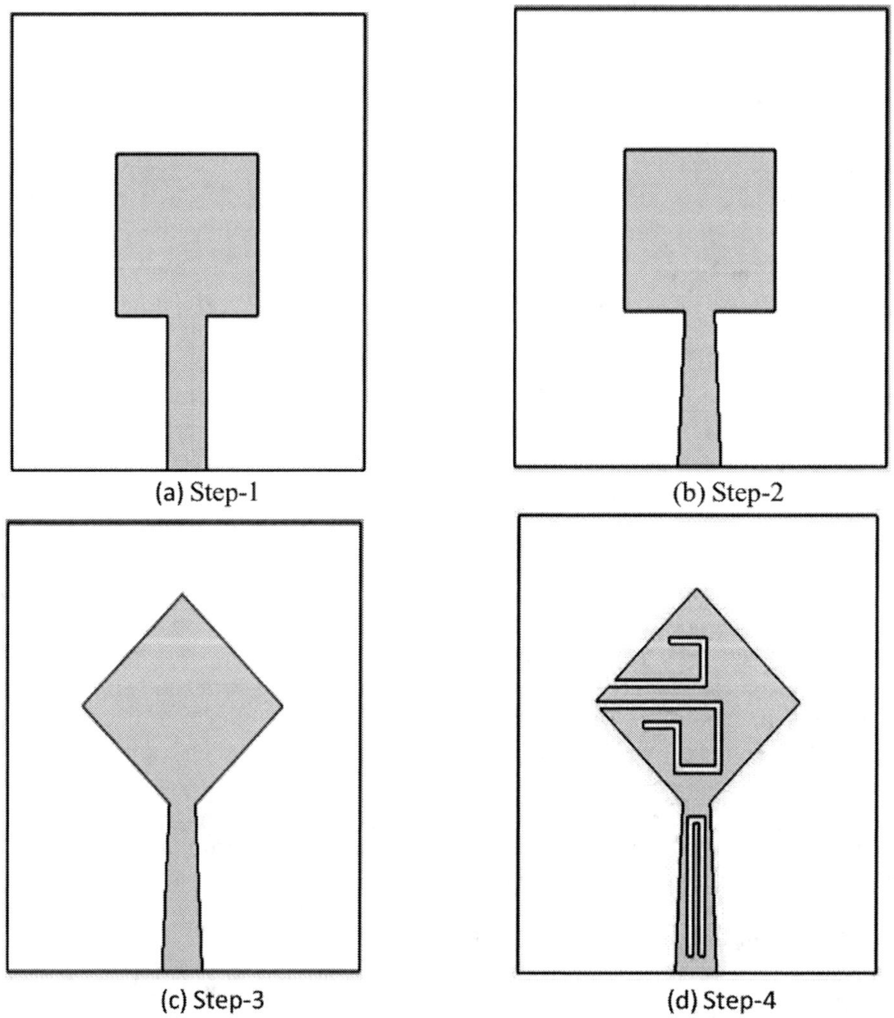

Figure 1. Geometry of the UWB antenna step -1,2,3,4 Evolution steps of the structure of the UWB antenna element.

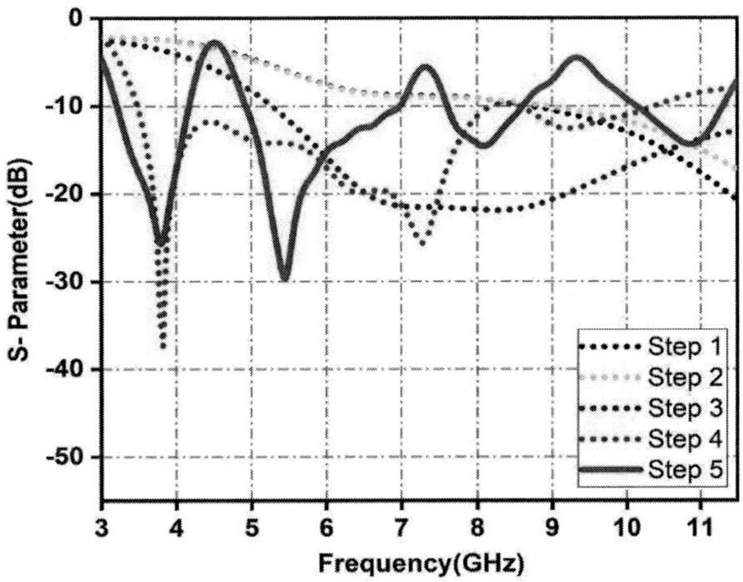

Figure 2. S-Parameter of the evolution of single element antenna.

2.2 Two-element UWB antenna

After optimizing the single element of the antenna is easily extended into two element MIMO antenna. The optimized geometrical arrangement of the proposed ultra-wide-band notched based antenna is described in figure.3, the spear-head shaped radiators with three slots are implanted on FR-4 substrate size 28*18*1.6 mm^3. The top and back view of proposed antenna is shown in figure 3.(a) and figure 3.(b) respectively. Additionally figure 3.(c) depicts the notch and inner dimensions of the patch which are taken into consideration to achieve this structure. The complete dimensions are discussed in Table 1. The partial connected ground is printed on the back side of the FR-4 substrate. In order to achieve the band-notched characteristics, the spoon-shaped slot is etched out from the radiating patch for the frequency band (4.17-4.9) GHz. Secondly, the L-shape slot from radiating patch gives the secondnotch having application satellite communication at (7.1-7.75) GHz and another inverted U- shaped slot is etched out from the feed line to achieve X-band application at the frequency of (8.5-9.6) GHz. These notches are described in simulated results of S-parameters. As observation from figure 4, the broad impedance bandwidth (3.1–11.3 GHz) is acquired with three notches. These simulated results exhibit the performance of UWB MIMO antenna.

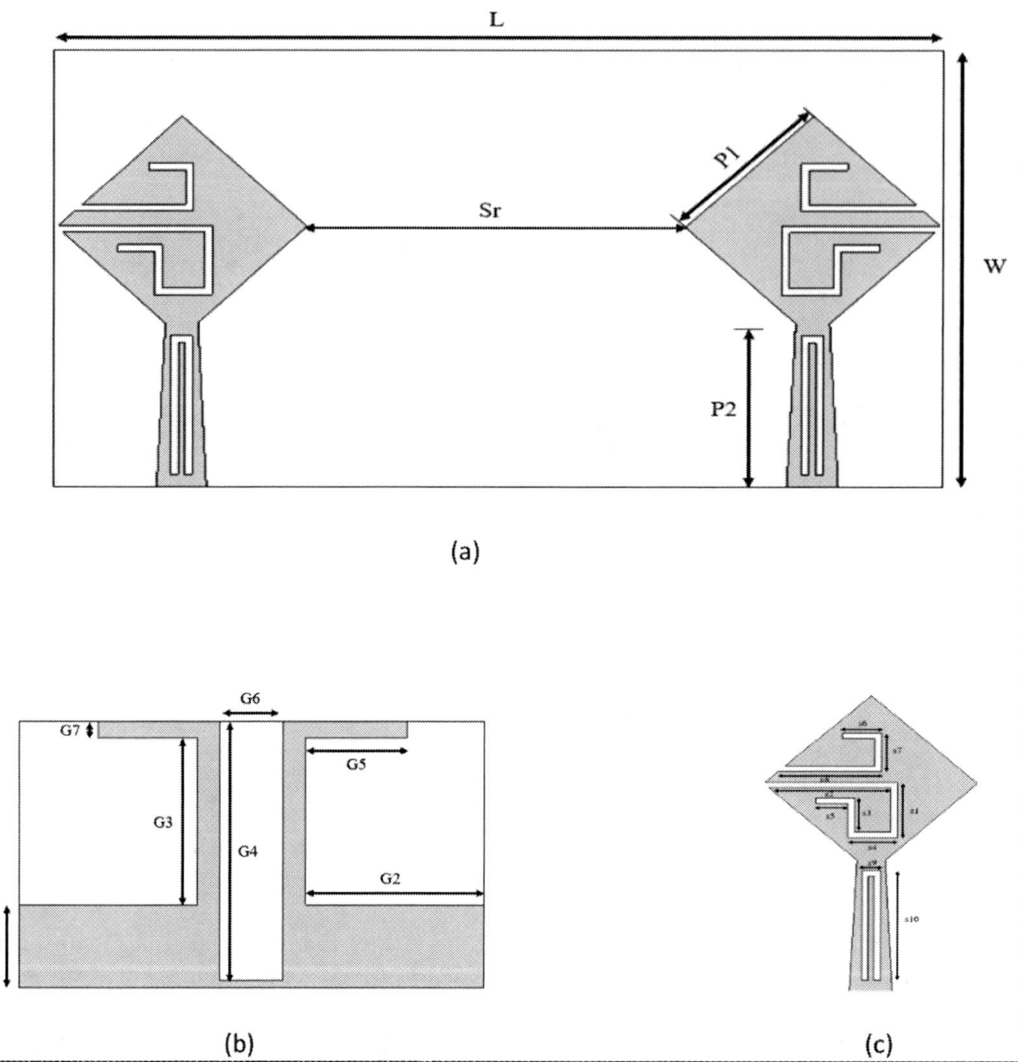

Figure 3. Configuration of the 2-element antenna (a) Top view, (b) Back view (c) Zoomed view.

Table 1. The dimensions of the suggested UWB-MIMO antenna's optimum specifications.

Parameters	Value (in mm)	Parameter	Value (in mm)
L	28	s1	2.5
W	18	s2	4.53
P1	5.66	s3	1.55
P2	5.98	s4	1.85
G1	5	s5	1.20
G2	10.75	s6	1.45
G3	10	s7	1.75
G4	15.5	s8	3.83
G5	6	s9	0.70
G6	3.8	s10	4.90
G7	1	Sr	11.85

3. Results and discussion

To validate the performance of proposed ultra-wideband-MIMO antenna with rejection capability, the simulated results S-parameter, radiation patterns, efficiency, and gain are discussed in this section. In Table 2 proposed antenna is compared with existing antennas and it is clearly observed that the presented antenna gives excellent impedance - agreeing, good Gain and ECC and is an excellent candidate for Ultrawide band applications.

Table 2. Performance comparison of the proposed antenna with existing antenna.

References	Size	Isolation	Gain	No. of Notches	ECC
[1]	33*15*0.5	82	5.7	0	<0.15
[2]	35*40*0.8	16	3.1	0	<0.01
[4]	58*58*1	17	4.3-13.5	0	<0.04
[6]	64*64*1.6	40	-	1	<0.06
[10]	35 * 23 *1.6	17	-	2	<0.01
[8]	80*80*1.6	25	5.8	2	<0.02
[17]	63*63*0.8	17	3.5, 5.5, 8.5	3	<0.01
This work	28*18*1.6	17	4	3	<0.01

3.1 S-Parameter

In figure.4. S-parameters (S11 and S22) results of the designed antenna has been described with isolation between ports 1 and 2. According to the figure4, the impedance bandwidth is acquired from
3.1 – 11.3 GHz and isolation is less than -19dB. As seen in the notched bands are obtained from 4.17-
4.9 GHz, 7.1-7.75 GHz and 8.5-9.6 GHz respectively. Moreover, an isolation dip (close to 50db) is observed at around 7.5GHz. Thus, it can be concluded that the isolation of the antenna produces good for MIMO characteristics.

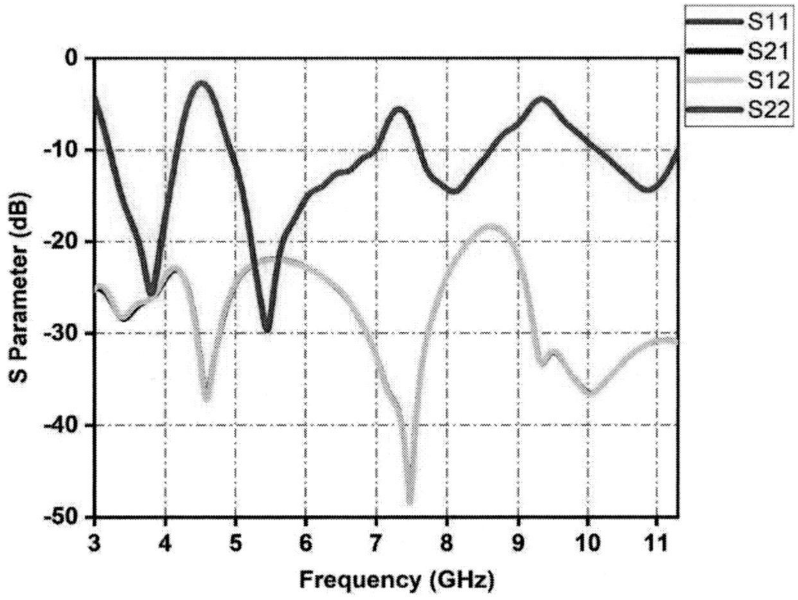

Figure 4. Simulated s-parameter against frequency of the proposed MIMO antenna.

3.2 Gain and efficiency

In figure 5. depicts the combine simulated result of gain(dB)pand efficiency of the proposed antenna. According to observation in figure 5, the maximum gain is 2.5dBi in throughout UWB spectrum, excluding for the notched bands at 4.17-4.9 GHz, 7.1-7.75 GHz and 8.5-9.6 GHz.Moreover, the gain achieved is less than zero at all rejected spectrums. This demonstrates that the suggested antenna has better gain for the UWB-applications. Furthermore, the radiation efficiency of proposed antenna for the UWB-MIMO spectrum is also discussed in figure 5 . The radiation efficiency of MIMO antenna is more than 70 % excluding the notched bands which is less than 50 percent at all notched bands. Therefore, it concluded that the gain and efficiency of the designed antenna are well under IEEE standards.

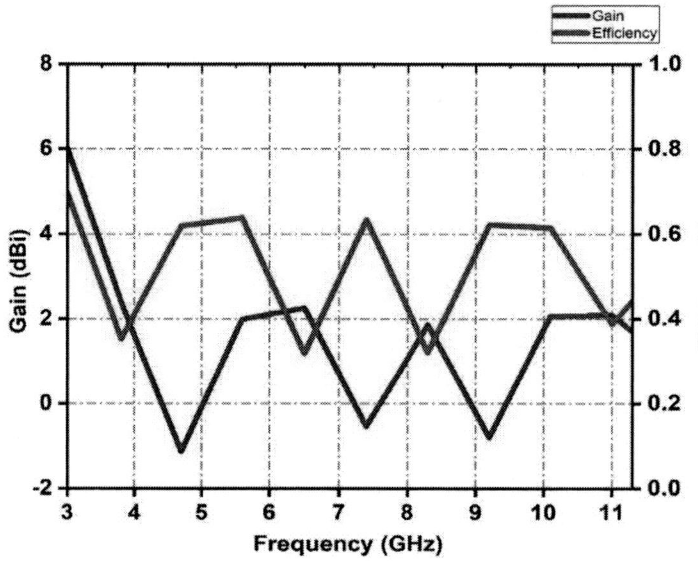

Figure 5. Gain and Efficiency

3.3. VSWR

The VSWR simulated results are described in Figure. 6 which can be seen below. It is understood fromthe figure that apart from the rejecting bands, VSWR shifts from 1 to 2 across the whole ultra-wideband.The VSWR value for all three rejection bands is more than 2, with a peak VSWR of 6.4 at 4.5GHz. These results suggests that the proposed antenna has better VSWR for UWB applications.

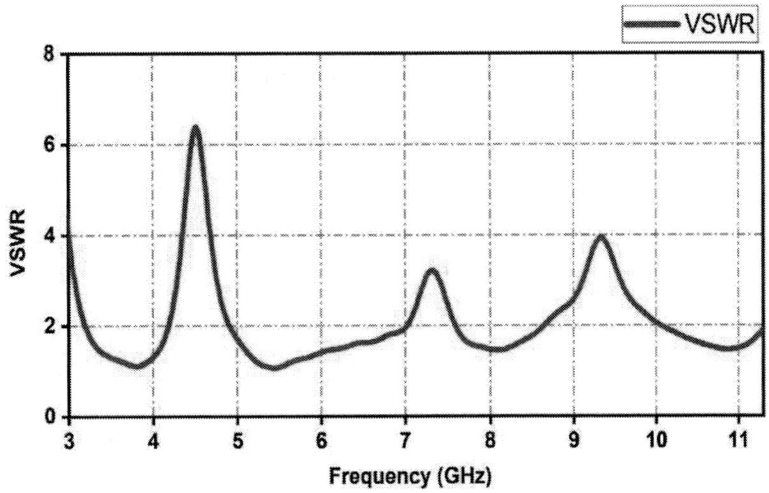

Figure 6. Simulated VSWR for proposed antenna.

3.4. ECC (Envelope Correlation Coefficient) and DG (Diversity Gain)

The ECC of antenna is depicted in Figure 7 and is calculated using equation (1). It is clearly understandable from Figure 7. that the simulated results of ECC (<0.01) are well under IEEE standards(<0.05) with the ECC is achieved to be 0.009 at 8.9GHz.

$$ECC = \frac{|S_{11}^* S_{12} + S_{21}^* S_{22}|^2}{(1-(|S_{11}|^2-|S_{21}|^2))(1-(|S_{22}|^2+|S_{12}|^2))} \quad (1)$$

The diversity gain of the proposed antenna is also depicted in Fig.7 and is calculated using equation (2)for both the ports. It is clearly understood from Figure. 7 that the simulated results of DG (>9.95) are well under IEEE standards(>9.95) with the minimum DG of 9.95 at 8.9GHz. Therefor, it can be concluded that proposed antenna has good MIMO diversity characteristics.

$$DG = 10\sqrt{1-|ECC|^2} \quad (2)$$

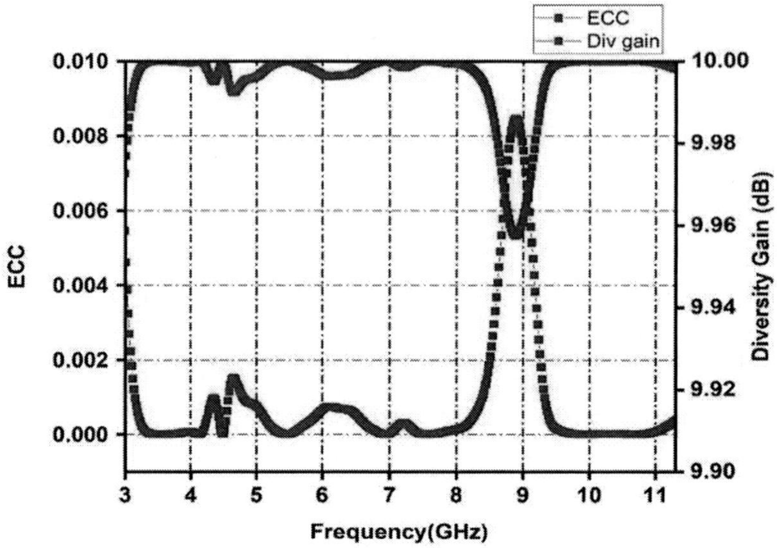

Figure 7. ECC and DG of the proposed antenna

3.5.Radiation Pattern

Radiation patterns of the proposed MIMO antenna are observed at 3.5GHz, 5.5 GHz, and 10.5 GHz in E-plane and H-plane which are shown in Figure 8. The radiation pattern in plane E-plane and H-Plane at different frequencies are stable and radiation patterns are bidirectional and nearly omnidirectional at different frequencies.

IOP Conf. Series: Materials Science and Engineering 1272 (2022) 012006 doi:10.1088/1757-899X/1272/1/012006

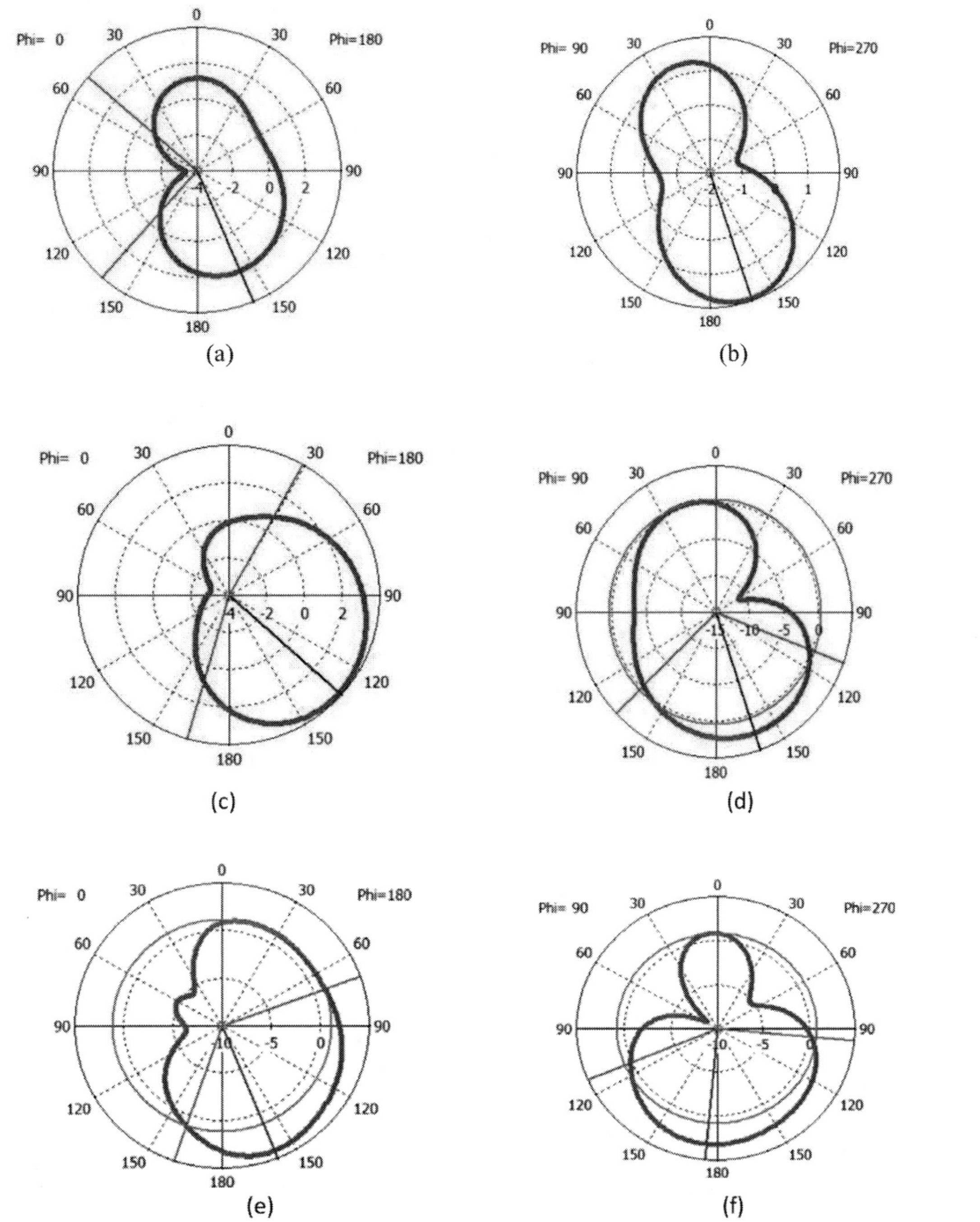

Figure 8. Radiation Pattern of the proposed MIMO antenna.

4. Conclusion

A novel dual-port UWB-MIMO antenna with tapered feed and triple band-notched characteristics has been successfully designed and simulated. The antenna design found to have an impedance bandwidth ranging from 3.1GHz to 11.3 GHz except for the three notched bands rejecting the interference for the application ARN, INSAT at (4.17-4.9), satellite communication at (7.1-7.75) GHz and X-band application (8.5-9.6) GHz . Furthermore, the isolation of the proposed antenna is < -19dB. The proposed antenna has maximum gain 2.5dB with radiation efficiency more than 60% throughout except notches bands in ultra-wide band spectrum. The MIMO diversity results of the proposed antenna are foundto be in good agreement with the IEEE standards. Hence, the proposed antenna can be utilized in ultrawide band applications.

5. References

[1] Rajmohan, R., Vishvaksenan, K.S. and Kalidoss, R., 2021. Compact four port slot based MIMO antenna with isolation enhancement. *International Journal of Electronics*, *108* (7), pp.1075-1088.
[2] Zhang, S., Ying, Z., Xiong, J. and He, S., 2009. Ultrawideband MIMO/diversity a ntennas with a tree-like structure to enhance wideband isolation. *IEEE Antennas and Wireless Propagation Letters*, *8*, pp.1279-1282.
[3] Chandel, R., Gautam, A.K. and Rambabu,K., 2018. Design packaging of an eye shaped multiple input multiple output antenna with high isolation for wireless U WB applications. *IEEE Transactionson Components, Packaging and Manufacturing Technology*, *8*(4), pp.635-642.
[4] Yu, C., Yang, S., Chen, Y., Wang, W., Zhang, L., Li, B. and Wang, L., 2020. A super- wideband and high isolation MIMO antenna system using a windmillshaped decoupling structure. *IEEE Access*, *8*, pp.115767-115777.
[5] Sipal, D., Abegaonkar, M.P. and Koul, S.K., 2017. Easily extendable compact planar UWB MIMO antenna array. *IEEE Antennas and Wireless Propagation Letters*, *16*, pp.2328-2331.
[6] Huang, H., Liu, Y., Zhang, S. and Gong, S., 2014. Uniplanar differentially driven ultrawideband polarization diversity antenna with band-notched characteristics. *IEEE Antennas and Wireless Propagation Letters*, *14*, pp.563-566.
[7] Chandel, R., Gautam, A.K. and Rambabu, K., 2018. Tapered fed compact UWB MIMO-diversity antenna with dual band-notched characteristics. *IEEE Transactions on Antennas and Propagation*, *66*(4), pp.1677-1684.
[8] Rekha, V.S.D., Pardhasaradhi, P., Madhav, B.T.P. and Devi, Y.U., 2020. Dual band notched orthogonal 4-element MIMO antenna with isolation for UWB applications. *IEEE Access*, *8*, pp.145871-145880.
[9] Singh, C. and Kumawat, G., 2020. A compact rectangular ultra-wideband microstrip patch antenna with double band notch feature at Wi-Max and WLAN. *Wireless Personal Communications*, *114*(3), pp.2063-2077.
[10] Li, J.F., Wu, D.L., Wu, Y.J. and Zhang, G., 2015, June. Dual band-notched UWB MIMO antenna.In *2015 IEEE 4th Asia-Pacific Conference on Antennas and Propagation (APCAP)* (pp. 25-26). IEEE.
[11] Tang, Z., Zhan, J., Wu, X., Xi, Z., Chen, L. and Hu, S., 2020. Design of a compact UWB-MIMO antenna with high isolation and dual band-notched characteristics. *Journal of Electromagnetic Waves and Applications*, *34*(4), pp.500-513.
[12] Zhu, J., Feng, B., Peng, B., Li, S. and Deng, L., 2016. Compact CPW UWB diversity slot antenna with dual band-notched characteristics. *Microwave and Optical Technology Letters*, *58*(4), pp.989- 994.
[13] Li, Z., Yin, C. and Zhu, X., 2019. Compact UWB MIMO Vivaldi antenna with dual band- notched characteristics. *IEEE Access*, *7*, pp.38696-38701.
[14] Kumar, P., Urooj, S. and Alrowais, F., 2020. Design of quad-port MIMO/Diversity antenna with triple-band elimination characteristics for super-wideband applications. *Sensors*, *20*(3), p.624.
[15] Jaglan, N., Gupta, S.D., Thakur, E., Kumar, D., Kanaujia, B.K. and Srivastava, S., 2018. Triple band notched mushroom and uniplanar EBG structures based UWB MIMO/Diversity antenna with enhanced wide band isolation. *AEU-International Journal of Electronics and Communications*, *90*, pp.36-44.
[16] Chen, Z., Zhou, W. and Hong, J., 2021. A miniaturized MIMO antenna with triple band- notched characteristics for UWB applications. *IEEE access*, *9*, pp.63646-63655.
[17] Raheja, D.K., Kanaujia, B.K. and Kumar, S., 2019.Low profile four-port super-wideband multiple-input-multiple-output antenna with triple band rejection characteristics. *International Journal of RF and Microwave Computer-Aided Engineering*, *29*(10), p.e21831.
[18] Lin, C.C., Jin, P. and Ziolkowski, R.W., 2011. Single, dual and tri-band-notched ultrawideband (UWB) antennas using capacitively loaded loop (CLL) resonators. *IEEE Transactions on Antennas and Propagation*, *60*(1), pp.102-109.
[19] Jaglan, N., Kanaujia, B., Gupta, S.D. and Srivastava, S., 2016. Triple band notched UWB antenna design

using electromagnetic band gap structures. *Progress In Electromagnetics Research C*, *66*, pp.139-147.
[20] Krishna, D., & Mathew, T. (2015). A planar UWB antenna design with triple band-notches for WiMAX, WLAN and downlink of X-band satellite communications system. *TENCON 2015 - 2015 IEEE Region 10 Conference*.

IOP Conf. Series: Materials Science and Engineering 1272 (2022) 012007 doi:10.1088/1757-899X/1272/1/012007

Design and Performance Improvement of 10T SRAM Using Sleepy Keeper and Drain Gating Techniques

Ranu Chaurasia[1], Brijesh Kumar[2], Sudhanshu Verma[3] and Akhilesh Kumar[4]

[1,3]Electronics and Communication Engineering, MMMUT, Gorakhpur, India
[2]Department of Information Technology, Indira Gandhi Delhi Technical University for Women, Delhi-110006
[4]NIT Silchar, Electronics and Communication Engineering, Assam, India
ranuchaurasia1999@gmail.com[1] and brijeshkr@igdtuw.ac.in[2]

Abstract. This paper presents the sleepy keeper and drain gating technique to improve the performance of 10T SRAM cell. The behavior of 10T SRAM cell is evaluated using benchmarked industry standard GPDK 45nm of the Cadence Virtuoso EDA tool. The performance is analyzed in terms of dynamic and static parameters of 10T SRAM cell and compared with 6T SRAM, where find the reduction in dynamic power and static power dissipation. Besides this, observed the reduction in leakage current using sleepy keeper and drain gating technique. The proposed modified topology works in single-ended write and differential read operation. The read delay product and the write delay product is decreased by 36.7 % and 67.5 %, respectively. The major goal of the suggested architecture is to provide the improved stability, reduction in delay, as well as reduction in leakage current.

Keywords: CMOS based SRAM Cell, Control Signal, Read-Sleepy Keeper, Drain gating technique, Power Consumption.

1. INTRODUCTION

Static Random-Access Memory (SRAM) is an unavoidable component of SoC design. This memory is widely employed in modern high-performance applications because of its good compatibility with logic design [1]. Many device characteristics, such as better performance, are made possible via technology scaling. With the device's size reduced, it consumes less power and takes up less space. Circuits are more sensitive to noise and instability when technology nodes diminish [1-2]. Furthermore, due to the energy limits of devices operated by the battery, research on low power dissipation memory elements has recently come to a lot of interest. Moreover, scaling the supply voltage is a popular option since it results in a quadratic degradation in dynamic power and a major reduction in leakage current. It may allow the evolution of applications of low power such as wearable health monitors, Internet of Things IoT) devices, wireless sensor networks etc. Furthermore, the author in [3] exhibits the cost of a reduction in static noise margin and performance.

Moreover, SRAM is commonly used for cache memory in microprocessors, mainframe computers, engineering workplaces, and memory in handheld devices because of its fast speed and low power consumption. Moreover, due to this, it has been quick difference detection and improved balance structure that takes up less space, the six-transistor SRAM cell has become an industrial standard. Furthermore, the cell ratio should be above one and pull up ratio is below one to achieve

IOP Conf. Series: Materials Science and Engineering 1272 (2022) 012007 doi:10.1088/1757-899X/1272/1/012007

continuous operation of read and write. The cell ratio (CR) and pull up ratio (PR) are given below-

$$CR = \left(\frac{pull-down-transistor-width}{access-\ transistor-width}\right)$$

$$PR = \left(\frac{pull-up-transistor-width}{access-transistor-width}\right)$$

Moreover, cell ratio and pullup ratio create a problem, that is identified the access transistor sizing problem (ATS). Furthermore, when the supply voltage is high, the size of the transistor is necessary for the traditional 6T topology to assure proper operation of the static memory circuit. However, maintaining stability and access time while addressing low-power challenges, on the other hand, remains a challenge. Furthermore, cell current (I_{on}) should not be affected when resolving the ATS issue. Moreover, during a read operation, the cell current collected from the accessible cell is referred to as on current. Furthermore, the advantages of an upswing in I_{on} are twofold. Moreover, for starters it provides the designer to make a big array with each column of many bit cells. Another this reduces the taking time for that cell to be read.

However, for instance when the reduction in supply voltage, the write operation is limited by MOSFETs at the slow NMOS, fast PMOS corner. Furthermore, due to this NMOS is weaker at the SF corner than PMOS of the same size [4]. However, due to opposite dependence of current on the subthreshold slope, temperature variations enhance this effect. However, at higher temperatures, the slopes are nearly identical for PMOS and NMOS, but PMOS reduces more quickly as temperature drops [1]. Furthermore, due to this, PMOS current driving in the low voltage area improves. Further, as a results, the PMOS transistor's driving current exceeds that of the NMOS transistor, causing PR reduction.

2. Conventional 6T SRAM

The conventional simplified 6T SRAM circuit is depicted in Fig.1.Furthermore, this SRAM cell has 6 transistors and is connected back-to-back with two inverters, However, they have (N1, N2) two pull-down transistors (P1, P2) two pull-up transistors and (N3, N4) two pass transistors. Moreover, the word line inputs control the gate of pass transistors (WL). This cell has two-line, bit line (BL) and bit line bar (BLB) are linked from the cell whenever the word line is made high, allowing the cell to be read out or written into from the BL. Furthermore, dynamic power dissipation refers to the amount of power calculated in this condition due to output switching activity.

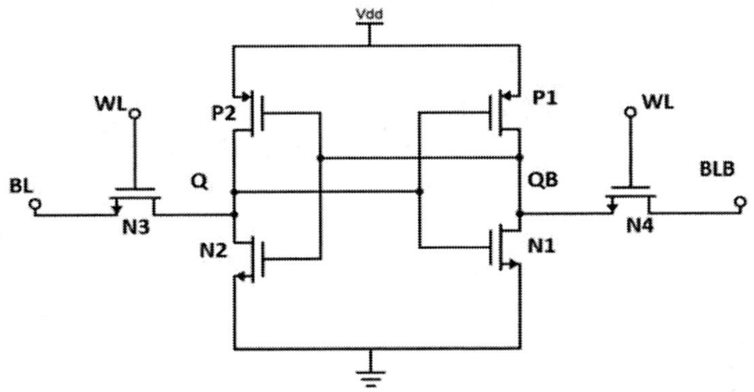

Fig. 1 Schematic diagram of conventional 6-T SRAM Cell.

IOP Conf. Series: Materials Science and Engineering 1272 (2022) 012007 doi:10.1088/1757-899X/1272/1/012007

Moreover, if the word line having zero, that time cell does not execute any reading or writing operations, hence it is known as the hold state. Further, due to this leakage power dissipation or static power calculated in this state. However, for successful writing in the cell, operation of write that checks the occurrence of value and permit the data to be written into the cell is required [12]. Again, due to this write operation is a basic AND gate with write enable and data as inputs. Furthermore, the write enables of the write driver is turned off when reading is to be done. The BL and BLB must be pre-charged to a specified level of voltage before reading occurs so that both bit lines have the same voltage.
Moreover, a pre-charge circuit conducts this pre-charge. Further, Followim pre-charging, the capacitor at one end drains according to the voltage difference between nodes of the access transistor, dependent on the value stored in the SRAM cell's Q and QB (output). However, due to these characteristics, to detects and amplifies the difference between the voltages in Bit Line and Bit Line Bar used the sense amplifier which is a differential amplifier. Furthermore, at the process of reading, the output lines Bit Lines and Bit Lines Bar are used. Furthermore, all the above characteristic which are explained about 6T SRAM.

3. Present Status of SRAM ranging from 6 to 12-Transistors

To resolve issues, a lot of new topologies ranging from 6 to 12 T have been proposed. Furthermore, differential read and differential write, differential read and single-ended write, single-ended read and single-ended write, and differential write single-ended read are some of the working operations in these papers. Moreover, several methodologies have been examined for reducing power dissipation while improving speed time and stability but maintaining other parameters under the regulator. The following papers show the contribution of different parameter read contribution circuits [4], write and read port decoupling [5], write contribution circuits performance [6], and for improving write and read noise margins a loop cutting technique [7] is used while also voltage scaling is increasing.

H.A.kusha [8] *at. al.*, presented to improve the bit cell's (RSNM) read static noise margin at a low supply voltage in a 7T SRAM cell. Moreover, in addition the author has been shown research on the impact of the process on the projected methodology which has different variations, but during a read operation, this cell absorbed more power. They introduced an 8T SRAM cell with improved read margin and lower read power by Farshad Moradi *at. al.* [9]. However, with a reduction in supply voltage, the circuit increases the read delay significantly larger than for a normal 6T cell. At process corners to increase read and write margin, there has been constructed an extra differential read and differential write 8T SRAM circuit [10] with configurable access. Further, due to this circuit was employed by the using source of buffer transistors connected with a common (word line).

Z. Volkan kursun Liu *at. al.* [11], has been introduce about 9T SRAM cell shown the noise margin in bit cell is increased when leakage power is decreased. This cell is used to separate access transistors to achieve a better noise margin in read mode and control leakage current using sleep transistors. Furthermore, a logic design of p-p-n based to drive the cell at low supply voltages and having less leakage current in the circuit has been explained in [12]. Again, Author in [12], used a logic design of p-p-n based to drive the cell at low supply voltages and having less leakage current in the circuit. In addition, the cell showed a good amount of versatility in terms of processing changes. V. Sharma *at. al.* in [13], created an additional improvement in write/read noise margin and proposed for low leakage operation in this circuit in Ref. [12]. when the use of the reading coupled approach, the redesigned circuit at process corners simultaneously addresses read/write stability difficulties in

Ref. [13].

Furthermore, Ref. [13] described the data-aware power supply methodology that allows circuits to be proposed for a reduction in hold power significantly. In addition, the half select issue overcomes in cell-like [12]. In Ref. [13] the revised version specifically mentions applications of IoT and significantly reduces leakage by employing the stack transistor. Moreover, a few other studied techniques are NMOS using sleep mode has been studied in [14-15]. Furthermore, author in [16-19] using read buffers, and monitoring of swing in voltage for maintaining constancy with reducing dissipation of power for developing topologies of SRAM. However, to reduce switching and leakage power while also reducing chip space, topologies that use read single-ended or write single ended techniques are used. Moreover, this method has a longer write/read access time. Further, in single-ended techniques, write and read sensing capacity decreases when supply voltage scales in the context of process changes [4-6, 7]. This happens due to the absence of a complementary bit line (BLB). Furthermore, to solve these problems, a few write singleended and read single-ended methodologies with various assist strategies have been proposed [4-6]. Moreover, to implement these aids, a silicon area is required, which is seen as a significant tradeoff in the present drive for greater memory and smaller device dimensions. Many researchers are also reported the latest unconventional devices and circuits for future flexible memory circuits [20-24].

4. Previous Work of 10T SRAM

Fig. 2 shows a schematic illustration of a 10T SRAM circuit. Table 2 shows the read/write/standby mode control signal position for a 10T circuit. Furthermore, during a read operation, the described topology operates in differential mode. However, for this architecture, P1 and P2 transistors, as well as N1 and N2 transistors, form the basic latch structure as shown in Fig. 2. Further, read (RD) and write (WR) signals to control the gate control of N3 and N4 access transistors. Furthermore, in addition, N5 and N6 transistors work as reading buffer transistors and are managed via the QB and Q hubs, respectively. The RD signal on the transistor N8's gate controls its operation as a buffer stack transistor. Furthermore, for this topology, transistor N8 is only active during reading operations and is turned off during write and standby operations.

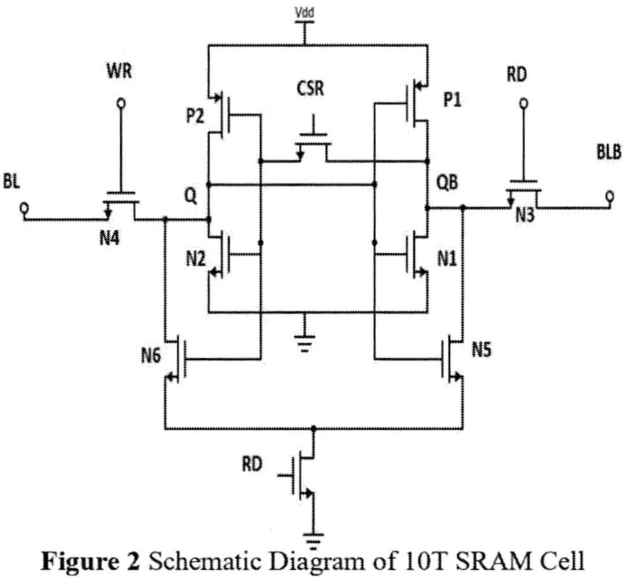

Figure 2 Schematic Diagram of 10T SRAM Cell

The control signal read is linked to the gate node of the (CSR) transistor N7. Again, for reading and hold activities the CSR signal remains logic '1' and its opposite stage is used during write operations. Moreover, when reading, this circuit operates in differential mode, and when writing, it operates in single-end mode. Table 2 indicates that throughout the read operation, all controller signals are put at logic '1'. Consider the case when logic '1' and '0' occur on nodes Q and QB, respectively.

Table 1. Different modes of 10T SRAM cell

Operation	Read Signal	Write Signal	Control Signal Read
Read-Mode	1	1	1
Write-Mode	0	1	0
Hold-State	0	0	1

The N1, N3, N4, N5, N7, N8, and P2 transistors are switched on in this condition. Furthermore, the N8 read stack transistor offers the discharge path to the transistor N5 during the operation of reading. The operation write is single-ended. No, write aid circuit is used in this topology. As a result, access transistor WR is put strong enough to achieve the intended write operation, and one of the feedback loops is disabled. To turn off the transistor N7 kept the CSR signal low, weakening the feedback loop.

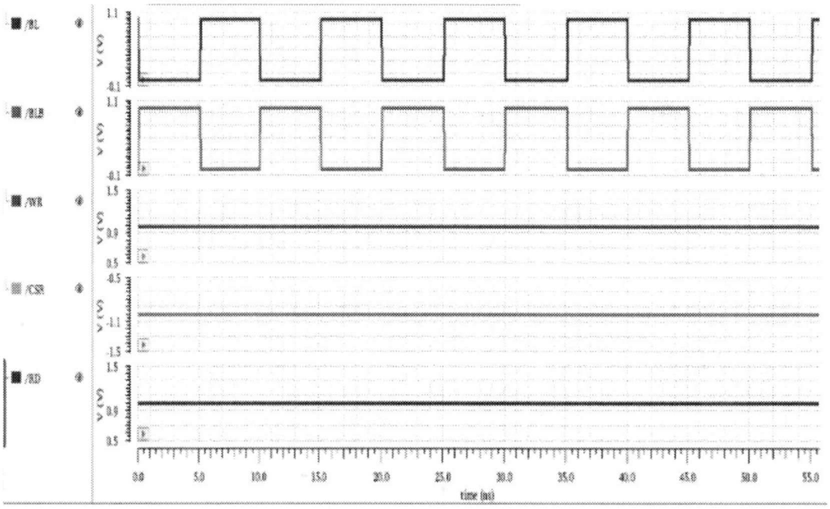

Figure 3. Wave-form for Read Operation of 10T SRAM Cell

In addition, to get a non-destructive write operation, through the data BL is straight linked to the feedback of the latch in this cell, while the other feedback is eliminated by disabling CSR. To enforce bit flipping, this provides an extra charge to the targeted. Moreover, in addition the disabled N8 transistor causes N5 and N6 which are the read buffer transistors to become inactive to sink current. At point Q directly stored the value from the bit line, whereas the opposite value is saved on QB. As an example, when the WR control signal is logic '1' and the BL signal is also logic '1', node Q change to logic '1'. As a result, transistor P1 goes off, whereas transistor N1 is switched active. As a result, any voltage higher than logic '0' at the QB point is sunk, transistor N1 to the ground, confirming logic 'low' at the QB node. The wave form for read operation and write operation are shown in Fig. 3 and Fig 4, respectively.

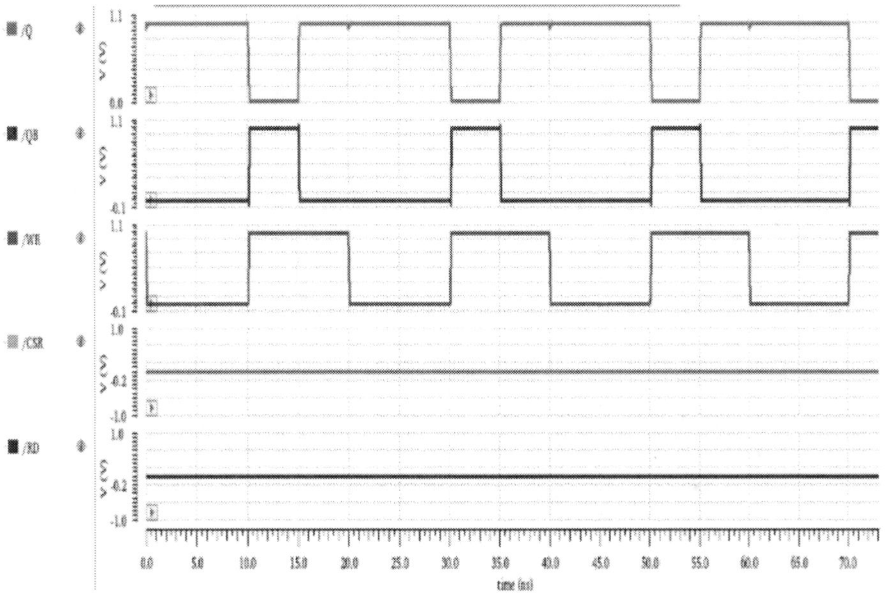

Figure 4. Wave form for Write operation of 10T SRAM Cell.

5. Drain Gating and Sleepy Keeper Technique

This section describes the in-depth about drain gating and sleepy keeper techniques along with proposed SRAM Cell configuration.

5.1. Drain Gating and Sleepy Keeper

In this method, we used Drain Gating and Sleepy Keeper techniques. However, by using these techniques to get reduced static power. Moreover, Sleepy Keeper is used as a combination of PMOS and NMOS linked between supply and circuit and the second combination of PMOS and NMOS is linked between Ground and circuit. P5 and P6 are two aid transistors are handled by output to prevent the data retention problematic. Drain gating has two sleep transistors linked between Pull-up and Pull-down. PMOS sleepy the transistor is connected between PUN (Pull Up Network) and output second NMOS combination of the sleepy transistor is linked between PDN (Pull Down Network) and output, together driven by externally.

5.2 Proposed SRAM Cell

The proposed circuit is analyzed in this section. The modified circuit works in a different mode. when we operate in Reading mode transistors P3, P4 and N8, and N9 are used as the Drain Gating technique. During reading mode control signal read transistor N7 And read signal N3 are made 'high'. N5 and N6 are driven as buffer stack transistors. At P3 and P4 transistors gives the low skd signal. Transistors N8 and N9 are having sbkd 'high' signals. Let us assume the logic 'High' at Q and logic 'Low' at QB. Transistors N2, N6, and P2 will be off during the read operation. P2, P3, P4, N1, N8, and N9 transistors conduct to allow the read operation. Proposed SRAM cell schematic diagram of SRAM cell is shown in Fig. 5

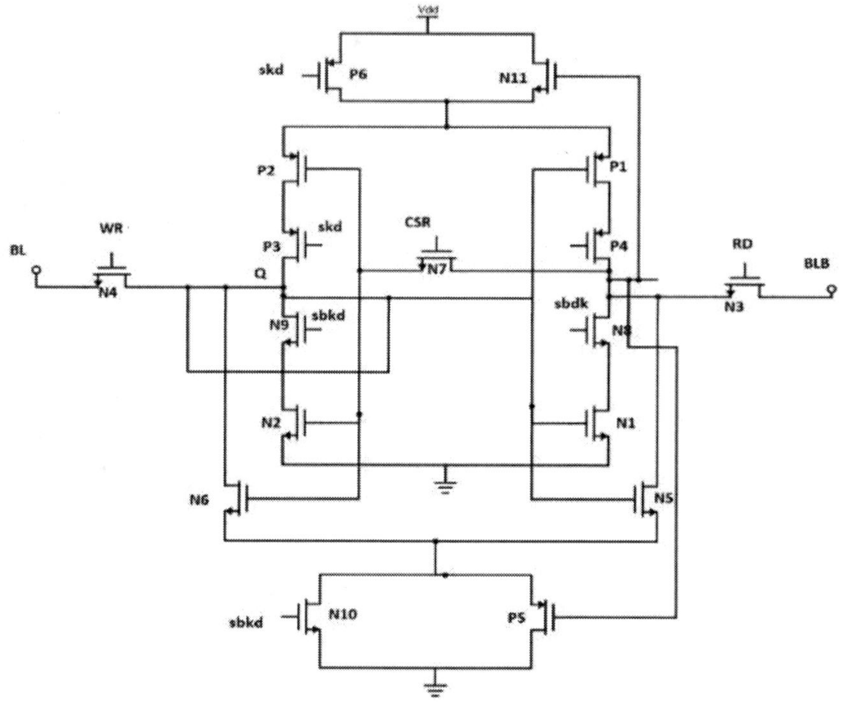

Figure 5. Proposed SRAM cell Drain Gating and Sleepy Keeper Schematic Diagram

In write mode operation according to Table.1 read signal and control signal read is made low. WR signal is kept sufficient Strong to get the desired operation. Due to CSR low feedback loop became weak. P1 will be disabled and N1 gets turned on. Transistors N1, N2, P2, P3, P4, N5, N6, N8, and N9 are taking the participant in a write operation. The read buffer transistors N5 and N6 are unable to sink current. At point Q, the data from the BL is saved directly, whereas at QB, the opposite value is saved. The cell signaling scheme for a write operation is shown in Fig. 6 (b).

The wave form for Read operation of proposed circuit and Write operation of proposed circuit is shown in Fig 6 (a) and Fig. 6 (b). Furthermore, in proposed work reading buffer transistors, during write and read operations eliminate conflict in Access transistor sizing (ATS). However, this work also ensures that PR and CR are selected separately depending on the write operation required. Moreover, for the proposed topology, transistor size criteria are applied which are describe as P1 and P2 are the pull-up transistors which put weaker to reduce the Pull Ratio and therefore make the write operation easier, comparable to conventional 6T SRAM. The cell discharge path is formed by the read buffer transistors (N5 and N6) and the transistor N8(read stack), and thus must be extended to allow a bigger read current to flow through and hence a shorter access time during operation. At condition read and write operations, access transistors (N3 and N4) are fundamentally put strongly to provide a more conduction channel between the accessed cell and BL.

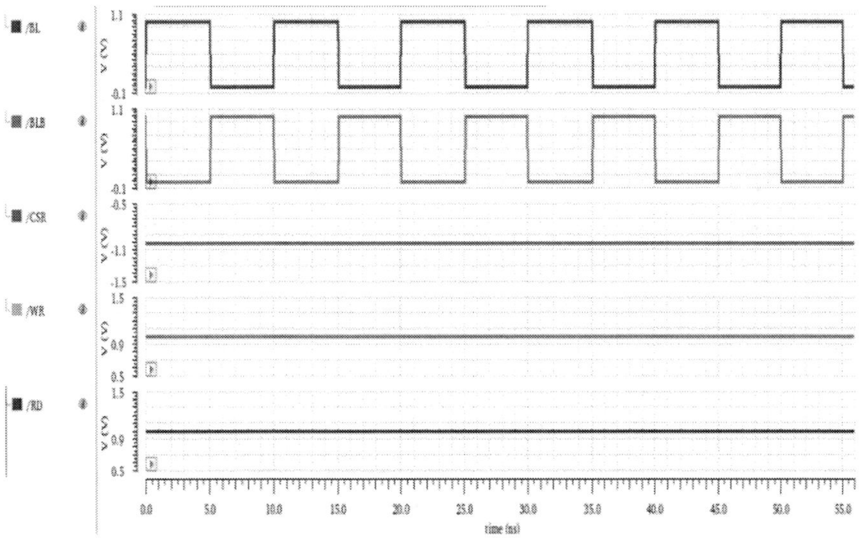

(a)

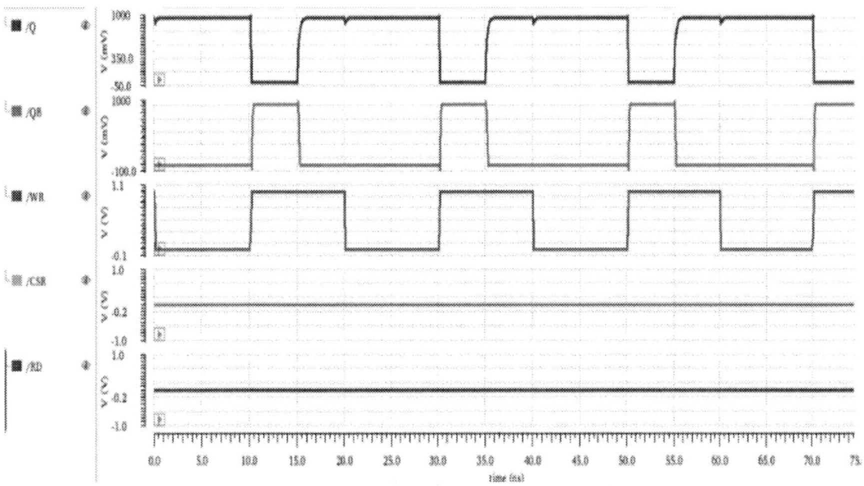

(b)

Figure 6. Wave form for (a) Read operation of proposed circuit and (b) Wave form Write operation of proposed circuit.

6. Results and Discussions

This section is described the result and discussion of simulated data of Drain Gating and Sleepy Keeper techniques. The Table 2 includes the results of the average power dissipation, delay, and power delay product (PDP). The performance parameters are extracted that includes the average power during the read and write operation of the proposed SRAM cell and compared with reported design and observed the overall better results. In this research paper, all result is extracted through the cadence virtuoso using 45nm technology node state of art industry standard tool. The comparison of read power and write power of 10T SRAM cell and proposed SRAM cell as characteristics curves are depicted in Fig. 7 (a) and (b), respectively. The comparison results of read PDP and write PDP are shown in Fig. 8 (a) and (b), respectively.

Table 2. Comparison of different parameter 10T SRAM and proposed SRAM Cell.

S.N.	Parameter	10T SRAM Cell	Proposed SRAM Cell
1.	Read power (μW)	14.44	2.823
2.	Write power (μW)	7.063	2.291
3.	Read Delay (pS)	18.55	60.04
4.	Write Delay (nS)	10.15	10.15
5.	Read PDP (aJ)	0.267	0.1695
6.	Write PDP (aJ)	71.69	23.25

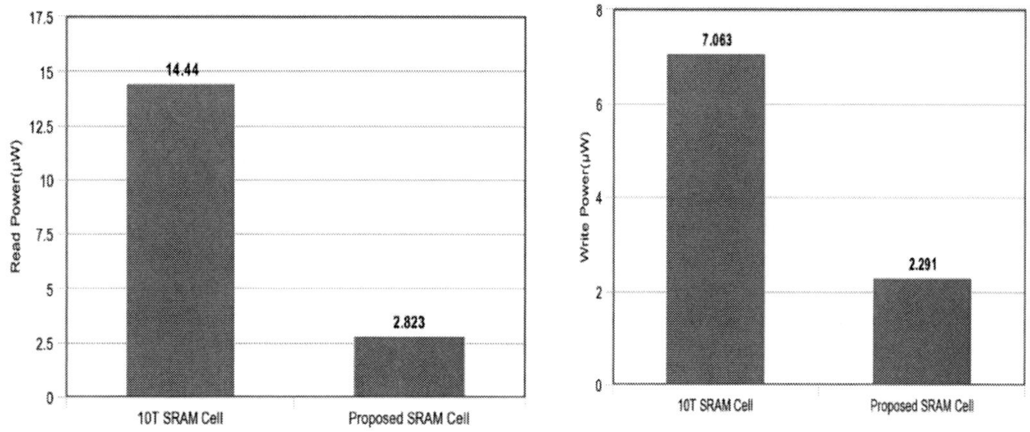

Figure 7. Comparison of the results for (a) Read Power and (b) Write Power for 10-T SRAM and Proposed Drain Gating and Sleepy Keeper SRAM Cell.

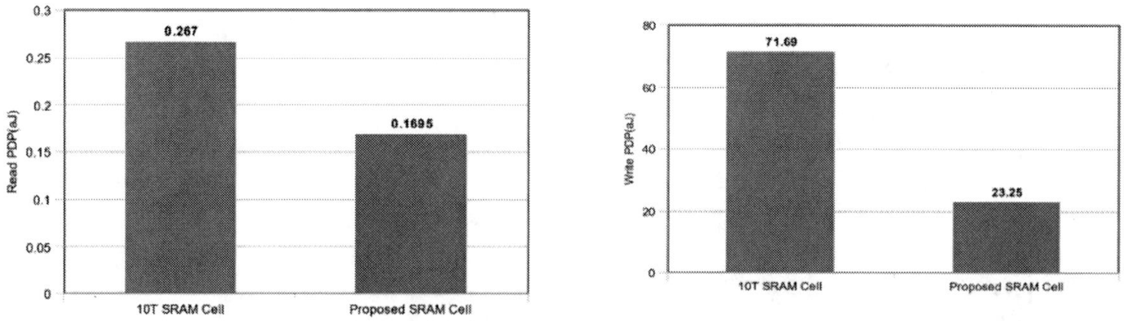

Figure 8. Comparison of the results for (a) Read PDP and (b) Write PDP for 10-T SRAM and Proposed Drain Gating and Sleepy Keeper SRAM Cell.

7. Conclusion

The low PDP and current leakage are two main issues that are addressed in this paper and that helped in improvement of SRAM design and overall performance. Where these techniques significantly decreased the power, delay, and PDP values. However, the area has been expanded due to the addition of a sleep transistor using the SK (Sleepy Keeper) and DG (Drain Gating) techniques, however, power consumption is decreased largely by use of this techniques. PDP has been lowered in this circuit when compared to conventional 10-T SRAM circuitry. The Read Delay Product and the Write Delay Product have been decreased by 36.7 % and 67.5 % respectively that is better for compact digital circuit designs.

References

1. Calhoun, B.H., Chandrakasan, A.P.: Static noise margin variation for sub-threshold SRAM in 65-nm CMOS. IEEE J. Solid State Circ. 41(7), 1673–1679 (2006).
2. Verma, N., Chandrakasan, A.P.: 256 kb 65 nm 8t subthreshold SRAM employing sense-amplifier redundancy. IEEE J. Solid State Circ. 43(1), 141–149 (2008).
3. Calhoun, B.H., Wang, A., Chandrakasan, A.: Modeling and sizing for minimum energy operation in subthreshold circuits. IEEE J. Solid State Circ. 40(9), 1778–1786 (2005).
4. Gupta, S., Gupta, K., Pandey, N.: A 32-nm subthreshold 7T SRAM bit cell with read assist. IEEE Trans. Very Large Scale Integr. Syst. 25(12), 3473–3483 (2017).
5. Verma, N., Chandrakasan, A.P.: 256 kb 65 nm 8t subthreshold SRAM employing sense-amplifier redundancy. IEEE J. Solid State Circ. 43(1), 141–149 (2008).
6. Farkhani, H., Peiravi, A., Moradi, F.: A new asymmetric 6T SRAM cell witha write assist technique in 65nm CMOS technology. Microelectron.J. 45(11), 1556–1565 (2014).
7. Wen, L., Li, Z., Li, Y.: Single-ended, robust 8T SRAM cell for low-voltage operation. Microelectron. J. 44(8), 718–728 (2013).
8. Hassan Akusha, Ansari, M., et al.: A near-threshold 7T SRAM cell with high write and read margins and low write time for sub-20 nm in FET technologies. Integration. 50, 91–106 (2015).
9. Farshad Moradi, Anh-Tuan, D., et al.: An 8T differential SRAM with improved noise margin for bit-interleaving in 65 nm CMOS. IEEE Trans. Circ. Syst. I: Reg. Papers. 58(6), 1252–1263 (2011).
10. Pasandi, G., Fakhraie, S.M.: An 8t low-voltage and low-leakage half- selection disturb-free SRAM using bulk-CMOS and FINFETS. IEEE Trans. Electron. Dev. 61(7), 2357–2363 (2014).
11. Upadhyay, P., et al.: A design of low swing and multi-threshold voltage based low power 12T SRAM cell. Comput. Electr. Eng. 45, 108–121 (2015).
12. Hung C, Lo, S.Y.: P-P-N based 10T SRAM cell for low-leakage and resilient subthreshold operation. IEEE J. Solid-State Circ. 46(3), 695–704 (2011).
13. Sharma, V., et al.: A robust, ultra-low-power, data-dependent-power-supplied 11T SRAM cell with expanded read/write stabilities for internet-of-things applications. Analog Integr. Circ. Signal Process. 98(2), 331–346 (2019).
14. Pasandi, G., Fakhraie, S.M.: An 8t low-voltage and low-leakage half- selection disturb-free SRAM using bulk-CMOS and FINFETS. IEEE Trans. Electron. Dev. 61(7), 2357–2363 (2014).
15. Upadhyay, P., et al.: A design of low swing and multi-threshold voltage based low power 12T SRAM cell. Comput. Electr. Eng. 45, 108–121 (2015).
16. CMR, P., Singh, A.K.: Novel eight-transistor SRAM cell for write power reduction. IEICE Electron Express. 7 (16), 1175–1181 (2010).
17. Limachia, M.J., Thakker, R.A., Kothari, N.J.: Characterization of a novel 10T SRAM cell with improved data stability and delay performance for 20-nm tri-gated FINFET technology. Circ World. 44(4), 187–194 (2018).
18. Rawat, B.; Mittal, P.: Single Bit Line Accessed High Performance Ultra Low Voltage Operating 7T SRAM Read Cell with Improved Read Stability. International journal of circuit theory and application. 49(5), 1435–1449, (2021).

IOP Conf. Series: Materials Science and Engineering 1272 (2022) 012007 doi:10.1088/1757-899X/1272/1/012007

19. Rawat, B.; Mittal, P.: A 32 nm single ended single port 7T SRAM for low power utilization. Semiconductor Science and Technology. 36(9), 095006, (2021).
20. Rathod, A.P.S.; Lakhera, P.; Baliga, A.K.; Mittal, P.; Kumar, B.: Performance comparison of pass transistor and CMOS logic configuration based de-multiplexers. International Conference on Computing, Communication & Automation. 1433-1437, (2015).
21. Negi, S.; Mittal, P.; Kumar, B.: Modeling and Analysis of High-Performance Triple Hole Block Layer Organic LED Based Light Sensor for Detection of Ovarian Cancer. IEEE Transactions on Circuits and Systems I: Regular Papers. 68(8), 3254-3264, (2021).
22. Kaushik, B.K.; Kumar, B.; Negi, Y.S.; Mittal, P.: Prospects and Limitations of Organic Thin Film Transistors (OTFTs). Advances in Computer Science, Engineering & Applications. Advances in Intelligent Systems and Computing, Vol. 167 (2), pp. 125- 139, 2012.
23. Kumar, B.; Kaushik, B.K.; Negi, Y.S.: Design and Analysis of Noise Margin, Write Ability and Read Stability of Organic and Hybrid 6-T SRAM Cell. Microelectronics Reliability, Vol. 54, No. 12, pp. 2801-2812, Dec 2014.
24. Kumar, B; Mittal, P.; Kaushik, B.K.; Negi, Y.S.; Mandal, A.: Organic thin film transistors characteristics parameters, structures, and their applications. Proc. IEEE, Recent Advances in Intelligent Computational Systems (RAICS-2011), pp. 706– 711, Trivandrum, Kerala, India, Sept. 22-24, 2011

Design of Two Stage OTA based Sixth Order Butter-worth Low Pass Filter using CMOS 180nm Technology

Anikesh Agrawal[1], D S Ajnar[1] and Rajesh Khatri[1]

[1] Department of Electronics and Instrumentation Engineering
Shri G. S. Institute of Technology and Science, Indore, India

anikesh.agrawal04@gmail.com, ajnards@gmail.com, rajeshkhatri1@gmail.com

Abstract. This paper shows the design of 6th order Butter-worth low pass filter based on operational trans-conductance amplifier (OTA). The designing and simulation results are obtained using Cadence Virtuoso tool with 180nm (0.18μm) technology. The simulation results of OTA have a gain of about 97.3dB. The simulation results of butter-worth low pass filter for 1.25V supply voltage and 20μA biasing current gives -96.4mdB gain and 2.9GHz bandwidth for a frequency range of 100Hz to 10GHz. The results are presented in this paper.
Keywords. CMOS IC Design, OTA, Op-Amp, BW, Cadence, 180nm, Gm-C, Butterworth filter, LPF

1. Introduction

As per the latest trends developers are trying to build devices that consume less power and have small size. Operational Trans-conductance amplifiers (OTAs) [1] are used in low power applications and are a basic component of analog circuits due to their linear input and output properties. In OTAs there's a trade-off between speed, gain, power and slew rate as these are contradicting parameters [2]. OTA is a basic component in the implementation of systems, such as filters (Gm-C) [7], and oscillators. The output of the active OTAs can alter the efficiency of active filters.
The main property of a Butter-worth filter is its highly smooth magnitude response, therefore it is also called Maximally Flat response filter and that is why it is used in the applications where monotonic amplitude is required [2]. Higher order filters are used in most of the communication related fields as it can reduce noise. In this paper, a 6th order butter-worth LPF (BLPF) based on a 2 stage operational transconductance amplifier is presented with bandwidth in the range of GHz.
The contents in this paper are organized in following way. Section 2 has a brief introduction about conventional design of butter-worth LPF. Section 3 has brief overview of operational trans-conductance amplifier structure. The proposed two stage OTA based 6th order butter-worth LPF is given in Section 4. Simulation results of proposed filter are discussed in Section 5. Section 6 handles the conclusion.

2. Conventional Butter-worth Filter Based On Two Stage OTA

The conventional 6th order butter-worth LPF consists of three telescopic OTAs cascaded with a combination of Resistors and Capacitors. Each conventional OTA has a biasing voltage of VDD = 1.25V and biasing current Idc = 15.6μA. The values of Resistor, R and Capacitor, C is R = 1KΩ and C = 1pF. These connections are repeated for other OTAs. Telescopic OTA based conventional BLPF has significantly higher bandwidth and lower power consumption.
The parameters of conventional telescopic OTA are given below in Table 1.

Table 1 – Various parameter values of conventional telescopic OTA.

Parameters	Values
Technology (nm)	180
M1,M2,M3,M4,M5,M6,M7,M8 (W/L) [μm]	10/01
M9,M10 (W/L) [μm]	15/01
M11,M12,M13,M14 (W/L) [μm]	05/01
Gain (dB)	81.70
UGB (MHz)	27.107

3. Operational Trans-Conductance Amplifier

An operational trans-conductance amplifier has differential inputs much like an operational amplifier (Op-amp) but output of an OTA is defined by current. The input - output relationship of the amplifier is given by

$$Io = gm\ Vin \qquad (1)$$

where Vin = Input Voltage, Io = Output Current and gm = Trans-conductance of OTA.

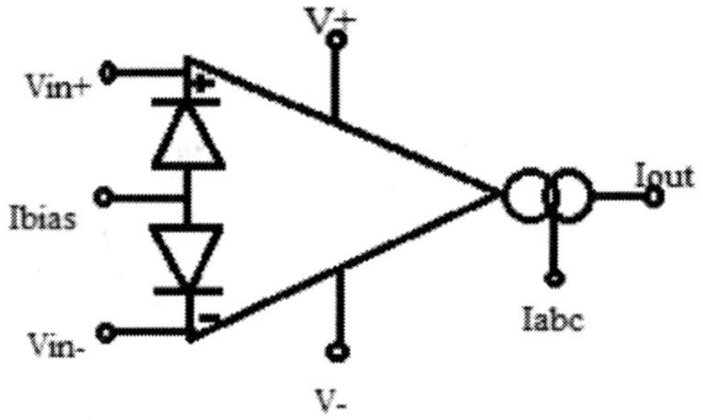

Figure 1. Figure showing OTA symbolic representation.

Figure 1 below shows the symbolic representation of the OTA. Gain(A_v), load capacitance(C), input voltages V_{in}(min) and V_{in}(max), etc. are the key parameters that characterize an operational trans-conductance amplifier. The performance of an OTA can be defined by the design parameters like I_{dc}, sizing of the transistors, etc.

In a two stage OTA first stage gives high gain and after that the second stage which provides the higher voltage swing at the output.

Figure 2 below represents the two stage OTA configuration.

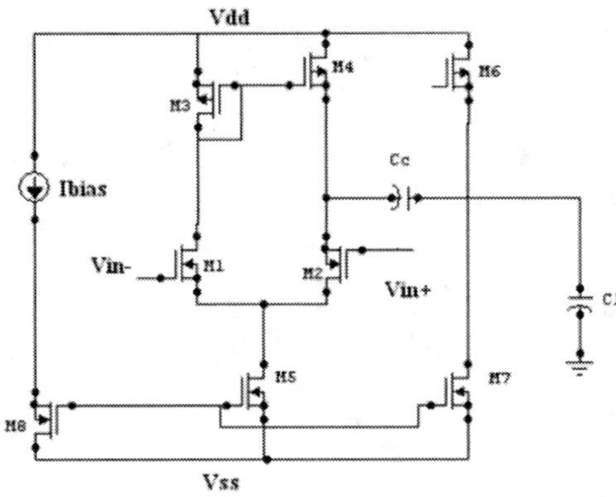

Figure 2. Figure depicting Two Stage Operational Transconductance amplifier.

4. Proposed Design Implementation

This section contains design of two stage OTA and based on it a sixth order butter-worth low pass filter. The proposed designs are implemented using Cadence design tools in 180nm technology.

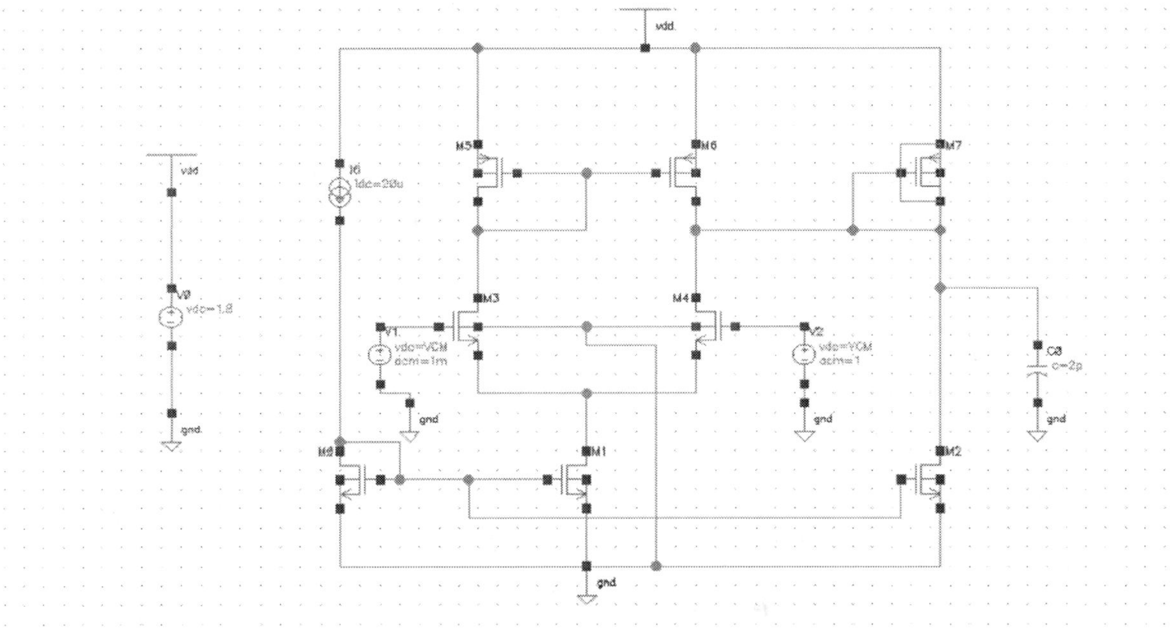

Figure 3. Schematic of the proposed 2 stage OTA.

Figure 3 given above shows the schematic of a 2 stage OTA. The above two stage OTA has 5 NMOS and 3 PMOS transistors. The applied voltage is VDD = 1.8V and biasing current Idc = 20µA and load capacitance C = 2pF.

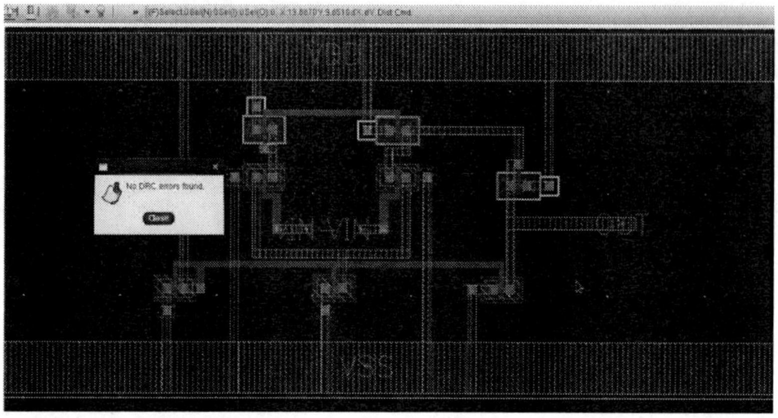

Figure 4. Layout of proposed Two stage OTA with no DRC error shown.

Figure 4 above depicts the layout of the 2 stage OTA. Layout of two stage OTA was made as per the design rules and it has no Design Rule Check (DRC) errors as shown in the figure.
Three two stage OTAs are cascaded with a sequence of resistors and capacitors to make the proposed 6th order butter-worth LPF. Each OTA in the filter has a supply voltage of VDD = 1.25V and current Idc = 20μA. At the input an AC source is applied. The output is taken at load resistor of 10KΩ which is connected to the last amplifier.

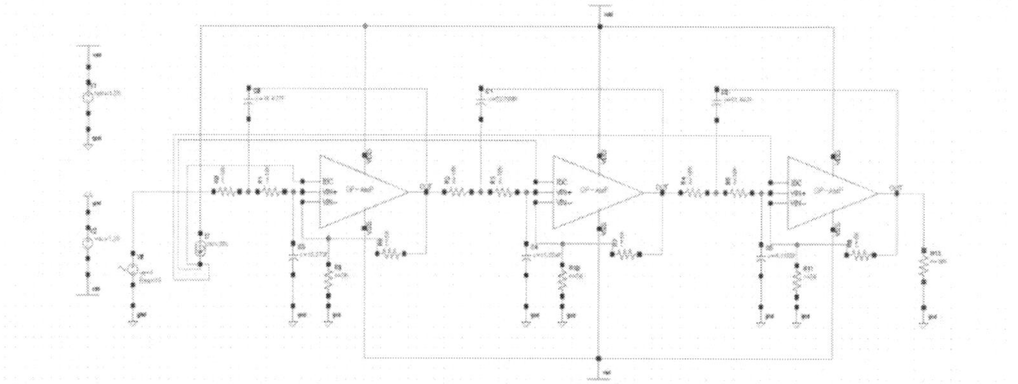

Figure 5. Schematic of 6th order Butterworth low pass filter.

Figure 5 above shows the schematic of proposed 6th order butter-worth LPF. The butter-worth filter is based upon Sallen - Key topology in which using a combination of resistors and capacitors of different values three OTAs are cascaded. Gain of each OTA is 97.3dB. Stage Gain of each stage is 2.

5. Simulation Results

The results of simulation of 6th order butter-worth low pass filter are given in Table II below. Figure 6 below gives the magnitude response and the phase response of the two stage OTA obtained using the given parameters.

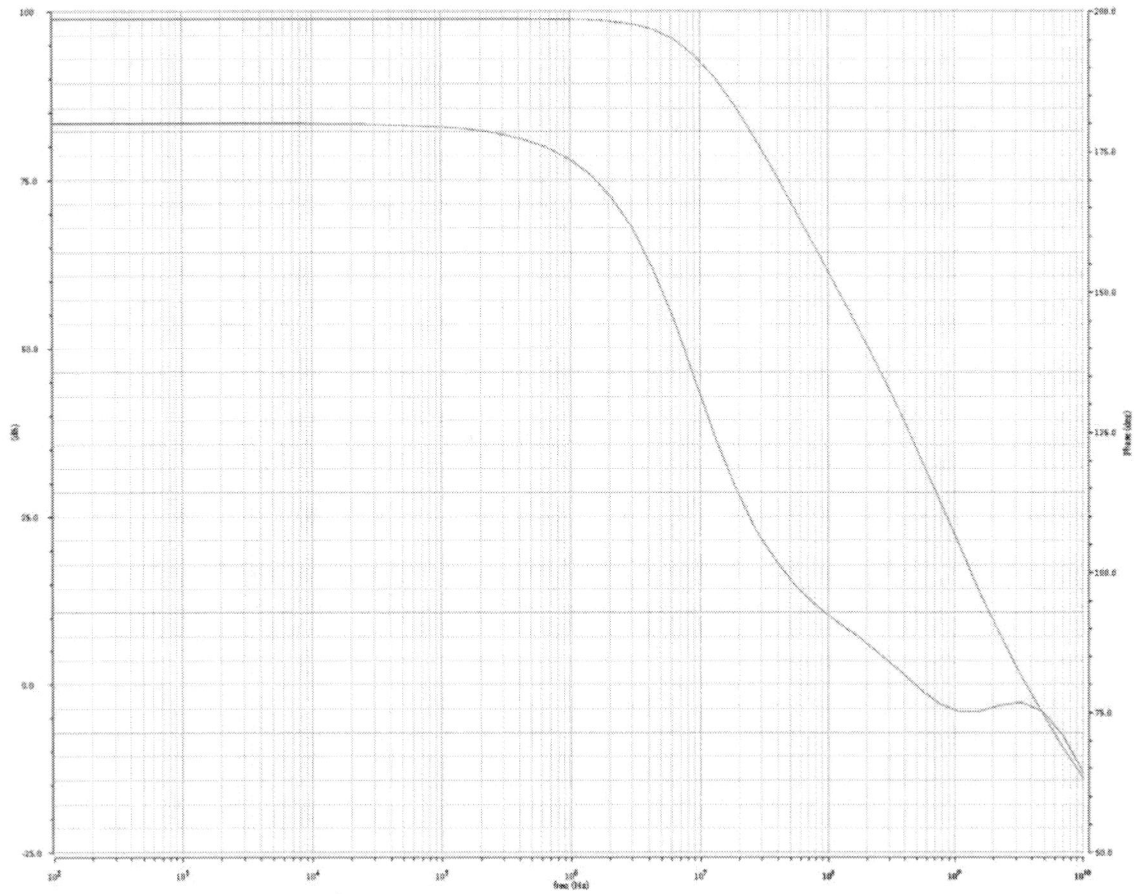

Figure 6. Magnitude response and phase response of the Two stage OTA

The magnitude response and the phase response for two stage OTA based sixth order BLPF is shown in Figure 7.

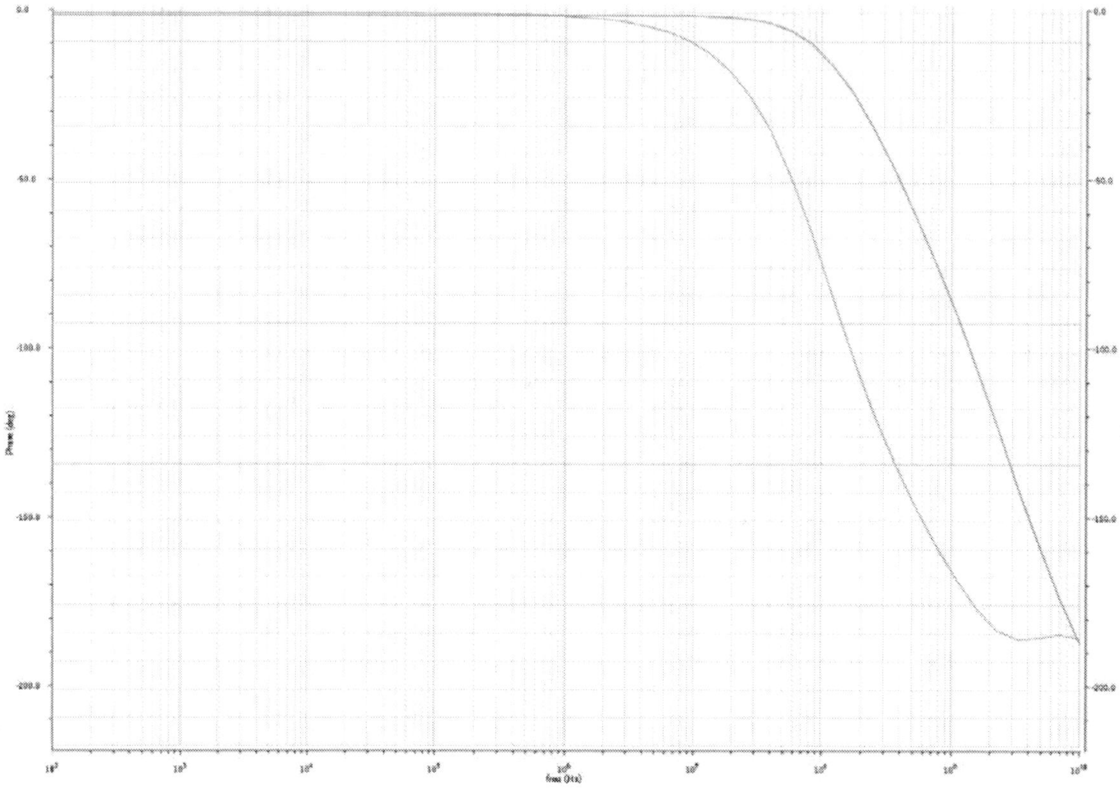

Figure 7. Magnitude response and phase response of Butterworth filter.

Figure 6 shows that the gain value obtained for the OTA is 97.3dB.
Figure 7 represents the gain-phase of butter-worth low pass filter and shows that at the stop-band frequency 2.9GHz of bandwidth is achieved.

Table 2- Conclusive table of parameters of proposed and conventional work.

Parameters	[8]	[2]	This Work
Topology	Gm-C	Gm-C	Gm-C
Technology (nm)	180	180	180
Order (n)	6	6	6
Supply Voltage VDD (V)	1.8	1.25	1.25
Bandwidth (MHz)	65	2200	2900
Area (mm2)	0.21	0.232	0.217

6. Conclusion

This paper shows the idea of a sixth order Butterworth LPF based on two stage OTA implemented using Cadence Virtuoso tool in 180nm CMOS technology. The proposed butter-worth can be used for high frequency applications.

7. References

[1] Pandya Kalpesh B. and Shah Kehul A. March 2013 Performance Evaluation of Different Types of CMOS Operational Transconductance Amplifier *Int. Journal of Science and Research (IJSR)* India Online ISSN: 2319-7064 Volume 2 Issue 3.

[2] Bhargava Gaurav and Majumdar Shubhankar 18-19 July 2020 Design of Telesopic OTA based 6th order Butter-worth Low Pass Filter using 0.18µm CMOS Technology *IEEE VLSI Device Circuits and System Conf. (VLSI-DCS)* pp 489-493 doi: 10.1109/VLSIDCS47293.2020.9179746.

[3] Geiger R. L. and Sánchez-sinencio E. March 1985 *Transconductance Amplifiers: A Tutorial* vol. 1 pp. 20–32.

[4] Lee S., Wang C. and Chu Y. Feb. 2019 Low-Voltage OTA–C Filter With an Area- and Power-Efficient OTA for Biosignal Sensor Applications *IEEE Trans. Biomed. Circuits Syst.* vol. 13 no. 1 pp. 56–67 doi: 10.1109/TBCAS.2018.2882521.

[5] Mathad D. R. S. 2014 Low Frequency Filter Design using Operational Transconductance Amplifier *IOSR J. Eng.* vol. 4 no. 4 pp. 21–28 doi: 10.9790/3021-04462128.

[6] Sun C. Y. and Lee S. Y. 2018 A Fifth-Order Butterworth OTA-C LPF with Multiple-Output Differential-Input OTA for ECG Applications *IEEE Trans. Circuits Syst. II Express Briefs* vol. 65 no. 4 pp. 421–425 doi: 10.1109/TCSII.2017.2695366.

[7] Hori S., Maeda T., Matsuno N. and Hida H. 2004 Low-power widely tunable Gm-C filter with an adaptive DC-blocking triode-biased MOSFET transconductor *30th European Solid-State Circuits Conf.* pp. 99–102 doi: 10.1109/ESSCIR.2004.1356627.

[8] Mincey J. S., Briseno-Vidrios C., Silva-Martinez J. and Rodenbeck C. T. 2017 Low-Power Gm-C Filter Employing Current-Reuse Differential Difference Amplifiers *IEEE Trans. Circuits Syst. II Express Briefs* vol. 64 no. 6, pp. 635–639 doi: 10.1109/TCSII.2016.2599027.

[9] Laouej D., Daoud H. and Loulou M. 2018 Design of sixth order butterworth Gm-C filter using Particle Swarm Optimization program for biomedical application *Proc. Int. Conf. Microelectron. ICM* vol. 2017-Decem no. Icm pp. 1–5 doi: 10.1109/ICM.2017.8268842.

Design of Improved Band Gap Reference Circuit for CS-VCO Application Using 180nm CMOS Technology

Amandeep Shrivastava[1], R. S. Gamad[1], R. C. Gurjar[1]

[1] Department of Electronics and Instrumentation Engineering
Shri G. S. Institute of Technology and Science, Indore, India

amandeep.shrivastavaa@gmail.com, rsgamad@gmail.com, rcgurjar94@gmail.com

Abstract. Electronics engineers are currently faced with the task of designing linear and broad scale voltage regulated oscillator for analogue and a varied signal applications for a shorter design period of time. The Band gap reference source is used here to compensate the noise which affects the Phase lock loop operation from power supplied noise. A cascode current mirror based band gap reference current starved voltage controlled Oscillator (BG-CSVCO) is discussed in detailed as a new design strategy for reducing power supplied noise in Phase lock loop (PLL) applications. A cascode current mirror based band gap reference CS-VCO design is simulated and direct employing in 180nm Cadence CMOS technology. A frequency range of band gap reference CS-VCO is 0.009 GHz up to the 2.07 GHz for adjusting range of 0.5V to 1.8V with the power dissipation is 0.564mW. The VDD as supplied voltage was 1.8volt.

Keywords. Cascode current mirror Band gap reference circuit, Band gap reference CS-VCO, analogue and mixed mode signal circuit, PTAT, CTAT.

1. Introduction

Many systems that are electronic, such as the VCO in frequency synthesizers or VCO in Phase lock loop design which reduces a power supplied noise, require precise reference voltage sources, as shown in Fig.1 [1]. Reference voltage and the current play an essential part in analogue IC design, with defined values based on application requirements. Bandgap reference circuit are often utilised to do this on-chip. These circuits facilitate a design of reference voltages independent of the supplied voltage alteration and temperature alteration to improve the noise performance [1-2]. The necessity for small area, low-voltage, and small power operation, on the other hand, becomes particularly apparent. For battery-powered portable devices, a high-precision BGR circuit with a lower power is expected. A primary BGR circuit in CMOS commonly utilize parasitic transistors or probably the diodes to obtain high-accuracy source voltage. An output voltage in Band Gap Reference is the sum of PTAT and CTAT voltages, first one is PTAT (Proportional to Absolute Temperature) and second is CTAT (Complementary to Absolute Temperature), indicated as [3]. To reduce power supply noise, this Current Mirror bandgap reference is used to the VCO.

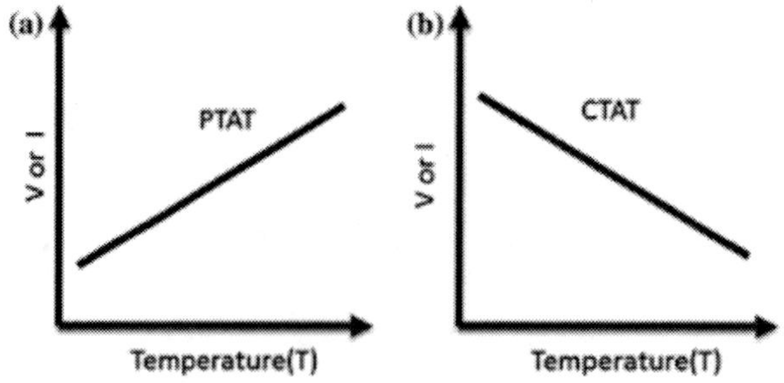

Figure 1 . Concept of BGR

IOP Conf. Series: Materials Science and Engineering 1272 (2022) 012009 doi:10.1088/1757-899X/1272/1/012009

The inclusion of PTAT and CTAT currents is the underlying concept behind the BGR circuit. With increasing temperature, a PTAT current exhibits positive slope. Their is CTAT current decreases as temperature rises. The sum will be independent of temperature if both two slopes is with uniform magnitude as well as sign.

2. Design of Cascode current mirror Bandgap reference circuit

There are various type of ways to designing the BGR circuit, some common method are by using conventional transistor, 1 or 2 stage based op amp BGR and etc. But in this paper cascode current mirror Band gap reference circuit is used to make constant voltage supply for the CS-VCO. The power supply rejection is adequate; however, cascode current mirrors can greatly improve it. MOSFETs M1 by way of M4 creates the feedback loop in this circuit, forcing a voltages in a sources M4 together with M5 should be identical. The current flowing in resistor R1 is therefore determined by a difference in the middle of VBE2 and VBE1. Now the PTAT current doubled by a ratio of M3:M2 that is ratio of current mirror and arrives at the output across resistor R2. The resultingreference voltage is

$$V_{REF} = V_{BE_3} + \left[\frac{R_2}{R_1}\frac{W_3}{L_3}\frac{L_2}{W_2}\right]\Delta V_{BE} \tag{1}$$

A width as well as a length of MOSFET M2 and M3, individually,are W3, L3, W2, and L2.

As we know,

$$V_{R2} = \frac{R_2}{R_1}\ln(N)VT \tag{2}$$

Where V_T is the PTAT Voltage and except V_T others term is constant.

$$\alpha_1 = 18.82 = \frac{R_2}{R_1}\ln(N) \tag{3}$$

Assume N=2 for PTAT current source BGR, where N- is the number of diodes or BJTs.

$$ID_5 = ID_4 = \frac{V_T \ln}{R_1} \tag{4}$$

$$V_T = 26mV \tag{5}$$

$$R1 = \frac{V_T l_n}{ID_4} \tag{6}$$

$$R1 = 3.62K\Omega \tag{7}$$

Therefore, From calculation (2)

$$R2 = 97.88K\Omega \tag{8}$$

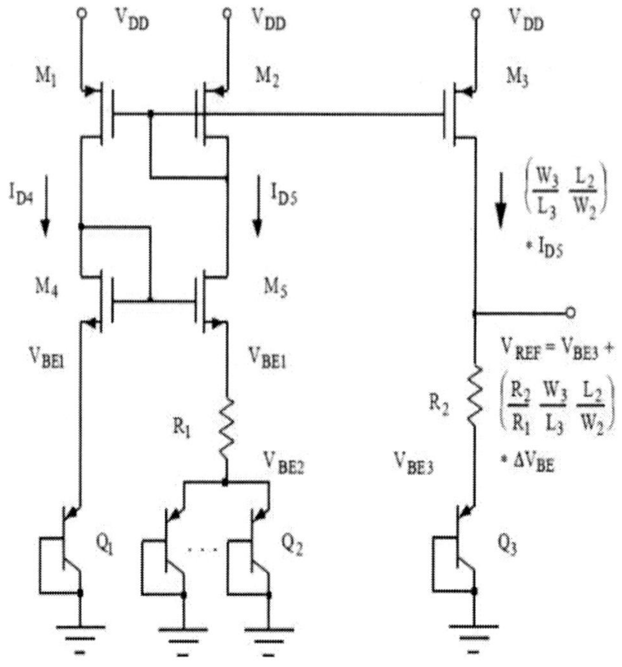

Figure 2. The Cascode Current Mirror Band gap Reference

3. Current starved VCO

A VCO is one in which the frequency of oscillation is controlled by the input voltage. A voltage controlled oscillator (VCO) is the crucial constituent while designing the high- frequency components that use Phase lock loop. The Block representation of current-starved VCO shown in Figure 3.

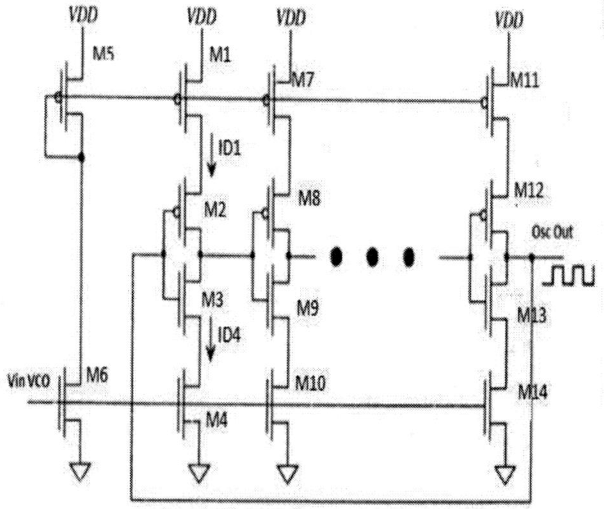

Figure 3. A simplified representation of a current starved VCO

IOP Conf. Series: Materials Science and Engineering 1272 (2022) 012009 doi:10.1088/1757-899X/1272/1/012009

A current starved voltage controlled oscillator is the most prevalent form of VCO circuit [2]. The inverter sections in this circuit is set to three. Both MOSFETs M2 and M3 function as inverters, whereas M1 as well as M4 perform like current sources. A MOSFETs Ml as well as M4 which perform like a current source that will restrict current approachable to the inverters M2 as well as M3; to put it simply, the inverter is current starved. There is identical drain current are flowing in MOS M5 and M6, which is controlled from the control input voltage. All inverter/current source step mirrors a currents in M5 as well as M6. On a drain of M2 and M3, the total capacitance is given by

$$C_{tot} = \frac{5}{2} C_{ox}\left(L_p W_p + L_n W_n\right) \tag{9}$$

A current starve VCO's oscillation frequency for the N stages are

$$F_{OSC} = \frac{I_D}{N C_{tot} V_{DD}} \tag{10}$$

$$I_D = N \times C_{tot} \times V_{DD} \times F_{OSC} \tag{11}$$

When VinVCO < V_{TN}, the VCO ceases oscillating. As a result, Vmin = V_{TN} and fmin = 0 can be defined. Finding I_D when VinVCO = VDD yields the maximum VCO oscillations frequency fmax, Vmax = VDD at the maximum frequency. A VCO's essential parameters are oscillations frequency, tuning scale, phase noise, and power dissipating [4].

4. Proposed work Simulation results and their comparison

Cadence software to simulate the voltage reference circuit along with CS-VCO in 0.18 μm standard UMC technology.

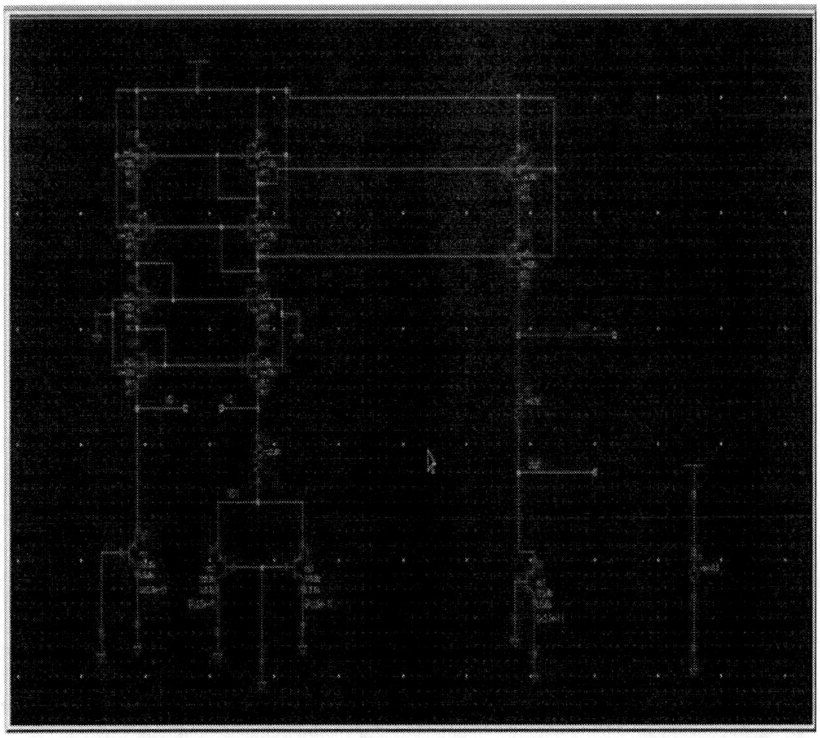

Figure 4. Schematic of proposed Cascode CM-BGR

Fig 4. shows a schematic representation of the cascode current mirror Bandgap reference Circuit, where the summation of the PTAT and CTAT of the circuit and resultant node of reference voltage.

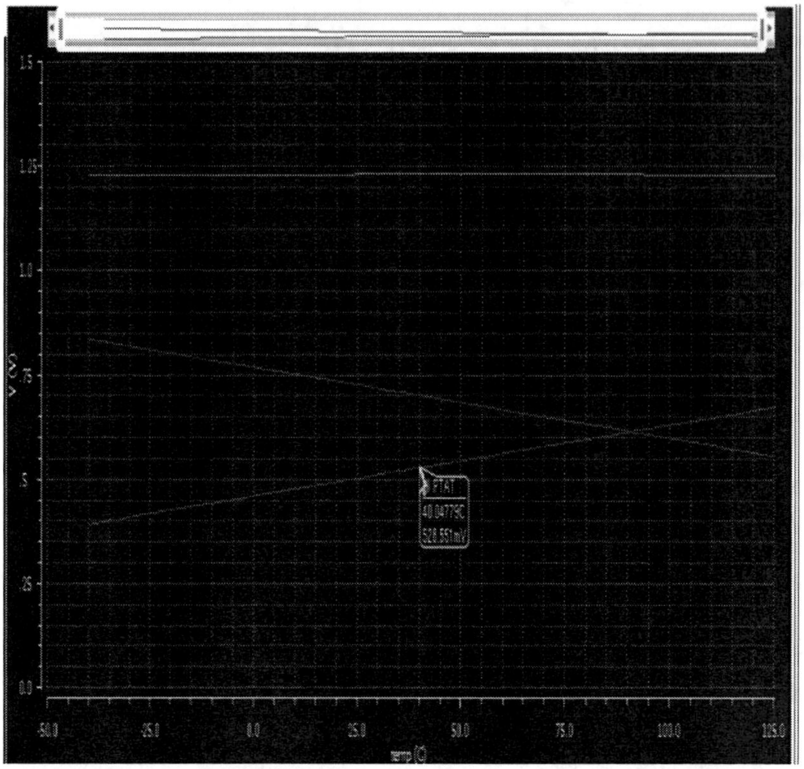

Figure 5. Schematic of Reference Voltage with temperature variation along with PTAT and CTAT

Fig 5. shows a schematic simulation of a resultant of PTAT and CTAT and Ref. voltage with respect to a temperature variation.

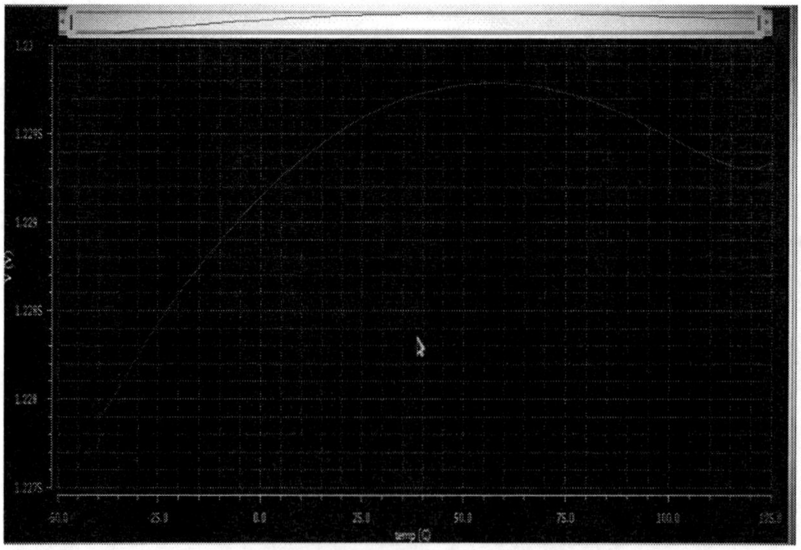

Figure 6. Extracted View of Reference Voltage with temperature variation

IOP Conf. Series: Materials Science and Engineering 1272 (2022) 012009 doi:10.1088/1757-899X/1272/1/012009

With the overall current consumption is 15.66 μA and the power consumption of 51.68μW, current is proportional to temperature.The temperatures vary from - 40 to 125°C, and the output reference voltage is approximately 1.22 V, as shown in Fig.6.

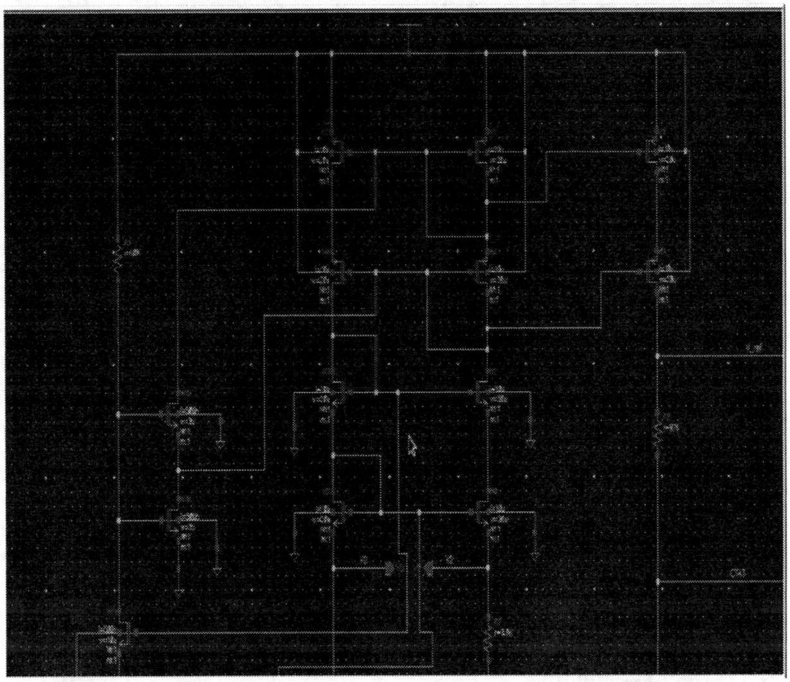

Figure 7. Cascode current mirror BGR with Startup Circuit

Fig 7. Shows the cascode current mirror BGR circuit with start up mechanism, the function of the startup circuit is to pump some current to the current mirror circuit so, that to make circuit work properly.

Table 1. shows comparison of the present work with previously published BGR work.

Parameter	[5]	[6]	[3]	This Work
Supply Voltage (Volt)	1	2.5	1.8	1.8
Current(μA)	33.50	38.00	33.79	15.66
Reference Voltage(mV)	632.9	617.7	789	1200
Process	0.18	0.35	0.18	0.18
Temp Range (°C)	-20 to 100	-15 to 150	-40 to 120	-40 to 125
Power Dissipation (μw)	-	-	60.82	51.68

IOP Conf. Series: Materials Science and Engineering 1272 (2022) 012009 doi:10.1088/1757-899X/1272/1/012009

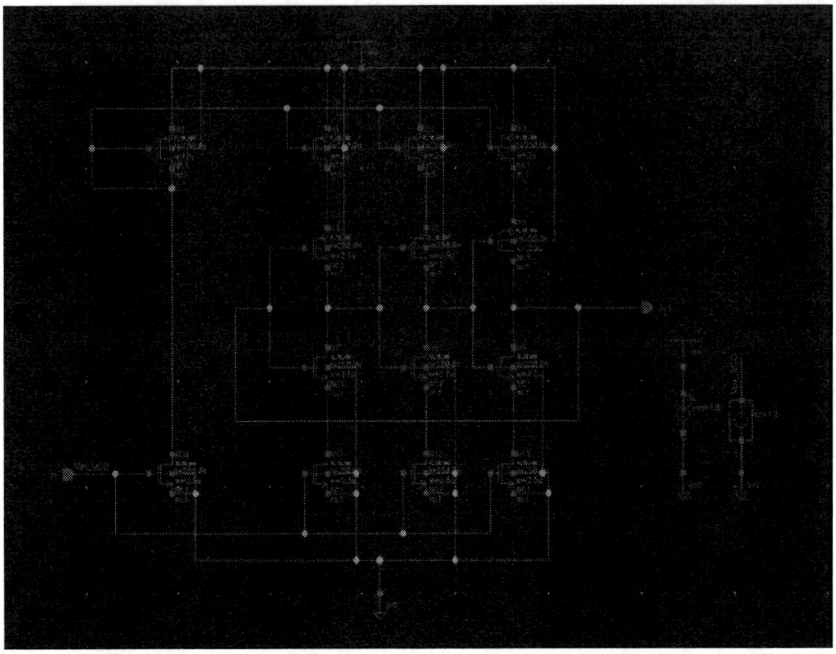

Figure 8. Schematic View of CS-VCO

Fig 8. Shows that, The identical drain currents are flowing in the MOSFETs M2 and M3 and are dominion by the Vin VCO. In all inverter current source step, a currents in M2 as well as M3 is mirrored. Both MOSFETs M2 and M3 function as inverters, whereas M1 and M4 function as current sources. The MOSFETs Ml as well as M4 which perform like a current source that will restrict a current available in to the inverter M2 as well as in M3; to put it simple way, the inverter is current starved.

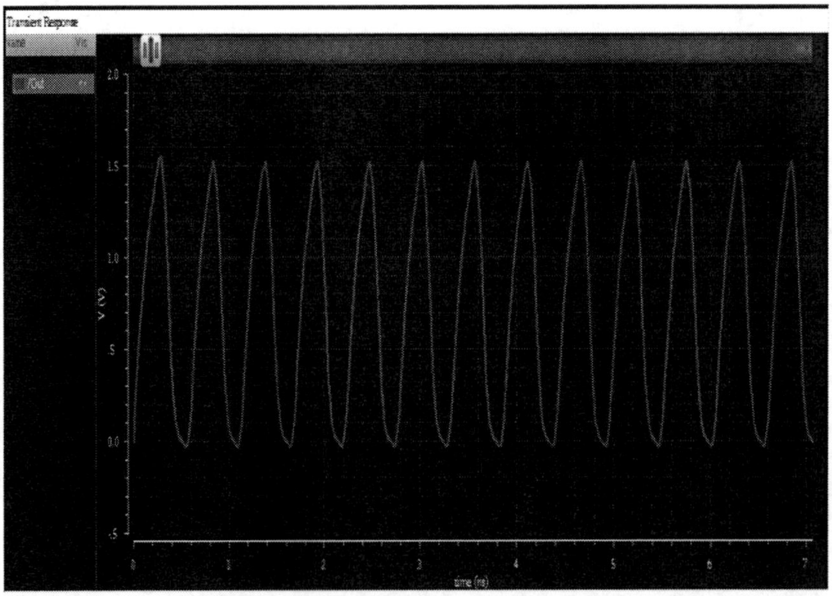

Figure 9. Simulation of CS-VCO

Fig 9. Shows the output waveform of a 3 stage CS-VCO, along with this the evaluation of a oscillating frequency comes out.

Table 2. Shows the aspect ratio of a schematic of 3 stage CS-VCO.

Device Name	Device Type	$W/_L$ *ratio* (μm/nm)
M1, M2, M3, M14	PMOS	2/250
M4, M5, M6	PMOS	2.9/250
M7, M8, M9	NMOS	2.5/250
M10, M11, M12, M13	NMOS	2.5/250

Table 3. shows comparison of the present work with previously published CS-VCO.

Specification	[2]	[4]	This work
Technology(nm)	180	180	180
Power Supply (Volt)	1.8	1.8	1.8
Power (mW)	9.5	0.6	0.56
Voltage Control (V)	-0.4–0.4	0.2-1.8	0.5-1.8
F_{OSC} (GHz)	2.6-4.4	0.002-2.15	0.009-2.07
Delay Elements	3	3	3

5. Conclusion

This research demonstrates how to create a cascode current mirror bandgap reference circuit with minimal voltage and temperature changes. Calculating a PTAT as well as CTAT currents over the resistor is a basic concept for obtaining a reference voltage. The power dissipation of 51.68 μW and current is 15.66 μA is improved in BGR from previous works.

To minimize power supply noise, exact control voltage sources is the important aspect of the VCO design. The frequency range of the BG-CSVCO is 0.009 GHz to 2.07 GHz, with a tuning range of 0.5V to 1.8V as well as a maximum power dissipation of 0.56mW at 2.07GHz. The supply voltage is 1.8 volts (VDD).

References

[1] V. G. Nasre, G. M. Asutkar, *Design of Area Efficient Bandgap Reference In 0.18μm CMOS Technology*, International Journal of Pure and Applied Research In Engineering And Technology, Vol.4 (7) pp. 64-71, March 2016.

[2] B. P. Panda, P. K. Rout, D. P. Acharya and G. Panda, *Design of a Novel Current Starved VCO via Constrained Geometric Programming*, International Symposium on Devices MEMS Intelligent Systems Communications 2011(ISSN 0975-8887) (ISBN 978-93-80747-80-2) April 12-14, 2011.

[3] Jasdeep Kaur, Roohie Kaushik, Akshita Srivastava, Sakshi Singh, *A Design Approach for Bandgap Reference Circuit*, 1st International Conference on Signal Processing, VLSI and , 2019.

[4] V. G. Nasre, G. M. Asutkar, *Design of CS-VCO with Band Gap Reference in 0.18μm CMOS Technology*, 2nd IEEE International Conference on Recent Trends in Electronics, 2017.

[5] Lee, Edward KF. *A low voltage low output impedance CMOS bandgap voltage reference*, In Circuits and Systems (ISCAS), 2013 IEEE International Symposium on, pp. 1480-1483. IEEE, 2013.

IOP Conf. Series: Materials Science and Engineering 1272 (2022) 012009 doi:10.1088/1757-899X/1272/1/012009

[6] Banba, Hironori, Hitoshi Shiga, Akira Umezawa, Takeshi Miyaba, Toru Tanzawa, Shigeru Atsumi, and Koji Sakui. *A CMOS bandgap reference circuit with sub-1-V operation*, IEEE Journal of Solid-State Circuits 34, no. 5 (1999): 670-674.
[7] Allen, Phillip E., (2002) and Douglas R. Holberg. *CMOS Analog Integrated Circuit Design.*
[8] Razavi, Design of Analog CMOS Integrated Circuit. New York: McGraw-Hill, 2001

IOP Conf. Series: Materials Science and Engineering 1272 (2022) 012010 doi:10.1088/1757-899X/1272/1/012010

Advanced Safety Programmable Logic Controller and Operational Programmable Logic Controller for Data Acquisition Control System of SADHANA Sodium Loop

M.Shanmugasundaram[1,6], D.Thirugnanamurthy[2,6], V.S.Srinivasan[3,6], T.SriDevi[4], B.K. Nashine[5,6].

[1] SO/G, IFS, SFG, RDTG, Indira Gandhi Centre for Atomic Research, Kalpakkam, India,
[2] Associate Director, EIG, Indira Gandhi Centre for Atomic Research, Kalpakkam, India
[3] Head, LIMS, SIRD, Indira Gandhi Centre for Atomic Research, Kalpakkam, India.
[4] Head, SIS, EIG, Indira Gandhi Centre for Atomic Research, Kalpakkam, India.
[5] Associate Director, SFG, Indira Gandhi Centre for Atomic Research, Kalpakkam, India
[6] Homi Bhabha National Institute, Training School Complex, Anushaktinagar, Mumbai - 400094

mss@igcar.gov.in

Abstract. The SAfety Decay Heat Analysis in NAtrium (SADHANA) sodium loop is designed with two PLC controllers. All safety control interlocks are monitored by a Safety programmable logic controller (SPLC), and temperature control, including averaging ON/OFF heaters, is managed by an Operational PLC (OPLC).The decay heat exchange system of the SADHANA sodium loop is evaluated using these combined safety and operational PLCs. In the case of reactor loss of offsite and onsite power, decay heat generated in the reactor core after shutdown is removed by four Safety Grade Decay Heat Removal (SGDHR) systems [1,2,3,4,5]. Each exchanger (AHX) is located outside the reactor containment building, and a sodium loop circulation is created by natural convection [6,7,8,9,10]. The purpose of construction of SADHANA loop, with simulated models of DHX and AHX to study the performance of the SGDHR system, has been planned [11,12,13,14,15]. The SADHANA sodium loop has been commissioned, operating at isothermal temperatures of 300°C and hot temperatures of 500°C/& 550°C. The smart instrumentation and control system comprising Loop Operational Data Acquisition and Control System (LODACS) and Experimental Data Acquisition System (EDAS) are used in the SADHANA sodium loop. During the tests, LODACS personal computer and Instrumentation panel monitor loop parameters such as sodium level, leak, and pressure. In SADHANA loop DHX and AHX temperature decay heat changes monitoring by EDAS. The temperature of the pool in Test vessel 4 (TV- 4) is controlled by immersion type sodium heaters via PID controllers. Currently, the loop consists of 64 surface heaters, 30 immersion type sodium heaters, and 205 thermocouples (84 Nos. of operational thermocouples and 121 Nos. of experimental thermocouples). According to the experimental thermocouples are being monitored by a separate fast data acquisition (scan rate of 10 seconds) by in-house developed software. This technical paper also describes the implementation of PLCs in different data acquisition control systems.

1. Temperature monitoring and control for sodium pipe lines and components.

The surface temperature of the sodium pipe line and SADHANA components is measured using a 3mm dia SS sheathed MgO insulated probe. All thermocouples needed for operation must be connected to an channel in OPDAS. These temperature readings are sent to the Large Component Test Rig (LCTR)

IOP Conf. Series: Materials Science and Engineering 1272 (2022) 012010 doi:10.1088/1757-899X/1272/1/012010

control room PC. The OPDAS will provide an ON-OFF control signal to SSR's in heater control panel in order to maintain the temperature of SADHANA's pipelines and components. Figure 1. show the location of surface heaters, leak detectors and thermocouples in the loop, pipe line and components. The SADHANA preheating system has 46 control thermocouples. Figure. 2. Shows the Flow sheet of SADHANA loop. Figure.3. Shows SADHANA Structure with AHX and Expansion Pot. Figure. 4 shows Test Vessel-4 with Decay Heat Exchanger. Figure 5 shows Experimental data acquisition system (AHX & TV-4 pool temperature monitoring)

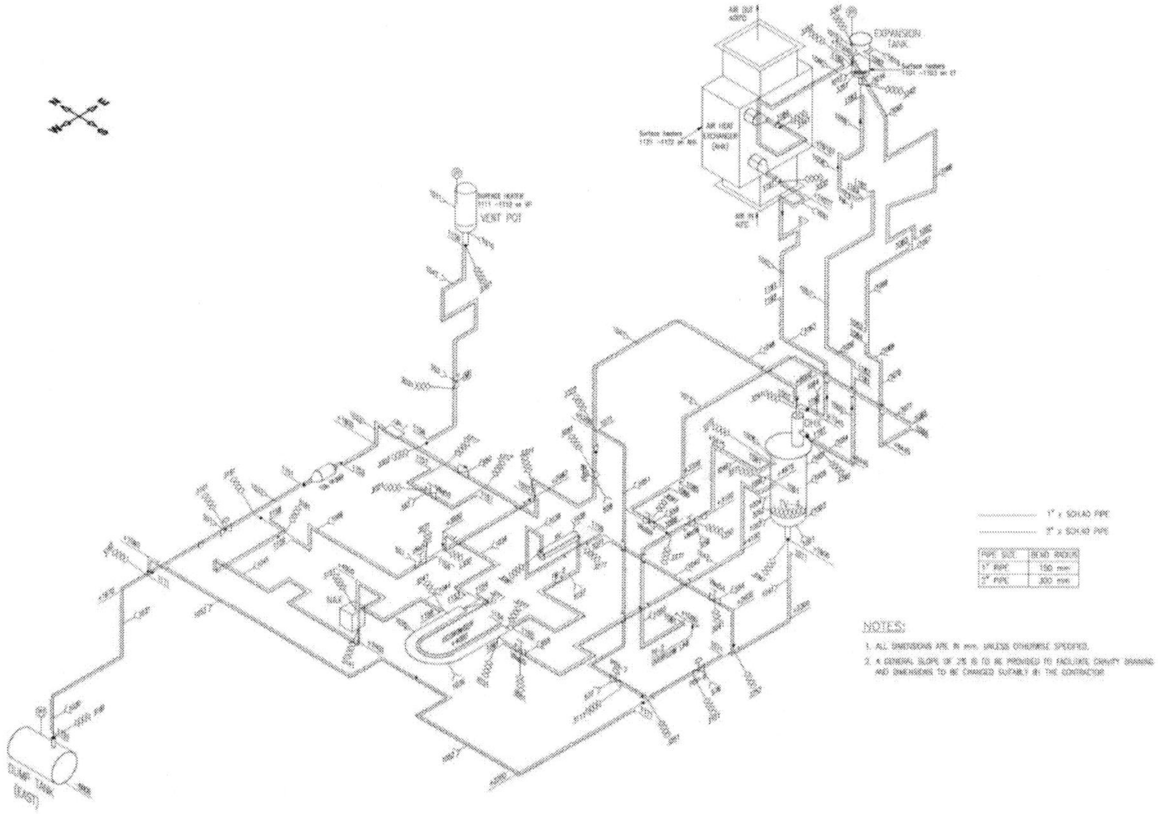

Figure 1. Surface heaters, leak detectors and thermocouples layout of SADHANA loop

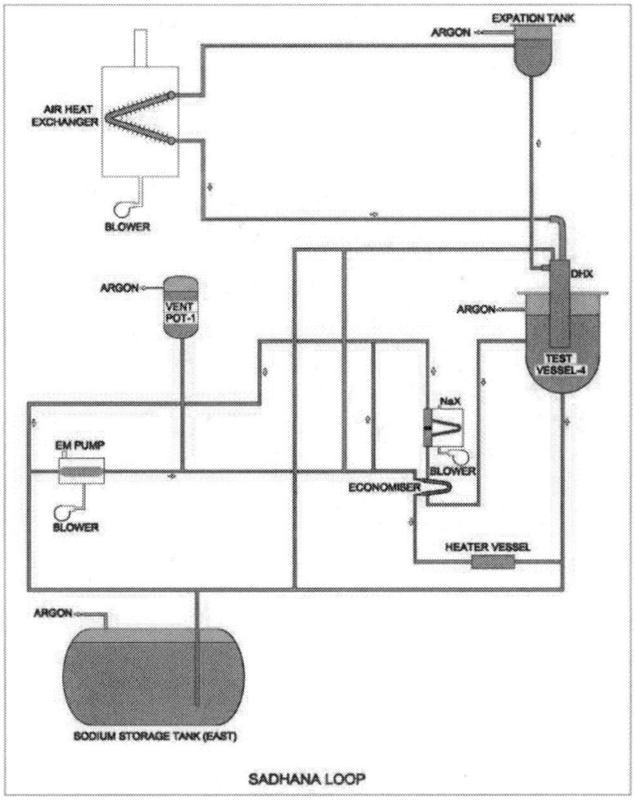

Figure. 2: Flow sheet of SADHANA loop

Figure. 3: SADHANA Structure with AHX and Expansion Pot

Figure. 4: Test Vessel-4 with Decay Heat Exchanger

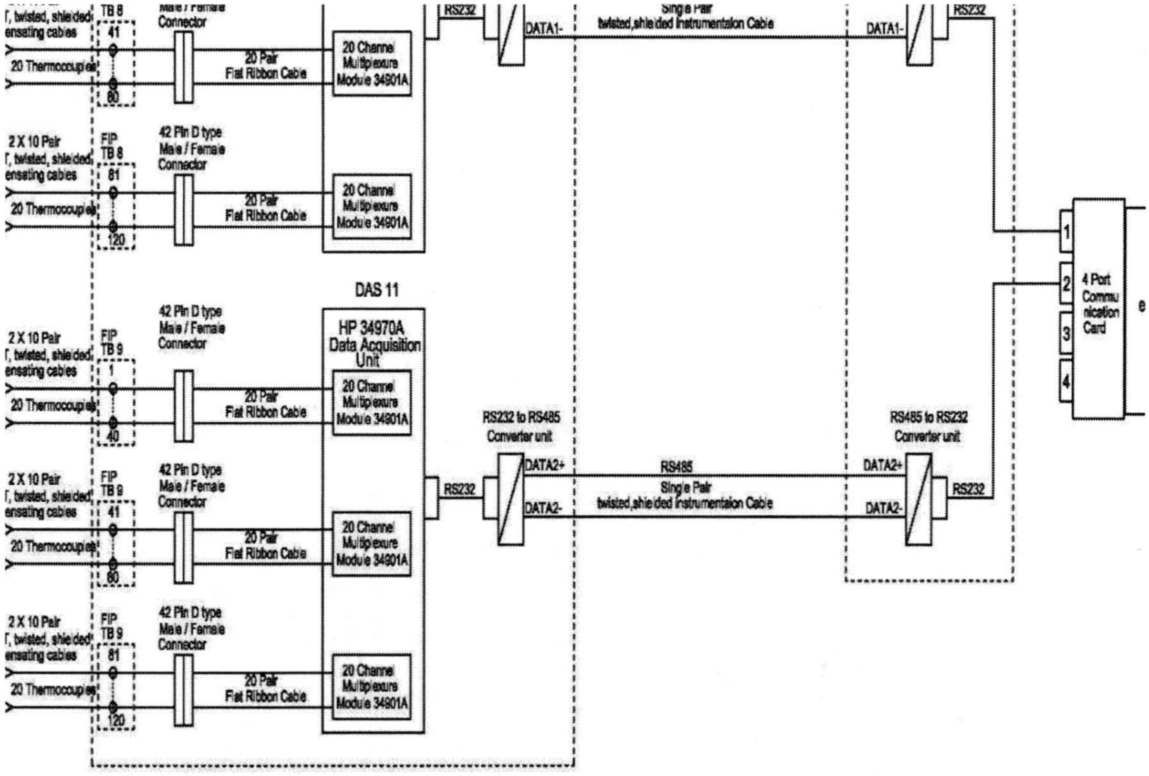

Figure 5. Experimental data acquisition system (AHX & TV-4 pool temperature monitoring)

2. Temperature Monitoring AHX and DHX temperature

The thermocouples installed on DHX and AHX are 1.5mm dia SS sheathed MgO insulated chromel - Alumel type that are installed during fabrication. These thermocouples are known as experimental thermocouples, and they are linked to two HP data loggers installed in the field instrument panel (FIP).Using three 20-channel multiplexer modules, each HP Data logger can receive 60 thermocouple signals. The thermocouple's single pair compensating cables will be connected to the EDAS10 & EDAS11 (HP data loggers) via 20 pair circular, male / female connector assemblies mounted on the field instrument panel and internal wiring. The experimental temperature data collected by the EDAS10 and EDAS11 (HP data loggers) will be sent to the control room PC via two independent serial communication link RS 232/ RS485 converters.

3. TV-4 temperature control

Electrical heaters are installed in Test Vessel -4 (TV-4) with a capacity of 355kW to simulate decay heat produced in the reactor core. The heaters are divided into six banks, four of which are controlled by an ON/OFF controller (TC2) and the remaining two by PID controllers (TC3 &TC4). The number of heaters in each bank is arranged in such a way that any amount of heating power required can be activated by operating one or more banks simultaneously. The surface temperature of TV-4 is monitored by an ON/OFF temperature controller (TC1), and if it exceeds 575° C, all six heater banks are automatically tripped. PLC executes the trip logic based on the ON/OFF temperature controller inputs. Figure .6.a,b,c,d. shows Annunciation System of SADHANA loop.

4. Heater vessel Temperature control

In heater vessel 80kW capacity, U type, immersion, electrical heaters are installed to raise the inlet sodium temperature to TV-4 to the required valve, PID controller(TC6) controls the heater vessel outlet temperature through SSRs mounted in the SSR box. The surface temperature of heater vessel is monitored if it exceeds 575°Cthe power supply to heaters will be cut off. If the flow rate of liquid sodium through the heater vessel is less than 1 cubic meter per hour all the heaters will be trip automatically. The trip logic is executed by PLC based on the inputs from ON/OFF temperature controller (TC5), and flow indicating controller (FC2).

IOP Conf. Series: Materials Science and Engineering 1272 (2022) 012010 doi:10.1088/1757-899X/1272/1/012010

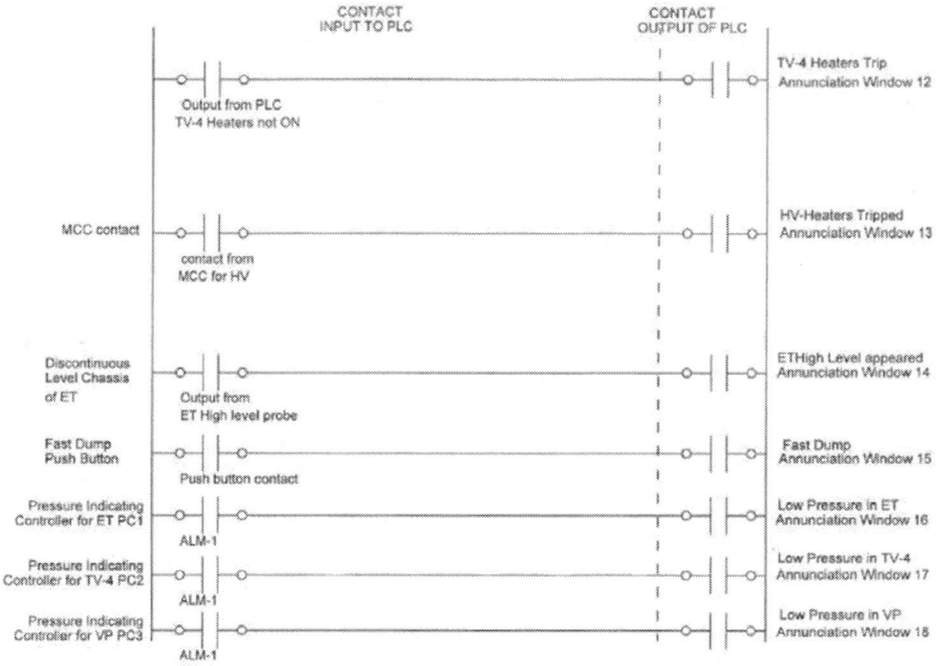

Figure. 6(a):

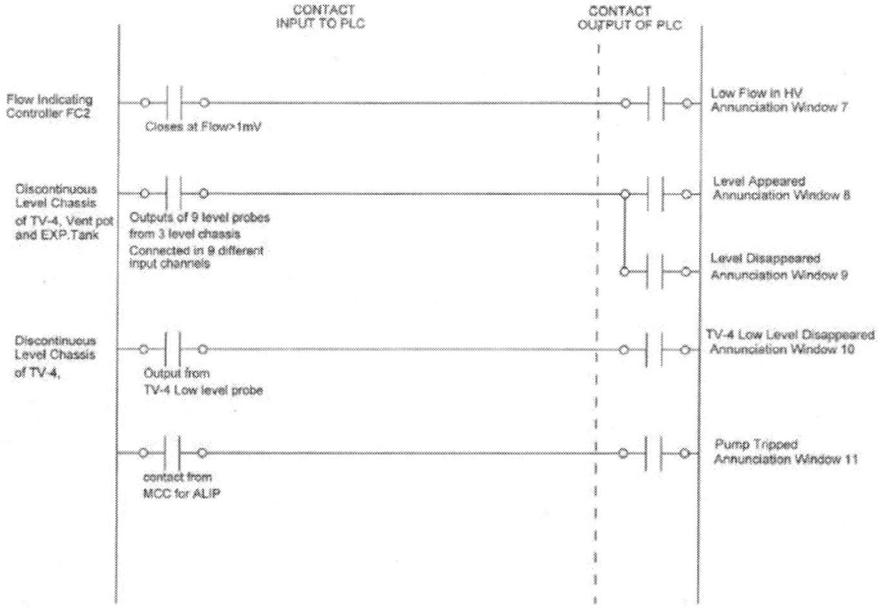

Figure. 6(b):

IOP Conf. Series: Materials Science and Engineering 1272 (2022) 012010 doi:10.1088/1757-899X/1272/1/012010

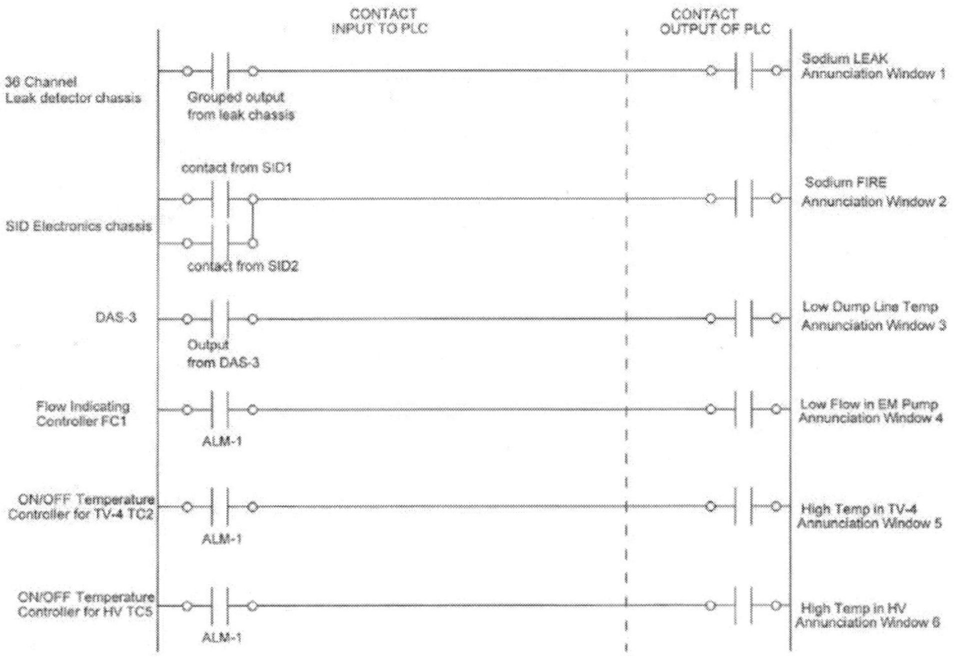

Figure.6(c):

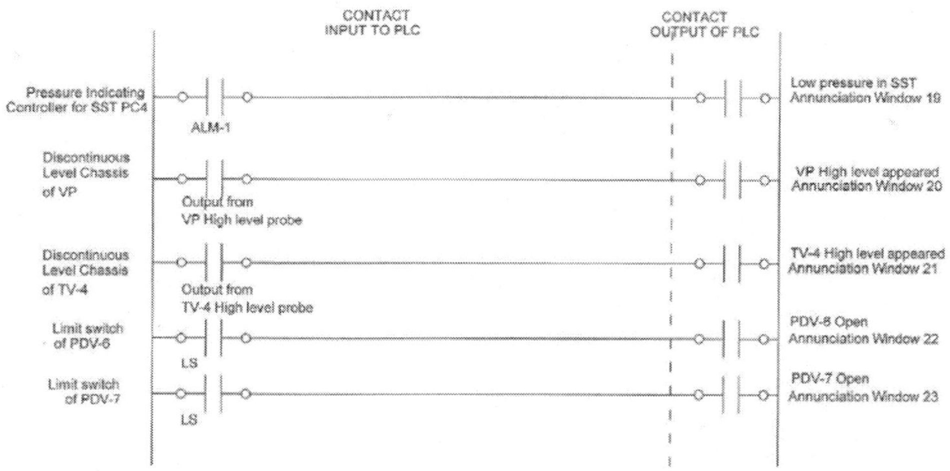

Figure.6(d): Annunciation System of SADHANA loop

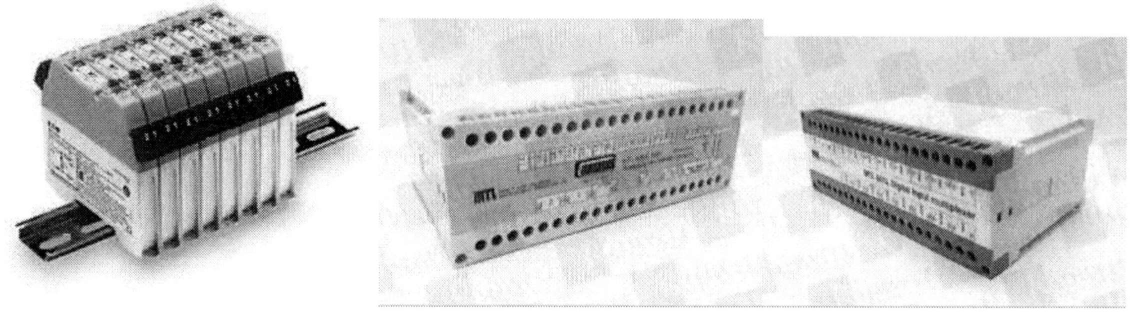

MTL PLC MTL 838B MBF RX MTL 831B MBF TR

Processor used as 8521 in MTL PLC, MTL838B-MBF RX multiplexer receiver, MTL831B- MBF TR multiplexer transmitter.

Figure 7. SADHANA Loop operational Data acquisition & control System

5. Commissioning of experimental Data acquisition system of SADHANA loop

SADHANA loop Surface temperatures of sodium pipe lines and components are measured using 300 mm long 3mm dia SS seethed MgO insulated Chromel -Alumel type thermocouples. Every pipe line and component surface heater will have at least one thermocouple to monitor the temperature. To monitor the temperature of all sodium valves have separate thermocouples. All of these thermocouples are required for temperature monitoring and control of surface heaters via a operational PLC based system. They are linked to OPDAS. These temperature readings are sent to the control room PC. Figure 7. Shows SADHANA Loop operational Data acquisition & control System. Figure.8.a, b & c. shows SADHANA Operational data acquisition and control system.

6. Sodium pool air and Thermowell temperature measurement

A 1 mm dia K type thermocouple assembled in a rack measures the temperature of the liquid sodium pool in TV-4. The rack has 19 thermocouples fixed at various elevations in the pool vertically up to the operating level. Inlet air is measured by seven thermocouples, while outlet air is measured by two thermocouples at each location. The Thermowell has 5 thermocouples installed at the DHX inlet, DHX outlet, AHX inlet, AHX outlet, and TV-4 inlet. These thermocouples are linked to the HP data logger EXDAS, which is installed in the field instrument panel (FIP). EXDAS has two 20-channel general purpose multiplexer modules installed. In channel numbers 101 to 120 and 201 to 215, 35 thermocouple signals are connected. In channels 216, 217, and 218, three millivolt signals are connected, two from flow metres F-7 and F9 and one from a continuous level probe. For connecting thermocouple to EXDAS, 10 pair and single pair, twisted, shielded compensating cable are used. For air velocity and differential air pressure measurements, the AHX air circuit is equipped with a hot wire anemometer, a pitot tube, and a differential pressure transmitter. The 4-20 ma current signal from the above instrument is connected to channels 121,122, and 221 of EXDAS, respectively. Copper instrumentation cables, single pair, twisted, shielded, are used to carry millivolt and 9 (mA) signals.

IOP Conf. Series: Materials Science and Engineering 1272 (2022) 012010 doi:10.1088/1757-899X/1272/1/012010

7. Air heat exchanger (AHX) temperature measurement

The Air Heat Exchanger (AHX) made up of finned tube inside a duct with a heating and cooling arrangement. The surface temperature of 21 AHX tubes is measured at four different locations (84 thermocouples) using a 3mm dia SS sheathed MgO insulated Chromel -Alumel type thermocouple. These are linked to two HP data loggers, EX DAS, which are installed in the field instrument panel (FIP). The HP data logger EXDAS is linked to sets of 60 thermocouple signals. EXDAS has three 20-channel general purpose multiplexer modules installed to receive 60 thermocouple signals. Thermocouple numbers 1331 to 1384, 54, and 1301 to 1306,6 are linked to EXDAS channels 101 to 102, 201 to 220, and 301 to 320. The next set remaining 24 AHX thermocouples 1307 to 1330 are linked to channels 117 to 120, 4 no's., and 201 to 220, 20 no's. of EXDAS. To connect thermocouples to EX DAS and EX DAS, 10 pair and single pair twisted, shielded type compensating cable are used and linked.

8. Decay heat exchanger (DHX) temperature measurement

The temperature of DHX is measured at 16 different elevations using 1 mm dia SS sheathed MgO insulated Chromel - Alumel type thermocouples of varying lengths[15,16,17,18]. These thermocouple signals are connected to the HP data logger EXDAS channel 101 to 116, with a total of 20 signals. To connect thermocouple to EXDAS, 10 pair and single pair twisted, shielded type, compensating cables are used. The field instrument panel FIP was fitted with an external air conditioning unit to keep the data acquisition systems and other electronic chassis at a controlled temperature. The FIP air conditioning unit is running continuously.

9. Data collection and storage

The experimental data required by the EXDAS is transmitted to a separate PC in the control room via an independent serial communication link using RS 232 / RS 485 converters. The installed HP-Bench link software can collect data from EXDAS every 3 seconds. Other software applications, such as MS Excel, require manual saving and exporting of data. EXDAS requires experimental temperature data, which is sent to another PC in the control room via two independent serial communication links that use RS232/RS485 converters. This PC is equipped with in-house developed software that can scan and automatically store the next set of 100 EXDAS channel data at 1 minute intervals. To estimate the amount of heat removed from the TV-4 sodium loop, process parameters of natural circulation of Na flow rate, inlet and outlet temperatures of AHX, DHX, air inlet and outlet temperatures are used. In all of these calculations, the accuracy of the online measurement during the SADHANA loop experimental campaign is used. To ensure accuracy, the experimental data acquisition system should be calibrated, and any errors discovered should be documented. Calibration of the EX DAS temperature channels was accomplished by unplugging the thermocouple connector at the thermocouple location and feeding the temperature value from the calibrator. Temperature readings from the EXDAS corresponding channel were recorded. The comparison of calibrator set values and measured values at the EXDAS channel resulted in a +3° C offset error on all channels.

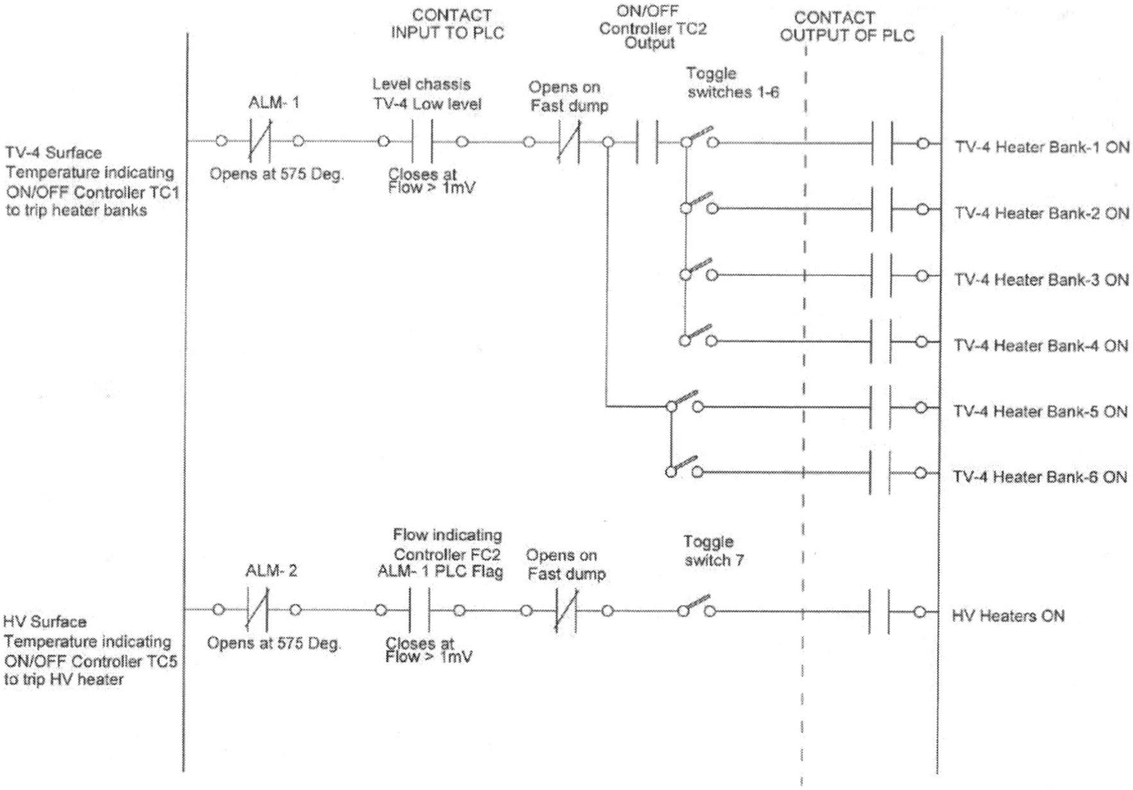

Figure.8(a): SADHANA Operational data acquisition and control system

IOP Conf. Series: Materials Science and Engineering 1272 (2022) 012010 doi:10.1088/1757-899X/1272/1/012010

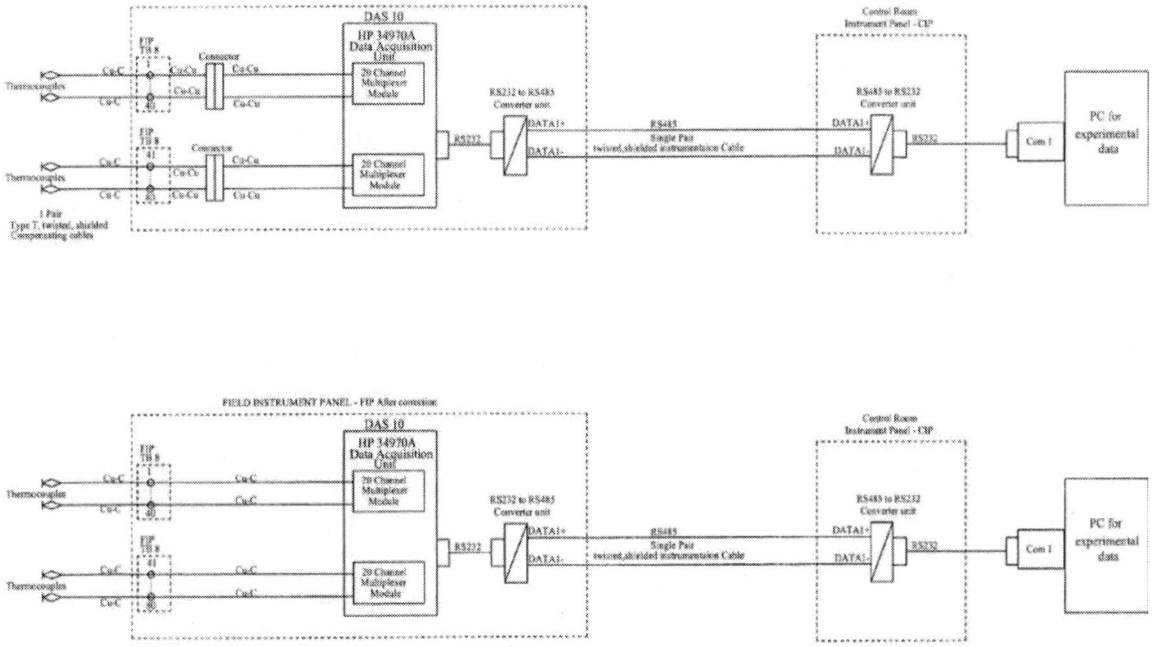

Figure. 8(b): SADHANA Operational data acquisition and control system

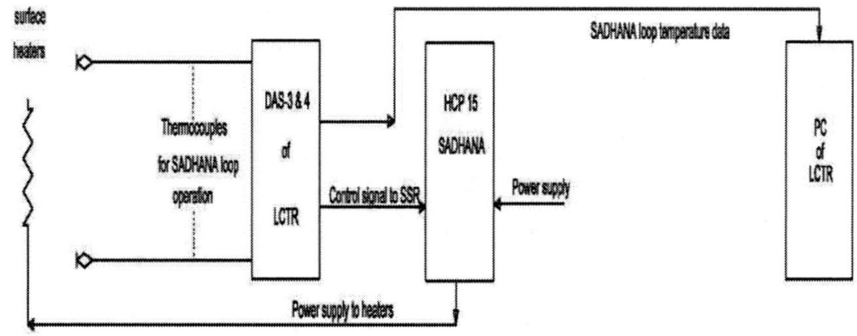

Figure.8(c): SADHANA Operational data acquisition and control system

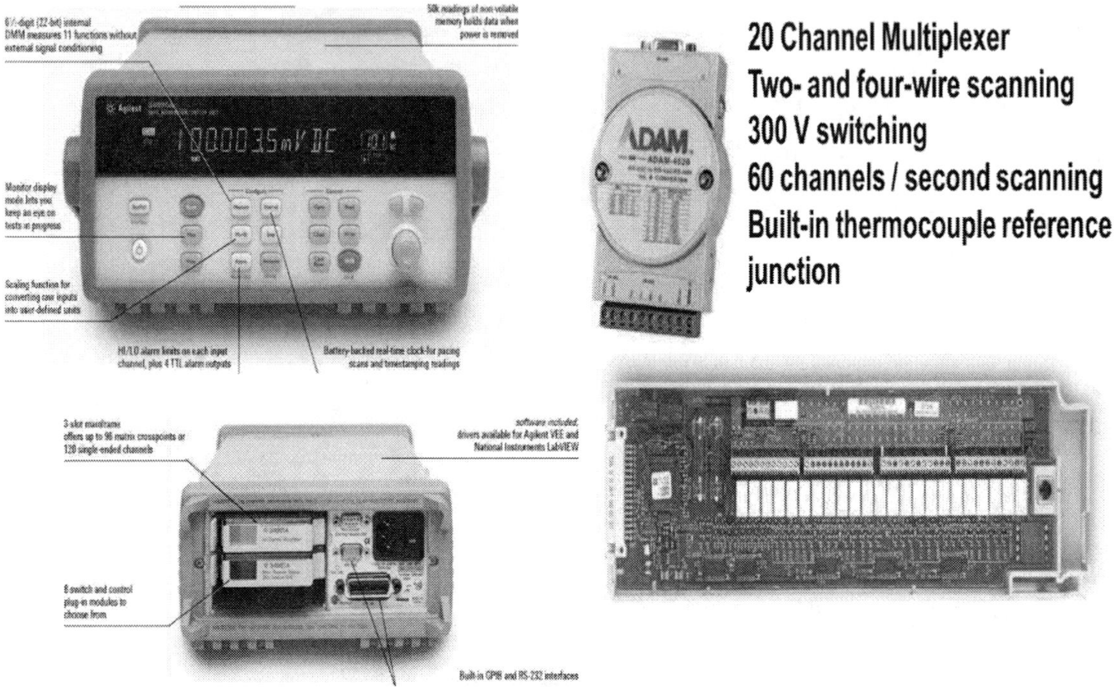

Figure.8(d): SADHANA Experimental data acquisition system

10. Leak detection system of SADHANA sodium loop

The insulator is made of star-shaped ceramic beads that can withstand temperatures of up to 800 °C. The lead wire is 0.7 mm nickel wire, and the detector chain is formed. By threading two lead wires through the holes in the ceramic beads, which are held in place by spot welding the support nickel wire at 15cm intervals. Wire type leak detectors are installed at various locations along the pipe line, nozzle weld joint, and critical weld joint of the component where stress concentration is high. On all bellow sealed valves SADHANA loops, a spark plug type leak detector is provided to detect bellows failure and subsequent sodium leak. Spark leak detectors are also included in AHX and NAX. Leak detectors in a group are connected in such a way that any discontinuity in the lead wire will be detected from the control room. This ensure from control room panel has a test button for simulation testing. The leak detection system is powered by a Class-II power supply. The leak detector is linked in series or terminated in the appropriate subgroup.

11. Integration of level measurement system

In the SADHANA field instrumentation panel, three Resistance Type Level Probe (RTLP) chassis and one continuous level probe chassis have independent indications. The chassis output of the discontinuous level probe electronics is interfaced with an annunciator to provide indication and alarm during level change. The wall mimic panel also includes an LED level indicator. From the control room instrumentation panel and the electronics chassis mounted in the field instrumentation panel, simulation of level to no level and no level to level is possible. The electronics for the TV4 continuous level probe are housed in the field instrumentation panel. The TV4 continuous level probe electronic chassis output is linked to a wall mimic panel level indicator to display the TV4's continuous sodium level.

12. Valve position Indication and control

In the SADHANA loop, four types of valves are used. Pneumatically operated air to open spring to close bellows sealed sodium valves, manually operated bellows sealed sodium valves, pneumatically operated argon equalising diaphragm valves, and a motor operated bellows sealed sodium valve are among the options. SADHANA has three pneumatic valves (PDV-6, PDV-7, and VNa-51). These three valves are spring-loaded and air-operated. The valves are opened by allowing compressed air into the air chamber of the valve via a solenoid valve. When the air in the chamber is vented, the valve closes automatically due to spring force acting on the valve stem. In the control room instrumentation panel, there are two self-illuminated pushbuttons for operating the valve. By pressing the CLOSE push button, 24 Volt DC is supplied to energise a relay RL1 & RL2 in the valve relay patch card (VRPC) for PDV-6 & PDV-7. The relay contact will activate the 5 way solenoid valve for PDV-6 and PDV-7, which shuts off compressed air to the valve chamber while venting existing air pressure from the valve and closing the valve by spring force. By pressing the OPEN push button, the relay and solenoid valve are de-energized, and compressed air is admitted to open PDV-6 and PDV-7. The NO contact of RL1 and RL2 serves as a latch to keep the power supply to the respective relays turned on. When DC supply fails then also the valves open automatically in the fail safe manner. 1.5 Sqmm, 2 core, PVC insulated cable used for powering the solenoid from Control Instrumentation Panel (CIP.). For VNa 51 a 3 way solenoid valve is used. CLOSE push button will de-energise the relay RL-3 and solenoid valve The normally closed port shuts the compressed air to VNa51, vents the air from the valve and valve closes by spring force. Pressing the OPEN push button the RL3 and solenoid valve is energised, compressed air admitted to open the valve.

In the pneumatic valve, the Open/close limit switches are wired to the indication lamps in Wall mimic panel and self illuminated OPEN/CLOSE push buttons in control panel by using 1.5 Sqmm, 4 core, PVC insulated control cable for each valve. The open limit switch contacts are multiplied and connected to PLC logic to trip the pump and heaters when the dump valves open. Schematic diagram of the valve operation, control circuit and indications are shown in Figure 9 (a),(b)&(c) for PDV-6, PDV -7 & VNa-51 respectively. The wiring diagram of valve relay patch card used for the control of pneumatically operated valves and equalising solenoid valves

The motor operated valve VNa-53 is operated from CIP through OPEN/CLOSE push button. The power and control circuits are available in motorized valve junction box of SADHANA. Open and close limit switches are torque switches are provided in the valve for indication and control purposes. The limit switches of motorised valve are wired for lamp indication at wall mimic panel for open/close indication by using 1.5 sq.mm, 4 core, PVC insulated control cable. A potentiometer sensed, continuous valve position indication in percentage (Open) is also available in the control room instrument panel of SADHANA. Schematic diagram of power supply, control circuit and indications of valve VNa-53 . Figure 8. (c) shows Schematic diagram of VNa -51 and control circuit for motorised valve.

All the seven manual valves (VNa52, VNa54, VNa59) are mounted with OPEN and close limit switches. The limit switches of manual valves wired for lamp indication at wall mimic panel OPEN/CLOSE indication. Wiring diagram of valve position indications of valves. Figure 9. (a) shows Schematic diagram of PDV6.

The solenoid valve SV1, SV2 & SV3 are used to operate the pneumatically operated diaphragm valves V74, V84 &V104 which are i9nturn equalise the pressure between sodium storage tank |(SST) & Expansion tank(ET), test vessel TV4 & vent pot (VP) respectively during operation. These valves are air to close and spring to open type and normally kept close condition by energising 3way solenoid valve normally closed port which admits air to close the valve. Toggle switches are provided in control room instrument panel (CIP) for individual solenoid valves operation. These valves will open automatically when fast dump key actuated. 1.5 sq.mm, 2 core cable is used for powering each solenoid valve from control room instrument panel. Figure 9. (b) shows Schematic diagram of PDV7.

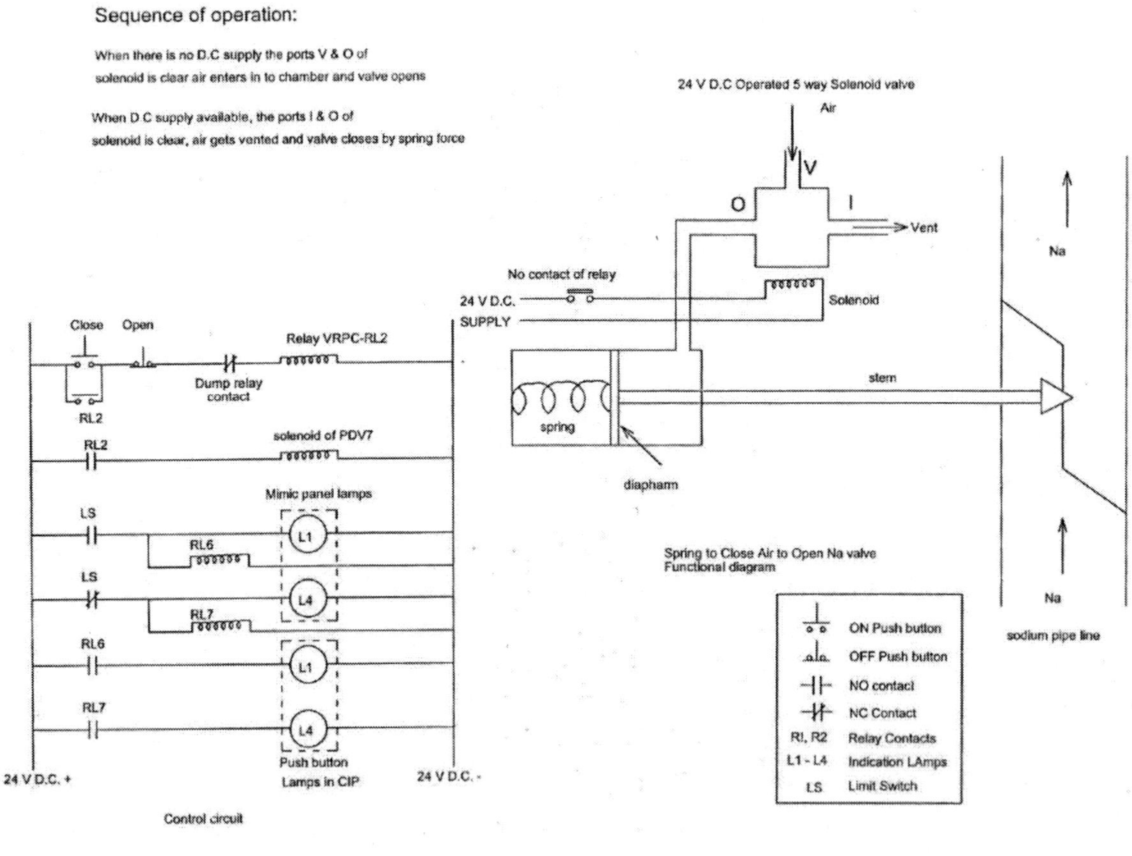

Figure. 9(a): Schematic diagram of PDV6

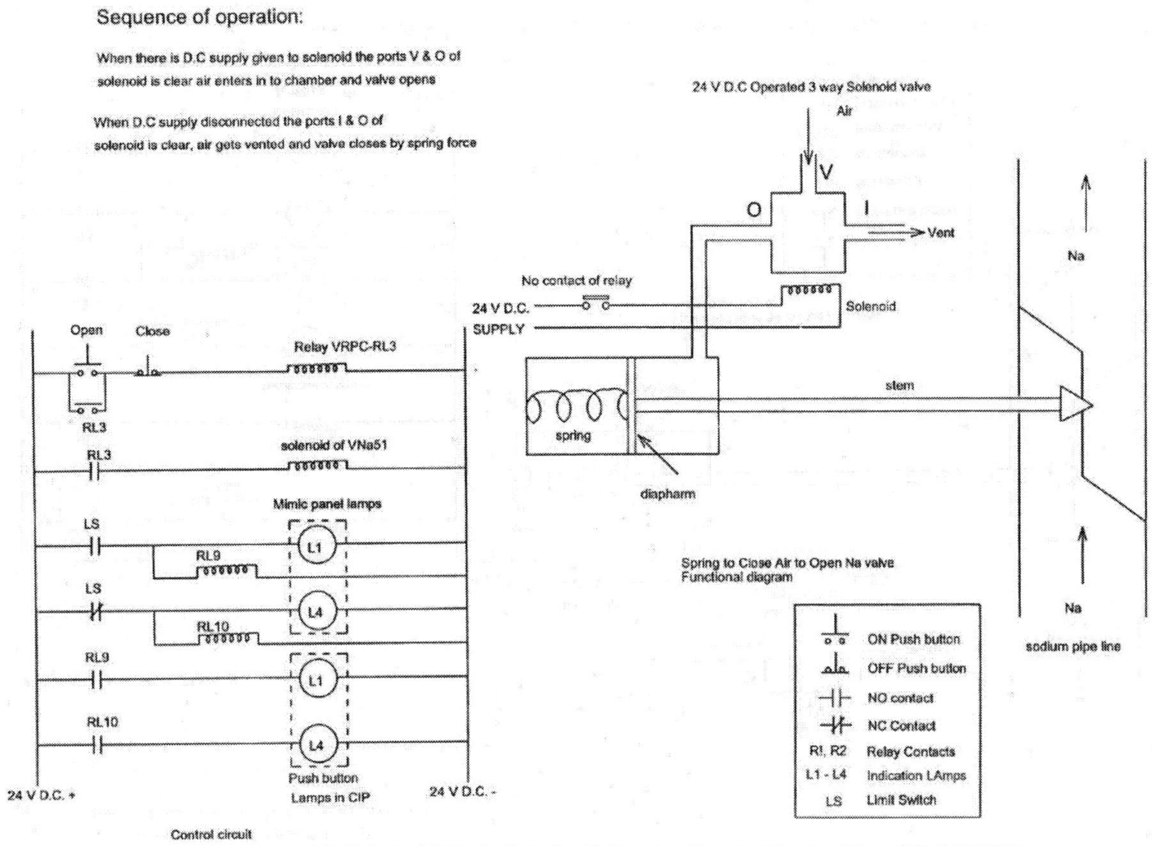

Figure. 9(b): Schematic diagram of PDV7

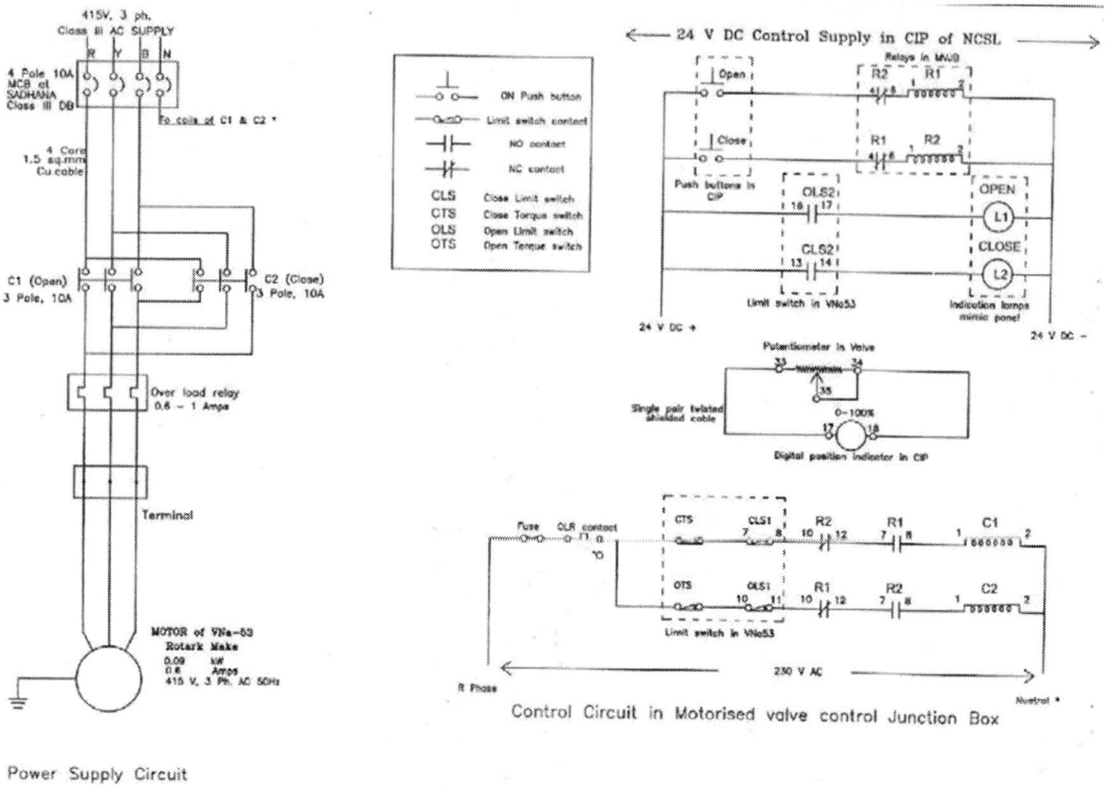

Figure. 9(c): Scehematic diagram of VNa -51 and control circuit for motorised valve

13. AHX Damper and Actuator system

The air flow in the AHX is being controlled by 2 dampers. Dampers are actuated by 2 independent belimo make model number GRVU24-SR Butterfly valve actuators. The actuators work on 24V Dc supply. The position control o0f the actuators is by 2-10 V Dc controlled signal applied on its control terminal which corresponds to 0-90degree mechanical deflection of the actuator(@4.5N-m torque). The actuator also delivers a position signal feedback of 2-10V DC. At 0 degree it gives 2V DC and at 90 degree it gives 10V DC at position terminal. The actuator is provided with end stops and it is overload proof requires no limit switches and automatically stops when the end stop is reached. System requirement for control and monitoring the position of the damper . It is required to design and construct a circuit to deliver 2-10V DC to the actuator and and indicator to accept 2-10V DC from actuator and to display it as 0-100 % opening of damper.

14. Positioner for the actuators.

As mentioned above, a damper actuating control signal of 2-10V DC corresponds to 0 degree to 90 degree angle of deflection. The positioned has been designed to give 1.25 -10.37V DC. The circuit uses LM3171 an adjustable 3 terminal positive voltage regulator which can supply more than 1.5 A of

IOP Conf. Series: Materials Science and Engineering 1272 (2022) 012010 doi:10.1088/1757-899X/1272/1/012010

load current with an output voltage adjustable from 1.2-37V. Output voltage is determined by 2 resistance R1 and R2 connected. R2 is parallel combination of 2.7K ohms resistor and 1 K ohms variable resistor (Potentiometer). The circuit gets 24V from class I supply. The Output voltage is given by

Vout= 1.25*(1+(R2/R1)) Volts

Thus circuit can provide 1.25 to10.37 V by tuning 1Kohm potentiometer mounted in CIP. Two similar circuits are used for individual control of 2 actuators. The outputs are protected by 100milli amps Fuses.

15. Damper position Indications

Two numbers of TOKY make Universal indicators have been mounted in CIP. They are wired to accept the 2-10V DC signal from the actuators. These indicators will display the instantaneous position of 2 dampers in percentage opening. ie, 0-100%. The indicators are working on 230V AC,50HZ power supply available in CIP. Figure 10. Shows Damper control and position indication system.

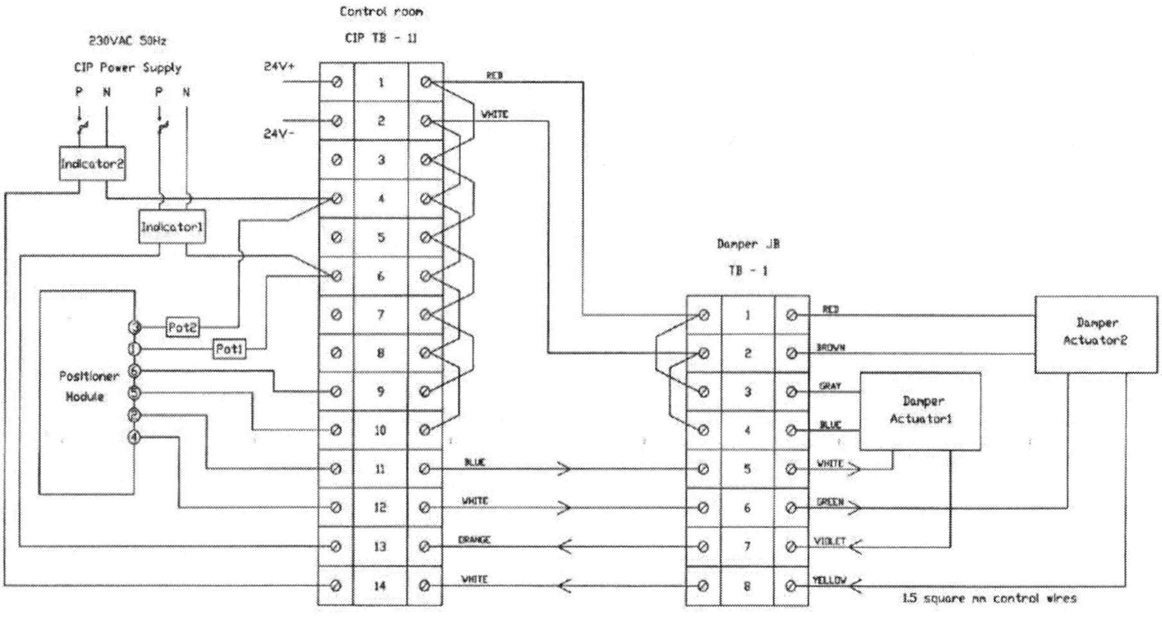

Figure. 10: AHX Damper control wiring Diagram

16. Control logics and annunciation system for SADHANA Sodium loop

The annunciation and indications in SADHANA loop are made available by wall mimic panel indications, flow indicators, temperature controllers Pressure indicators, flow indicators and 24 window alarm annunciator provided in control room instrument panel(CIP) All the annunciation windows in the annunciator driven by contact output from PLC in CIP. Control and safety interlock for safe operation of SADHANA loop are being implemented by using PLC. The PLC has 32 digital i8nputs and 64 digital outputs. The analog values required for the control interlock are measured by hardwired instruments like

IOP Conf. Series: Materials Science and Engineering 1272 (2022) 012010 doi:10.1088/1757-899X/1272/1/012010

temperature controllers , flow indicating controllers etc., depending on the set value in the controllers the contact output is derived from controllers and given a logical input to the PLC. Figure 11. Shows Control Interlocks of SADHANA loop

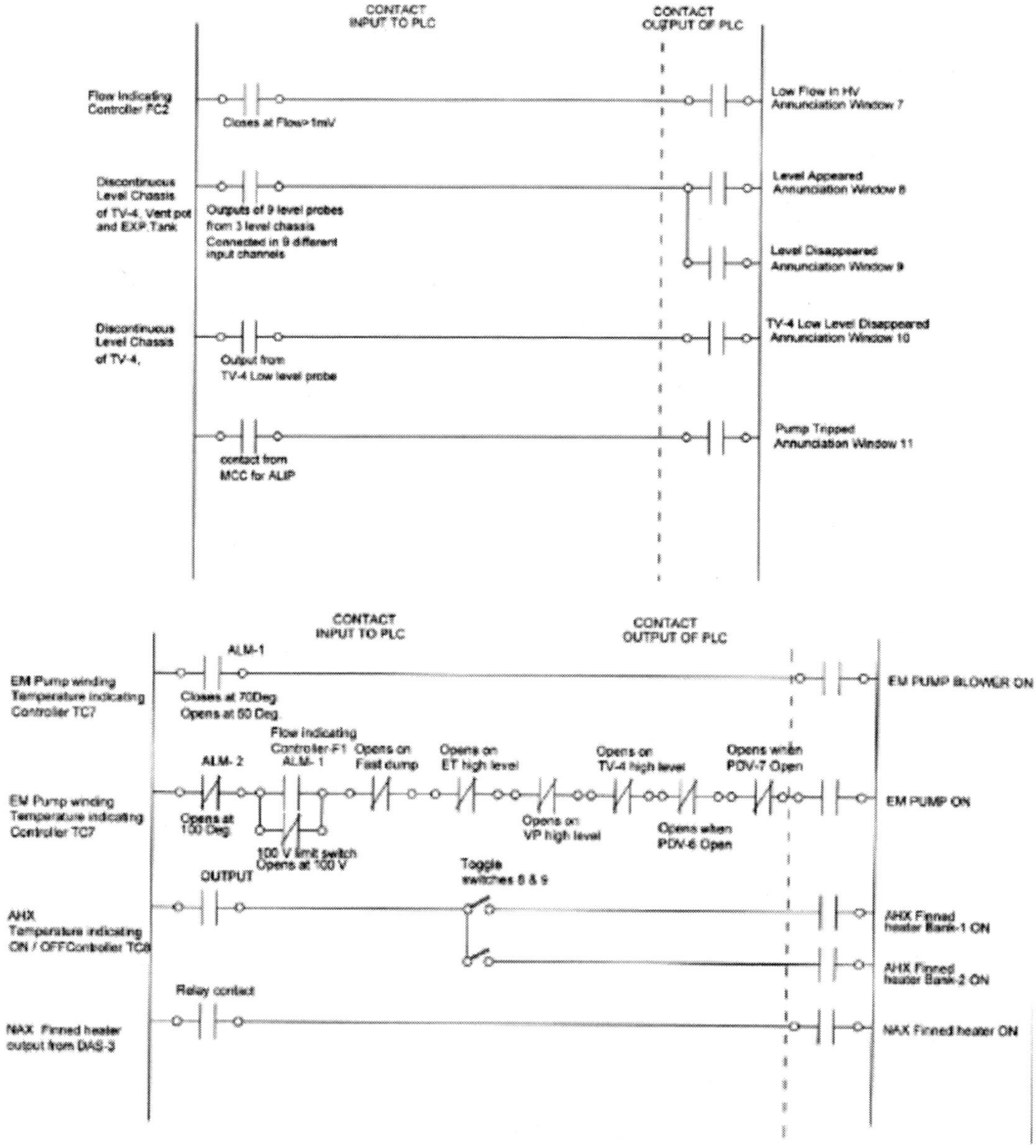

Figure. 11(a) & (b): Control Interlocks of SADHANA loop

IOP Conf. Series: Materials Science and Engineering 1272 (2022) 012010 doi:10.1088/1757-899X/1272/1/012010

17. Leak detection and Indication logics

All the leak detectors are grouped and terminated in a 36 channel leak detector chassis provided in the field instrumentation panel. Healthy, leak and open status of leak detectors are detected and corresponding indications will be available in the leak chassis. Grouped leak signal from leak chassis as contact output is connected in parallel to PLC input and " sodium leak" indication lamp in control room instrument panel. PLC will give output for annunciation window. Appearance of any leak will give "sodium leak" lamp indication window annunciation and hooter in control room instrument panel.

18. Sodium Ionisation Detector logics

There are two Sodium Ionisation Detector (SID) systems (SID1 &SID2) are in SADHANA . one for loop area and other for AHX air duct. If the SID ion current value exceeds the threshold value a contact output is actuated from SID chassis which is connected to two inputs of PLC in control room instruments panel (CIP) by control cable. There is an annunciation window named " Sodium Fire" driven by PLC output when any of the SID chassis gives sodium fire alarm.

19. Level Indication logics

Resistance type discontinuous level probes will be used for detecting sodium levels at three different levels in TV-4, Vent pot and Expansion tank. Three level probe electronic chassis are installed in field instrumentation panel and the outputs are connected in PLC in control room Instrument panel (CIP) using 1.5 sq.mm control cable. Lamp indication are provided in wall mimic panel (parallel connection to PLC inputs) for corresponding discontinuous levels. If there is a change in the status of any of the above 9 levels it will be annunciated by " level Appeared "and " Level Vanished" annunciation window driven by PLC output. Appearance of High level in ET,TV-4 & VP and vanishing of low level in TV-4 are annunciated by separate annunciation window.

20. Temperature Indication and control logics

All the loop temperature readings are available on DAS-3 page of LCTR PC screen. Experimental temperature readings I.e, AHX &DHX temperatures are brought to be separate PC for SADHANA in control room. There are five ON/OFF temperature controller (TC1, TC2, TC5,TC7 & TC8) and three PID temperature controllers(TC3,TC4 & TC6) fixed on control room instrument panel (CIP). These controllers are used for temperature indication and control the test vessel -4, heater vessel, EM pump winding and AHX finned heater bank 1&2 . Schematic diagram of control logic system for SADHANA loop

21. Test vessel TV-4 Temperature control and Indication logics

ON/OFF temperature controller TC1 &TC2 and PID controller TC3 &TC4 are provided in the control room instrument panel (CIP) for TV-4 temperature measurement and control. One ON/OFF controller TC1 used for measuring the surface temperature of TV-4 and its output contact is connected to PLC input. PLC trips as all the six heaters bank in Test Vessel-4 when the surface heater exceeds as 575°C , low level in TV-4 vanish and dump command is actuated. The second ON/OFF controller (TC2) measures Thermowell temperature of TV4 i.e, TV4 sodium temperature and gives ON/OFF control output to PLC. PLC will give six different outputs through six toggle switches (1 to 6) to switch ON/OFF the MCC module of heater banks 1-6 respectively. With the toggle switches heater banks can be selectively switched ON for required heater power. The PID controller TC3 & TC4 are used to control heater bank 5&6 of TV-4. They measure Thermowell temperature of TV-4. Time proportional PID control output signal from these controller will be driven the solid state relay (SSR) in SSR box of heater bank 5&6. PLC also drives the high temperature in TV-4 and TV-4 heaters trip annunciation window and heater in control room instrumentation Panel.

22. Heater vessel (HV) temperature control and indication logics

ON/OFF temperature controller TC5 &PID controller TC6 are provided in the control room Instrumentation Panel (CIP) for heater vessel (HV) temperature measurement and control. The ON/OFF controller (TC5) used for measuring the surface temperature of HV and its output contact is connected to PLC input. PLC will trip the heaters in HV when the surface temperature exceeds 575" C. The PID controller TC6 is used to control the heater vessel heaters. It measures the output temperature of HV. Time proportional PID control signal from TC6will drive the solid state relays (SSR) in SSR box of HV heaters. PLC also drives the high temperature in HV and "HV heaters tripped "annunciation windows and hooter in control room in Instrumentation panel

23. EM pump winding temperature control and Indication logics

In the control room, there is an ON/OFF temperature controller (TC7). Instrument panel (CIP) for measuring and controlling the winding temperature of an EM Pump (ALIP). The ON/OFF controller measures the winding temperature of the EM Pump and connects its alarm-1 contact to the PLC input; the alarm is set at 70°C with a dead band of 20°C. The EM Pump blower is started and stopped by the PLC at 70"C and 50"C, respectively. The NO contact of a 100V limit switch mounted on the ALIP auto transformer is wired to the PLC input for low flow trip. The PLC input is also connected to the open limit switch contact on the dump values. When the PLC trips the EM pump, the winding temperature exceeds 100"C , low flow appearance , high level in TV-4, ET& VP, dump command actuated, and dump valves are opened. The PLC also controls the ALIP high winding temperature, annunciation windows and hooters in the control room instrumentation panel.

24. Air heat exchanger (AHX) temperature control, and indications logics

The control room instrument panel (CIP) includes an ON/OFF temperature controller TC8 for AHX temperature measurement and control. The air temperature is kept stable by a manually operated centrifugal air blower and finned heater banks mounted at the AHX's air inlet. The ON/OFF temperature controller measures the air temperature inside the AHX duct and connects its output contact to the PLC input. PLC Turn AHX's finned air heater banks 1&2 on and off. Toggle switches 7&8 are provided to selectively turn on the finned heater banks. Temperature control and indications for a sodium-to-air heat exchanger (NAX). A manually operated centrifugal air blower and finned air heaters will keep the NAX temperature stable. The NAX finned heaters are controlled using an output from the LCTR's OPDAS. According to the OPDAS settings, a relay is activated by the OPDAS output, and the relay contact is wired to turn on the NAX finned heater.

25. Flow indication logics

The output of three flow meters connected to flow indicator (FC1, FC2, and FC3) fixed in the control room instrument panel. The mV readings of the respective flow meters in the CIP & wall mimic panel will be displayed by these controllers and the display unit in the wall mimic panel. The contact outputs of FC1 and FC2 controllers are wired to the PLC input for low flow trip, annunciation of the EM Pump (ALIP) and heater vessel, respectively.

26. Pressure indication logics

The pressure transducers' 4-20 mA outputs are connected to pressure indicating controllers (PC1, PC2, PC3, and PC4) mounted in the control room instrument panel. The pressure reading from the respective pressure transducer will be displayed by these controllers. The pressure indicating controllers' contact outputs are wired for pressure low annunciation in the control room for the respective components.

27. Equipment status indication logics

The working status of the SADHANA's equipment is readily available in the control room's wall mimic panel. The NO contact from the MCC module of the respective equipment/device is wired for ON/OFF indications in the control room wall mimic panel and self-illuminated push button in the control room electrical panel.

28. Summary

Advanced one Safety Programmable Logic Controller and one Operational Programmable Logic Controller for Data Acquisition Control System of SADHANA Sodium Loop as implemented. The performance of experimental study on the transient response of passive decay heat removal system also evaluated in the 1:22 model of PFBR reactor.

29. Conclusion

Two programmable logic controller applied Data Acquisition Control System of SADHANA loop. The above methodology verification and validation of safety interlocks and operational interlocks are easy. In this methodology dedicated and reliable PLC for safety of the system. The performance of decay heat removal system in TV4 vessel, DHX sodium loop and AHX air natural circulations are evaluated by opening of various position (+90 to -90°C) of damper in AHX inlet.1: 22 model heat removal capacity in line with design datas also verified.

Acknowledgment

The Authors thank Shri S. Raghupathy, Dr. B.K.Nashine, and Dr.B.Venkatraman for the encouragement and support.

References

[1] V.Vinod, S.Chandramouli, G.Padmakumar, K.K.Rajan. "Experimental study on the transient response of passive decay heat removal system 2014 Nuclear Engineering Design,280564-569.

[2] V.Vinod, T.Chandran, G.Padmakumar, K.K.Rajan "Calibration of an averaging pitot tube by numerical simulations; Flow Measurement and Instrumentation"; Vol – 24, (2012).pp. 26-28.

[3]G.Padmakumar,V.Vinod,G.K.Pandey,S.Krishnakumar,S.Chandramouli,G.Vijaykumar,R.RajendraPrasad,R.K.Mourya,P.Madankumar,M.Shanmugasundaram,V.Ramakrishnan,C.Meikandamurthy,K.K. Rajan" SADHANA facility for simulation of natural convection in the SGDHR system of PFBR", 2013.Progress in Nuclear Energy, 66

[4]V.Vinod,S.Krishnakumar,R.Punniyamoorthy,G.Vijayakumar,S.Chandramouli,G.Padmakumar ,B.K.Nashine, K.K.Rajan "Experimental evaluation of safety grade decay heat removal in prototype fast breeder reactor, 2013 Nuclear Engineering and design", 265, 1057– 1065.

[5] V.Vinod,S.Chandramouli,G.Padmakumar,B.K.Nashine,K.K.Rajan"Experimental study on the transient response of passive decay heat removal system, 2014Nuclear Engineering and Design", 280, 564-569.

[6] V. Vinod ∗, S. Chandramouli, G. Padmakumar, B.K. Nashine, K.K. Rajan Performance evaluation of decay heat removal system of PFBR with partial availability of air side dampers, 2017Nuclear Engineering and Design, 318, 174-181.

[7] Astegiano, J.C., Martin, R., Gesi, E., Theoretical and experimental analysis of Super phenix 1 thermohydraulique problems in natural convection. April 1985. In: ANS Topical Meeting on Decay Heat Removal and Natural Convection in FBR. Upton.

[8] Heisler, M.P., 1982. Development of scaling requirements for natural convection liquid metal fast breeder reactor shutdown heat removal test facilities. Nuclear Science and Engineering 80, 347-359.

[9] Hoffmann, F., et al., Investigation on natural convection decay heat removal for EFR status of the program. 1993. In: IAEA-IWGFR Specialist Meeting on Evaluation of Decay Heat Removal by Natural Convection in Fast Reactors, Feb-22-25. PNC, Mito, Japan.

[10] Ieda, Y., Kamide, H., Oshima, H., Sugawara, S., Ninokata, H.,. Strategy of experimental studies in PNC on natural convection decay heat removal. Feb 1993In: IAEA-IWGFR Specialist Meeting on Evaluation of Decay Heat Removal by Natural Convection. PNC, Japan.

[11] Lee, Tae-Ho, Eoh, Jae-Hyuk, Lee, Hyeong-Yeon, Lee, Jae-Han, Kim, Tae-joon, Jeong, Ji-Young, et al., 2009. Scientific design of large scale thermal-hydraulic test facilities. In: Proceedings of International Conference on Fast Reactors and Related Fuel Cycles, Challenges and Opportunities, (FR09).

[12] KAERI, Kyoto. Stehele, H., Damm, G., Jansing, W., Large scale experiments with a 5 MW sodium/air heat exchanger for decay heat removal. 1994. Nuclear Engineering and Design 146, 383e390.

[13] Tenchine, D., Grand, D., Onset of natural circulation in a sodium loop. 1989. In: Proc. Int. Conf. on Nuclear Reactor Thermal Hydraulics (NURETH-4), Oct 10-13. Karlsruhe, Germany.

[14] Ushakov, P.A., Sorokin, A.P., Role of the Reynolds number in modelling natural convection in liquid metals. 1998. Atomic Energy 84 (5), 314-319.

[15] Vijayan, P.K., et al., Scaling laws for natural Circulation loops. 1994 Nuclear Engineering and Design 152, 331-347.

[16] Weinberg, D., Suckow, D., et al., The transferability of reactor conditions of thermohydraulics model investigation of decay heat removal. 1990. In: Proc. Int. Fast Reactor Safety Meeting.

[17] Weinberg, D., Hoffmann, H., Ohira, H., Schnetgüke, G.,. The status of studies using RAMONA and NEPTUN models on decay heat removal by natural convection for the European Fast Reactor. 1993In: IAEA-IWGFR Specialist Meeting on Evaluation of Decay Heat Removal by Natural Convection. PNC, Japan.

Effect of Undersea Parameters on Reliability of Underwater Acoustic Sensor Network in Shallow Water

Padmavathy N[1] and Venkateswara Rao Ch

Department of ECE, Vishnu Institute of Technology, Bhimavaram, 534202, INDIA
padmavathy.n@vishnu.edu.in, venkateswararao.c@vishnu.edu.in

Abstract. Underwater acoustic sensor networks have attained intensifying significance within the research communality, since they are enabling technology for wide range of applications. Typically, these networks establish connectivity (single or multi-hop) by a means of acoustic links. Acoustic communications pose several challenges due to transmission losses, fading, multipath, and effect of medium characteristics (depth, salinity, temperature and pH) on sound speed. The variation in sound speed affects the link formation, which may lead to link failures among sensor nodes. It seems to be crucial to incorporate a channel model that reflect on effect of medium and channel characteristics (absorption and transmission losses) in order to accomplish reliable communication among the links in an underwater network. In this work, an evaluation method for the two-terminal reliability of underwater sensor networks deployed in shallow water scenarios has been proposed. The methodology takes into account an acoustic channel model that distinguishes the effect of frequency and the aforementioned parameters on link formation. Two-terminal reliability analysis has been carried out by considering the effect of underwater media properties. The experimental outcomes specify that, reliability is reaches its maximum value when the temperature and salinity discrepancy with respect to depth are insignificant at lesser frequencies than that of upper frequencies.

Keywords: absorption, acoustics links; reliability; sound speed; transmission loss

1. Introduction

The undersea acoustic sensor networks (UASN) are made up of sensors that are submerged and used to collect data on the areas of rivers and oceans that have not yet been studied. These networks are distributed over an area that has to be studied and include a configurable number of anchored and floating sensors as well as vehicles [1]. These nodes create a single hop or multiple hop communication channels between themselves. Underwater equipment can communicate with one another through optical, radio, electromagnetic, and acoustic waves [2]. Among these, acoustics are preferred by most of researchers because it offers long range. The network characteristics (node mobility, transmission range, power) and the underwater medium properties (temperature, pressure, salinity, and pH) of UASN introduce major research challenges (restricted bandwidth, multipath fading, constrained battery and partial information ability etc.) [3]. The main factors which are venerable in underwater environment are temperature, salinity, and sound speed with respect to depth [4]. In addition, numerous parameters such as; wave height; turbid currents; water pressure; chemical compositions of water; and wave speed causes underwater environment unpredictable in nature [5]. It

[1]Corresponding Author

is necessary that, the proposed channel model should identify the changes in underwater medium characteristics, and network properties in order to deploy a reliable network. Generally, the effectiveness of a channel model relies on the speed of the sound in underwater and operating frequency [6].

The sound speed in underwater is a function of temperature, salinity, depth, and pH. As the temperature and salinity varying with respect to depth in underwater, the sound speed alters accordingly [7]. The investigation of variation in temperature, sound speed and salinity has been carried out (see Figure.1 (a)), which illustrate the effect of depth on temperature, sound speed and salinity measurements. It is identified from Figure. 1(a); temperature is drastically decreasing with increase in scenario depth of the ocean. These temperature variations with respect to depth fluctuates the sound speed which inherently affects the connection between the sensor nodes. The salinity of the sea water changes with geographical region and with water depth. The salinity obtained by measuring concentrations of dissolved salts in ocean and expressed in parts per thousand (ppt).

The mean value of salinity lies between 31 to 37ppt, where as in Polar Regions this value is less than 30 ppt. It shows that, the salinity variations are abnormal at the lower depth when compared to the higher depths. This is due to the chemical compositions of the water and the temperatures are higher at the surface when compared to the ocean bottoms. These changes in salinity and temperature affect the sound speed in underwater acoustic communication [8]. Typically, sound travels with a speed of 1500 meters per second in seawater, however it varies with geographical regions, seasons, and with water depth [9]. The variation in the sound speed (see Figure.1 (a)) with respect to depth, where the sound speed decreases gradually with increase in depth in the upper layers, whereas it attains momentum in speed in the middle layers of the ocean. This is virtuously due to the dissimilarities in temperature, salinity, atmospheric pressure (depth), and chemical compositions of the sea water [10].

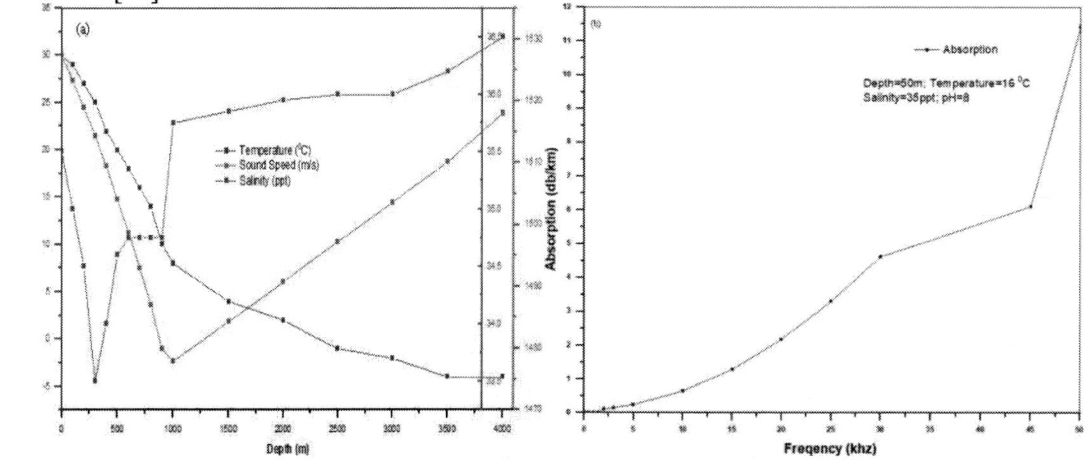

Figure. 1: (a) Variation of temperature, salinity and sound speed with depth (b) absorption of sound

On the other hand, the absorption of sound in sea water depends on salinity, temperature, transmission distance and frequency of operation [11]. Absorption coefficient articulated in dB per kilometre (see Figure.1 (b)) as a function of frequency calculated for a temperature of 16 ^{0}C, salinity of 35 ppt, and with pH of water taken as 8 in a shallow water scenario of depth 50 meter. It is clearly identified in Figure. 1(b); absorption is directly related with frequency of operation. As the frequency increases, the absorption increases rapidly which attenuates the acoustic signal and limits its range [12]. These unpredictable changes in sound speed (depends on temperature, salinity, water depth, pH) and absorption (depends on acoustic frequency, transmission distance) affect the connection between the sensor nodes. Moreover, the connectivity of a network depends on the successful communication link formation between the sensor nodes. Therefore, it is vital to integrate a viable acoustic channel model which deliberates the effect of sound speed and absorption losses to achieve a consistent linkage among the sensor nodes in an underwater network. Hence, this work is focused on developing a methodology which considers the effect of underwater parameters and network parameters for evaluating two-terminal reliability based on acoustic channel model.

2. Literature Review

The several of complications (transmission loss, multi-path, Doppler effect, absorption) in the underwater medium makes UWSNs a fascinating region for scientists to deal with. The development in sensor innovation and remote advances, UWSNs has pulled in a great deal of specialists and has contributed fundamentally to this field. An underwater network is said to be reliable, if there exist a perfect link between the source and terminal pair either by direct or by multi-hop [13]. The link existence between the sensor nodes is a responsibility of channel characteristics (pathway-loss, frequency, capacity), and medium characteristics (temperature, salinity, pH, depth). The UWSNs are deployed and modelled in order to achieve the acceptable connectivity rate among the sensor nodes, which enhances the reliability of network [14]. Ocean exhibit unpredictable environment which alters with the geographical regions, seasons and with time. The fundamental physical properties such as; temperature, salinity, pressure, pH of the ocean have an adverse effect on communication networks performance and their connectivity.

A mathematical model [15] that provides the conversion of atmospheric pressure to depth, and depth to atmospheric pressure has been proposed for reduce the errors in sound speed in oceans and sea. The sound propagation model in underwater can be attained from interface between acoustic wave and temperature, salinity, and pressure of the medium. An experimental set up has been carried out in [16], which illustrate the effect of salinity, temperature, and pressure on physical characteristics of the deployed environment and sound speed variations. Numerous localization strategies used in underwater are based on hyperbolic positioning [17]. Hyperbolic positioning assumes the underwater speed of sound to be constant. Hyperbolic positioning expects the submerged speed of sound to be steady. Notwithstanding, a few examinations have demonstrated that this speed changes with salt content temperature and pressure factor [18]. An algorithm [19] has been presented which estimates the sound speed at a particular location with time and improves the localization accuracy in underwater. Observing submerged conditions is a challenging task. Terrestrial and aeronautical reconnaissance innovation cannot be embraced directly since optical and electromagnetic waves have excessive attenuation in water. On the other hand, the terrestrial MANET reveals different network performances [20-29] in terms of transmission delay, topology, power requirements, movement of nodes, network life time, data rate etc.

The literature has provided an insight on how the sound speeds in underwater alters with salinity, temperature, and with depth. Based on these studies the authors have analyzed the channel characteristics in underwater. However, there is no literature that assesses the network dependability of UWSNs while taking into account the auditory channel characteristics. However, there is only a minute literature available [30–32] that provides an evaluation of dependability based on head nodes, throughput, and communication between the networks, respectively. As a result, a mathematical model that takes into account the underwater channel between channel characteristics in shallow water (temperature, depth, salinity, pH, absorption, spreading losses, and transmission losses) has been constructed in this work. By altering all underwater medium variables, such as temperature, salinity, and depth with set network parameters, the network's dependability has been assessed (number of nodes in the network, coverage area, communication range etc.).

3. Assumptions

- At the commencement of the mission, the UWSN is uniform and functioning.
- The buoyed source node is deployed at a specified depth SND.
- The transmission range of the nodes is higher than their sensing range ($R_t > R_s$).
- RWPM is used to describe the movement of nodes in underwater, with zero pause time, node velocity (V_{max}, V_{min}), and node orientation (0, 2ϕ).
- The failure of nodes follow a known distribution (Weibull) with scale parameter (θ) and

shape factor (β).

- UWSN node failure is statically self-sufficient, and if a node fails, the mission is considered unsuccessful for the length of the mission.
- Bidirectional communication links (L) are used throughout the system.
- There is no consideration of node interference or overlap.

4. Methodology

4.1. Deployment Model

The proposed network (see Figure.2) is made up of n sensor nodes, each of which has the same transmission range (R_t) and the sensing range (R_S) is taken exactly half of transmission range. In an underwater scenario, the sensor nodes are planted at random, while the source node and its buoy are deployed at a depth that may be adjusted. The sensor node sites were analyzed by considering the effect of *SD* and *SND* with respect to varying temperature and salinity in shallow water [33-35].

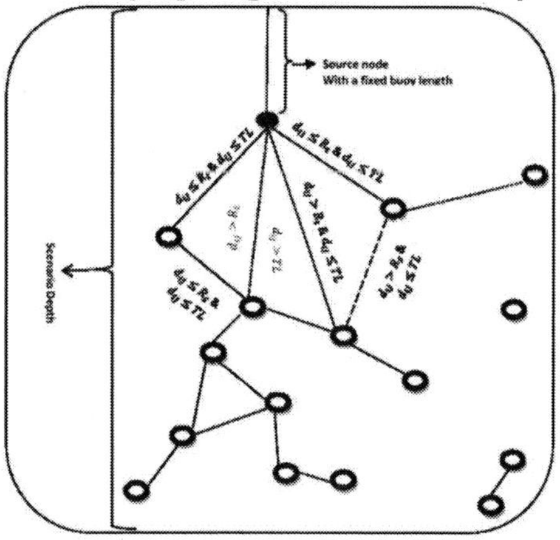

Figure.2: Deployment model of underwater sensor network

4.2. Network Model

The sensor network can be modelled with a superlative source-destination pair, which is ideal for entire simulation time. Hence, the reliability at a meticulous instant will be defined by (1).

$$R_G(\tau)=\left(\prod_{u_i\in p}R_{u_i}(\tau)\right)R_{\left(G/p\right)}(\tau),\text{ where } p=\left(u_1,u_2,\ldots\ldots,u_p\right) \tag{1}$$

4.3. Channel Model

An acoustic channel has been modelled by taking into the consideration of sound velocity in undersea environment. The speed of the sound depends on the medium properties such as; temperature, salinity, depth, and Ph. Hence, the sound speed in undersea [36] is calculated using (2).

$$c(T,S,z)=a_1+a_2T+a_3T^2+a_4T^3+a_5(S-35)+a_6z+a_7z^2+a_8T(S-35)a_9Tz^3 \tag{2}$$

4.3.1. Spreading Loss

Spreading loss (3) is the amount of signal intensity loss that occurs as a result of algebraic distribution as a sound wave travels away from its source.

$$L_{sp}=k\times 10\log_{10}\log_{10}R_t \tag{3}$$

4.3.2. Absorption Loss

Acoustic wave propagation's absorptive loss is frequency-dependent [36]. The coefficient of absorption is characterized using (4) and the total absorption is evaluated using (5).

$$\alpha = \frac{A_1P_1f_1f^2}{f^2+f_1^2} + \frac{A_2P_2f_2f^2}{f^2+f_2^2} + A_3P_3f^2 \tag{4}$$

$$L_{ab} = \alpha \times R_t \times 10^{-3} \tag{5}$$

4.3.3. Transmission Loss

When the sound pressure wave moves away from the sound source, there is a cumulative decrease in acoustic intensity known as transmission loss (TL) and it is describe using (6).

$$TL_{dp} = 10\log_{10} R_t + \alpha R_t \times 10^{-3} \tag{6}$$

4.4. Depth Calculation

The nodes which are deployed randomly are settled in underwater at different depths. Each deployed sensor node's unknown depths are assessed [36] using (7), the source node as a benchmark, and the Euclidean distance assessed by utilising (8).

$$Z_k = D - \left(SN_z + d_{ij}\right) \tag{7}$$

$$d_{ij} = \sqrt{(x_s - x_d)^2 - (y_s - y_d)^2} \tag{8}$$

4.5. Node Status

A vector U defines the nodes in the UWSN, and u_i defines their individual status. Due to the harsh undersea ambience and node movement, nodes in a UWSN are not always reliable or functioning. As a result of assumption # 5, the sensor node reliability is evaluated using (9).

$$R_{u_{(i)}} = \Pr\left(u_i(\tau) = 1\right) = e^{\left(\frac{-\tau}{\theta}\right)^{\beta}} \tag{9}$$

$$u_i(\tau) = \begin{cases} 1 \; if \; i^{th} \; node \; is \; operational \; at \; time \; \tau \\ 0 \qquad if \; i^{th} node \; is \; fails \; at \; time \; \tau \end{cases} \tag{10}$$

4.6. Link Status

(11) and (12) define the link existence as a function of distance between deployed nodes, respectively. Where L_{mn} denotes direct link construction between nodes, l_{mn} denotes intermediate link status between nodes, and d_{mn} signifies the distance among nodes, and it is denoted as (13).

$$L_{mn} = \begin{cases} 1 & if \; d_{mn} \le R_t \; and \; d_L \le TL_{R_t} \\ 0 & otherwise \end{cases} \tag{11}$$

$$l_{mn} = \begin{cases} 1 & if \; d_{mn} \le R_s \; and \; d_L \le TL_{R_s} \\ 0 & otherwise \end{cases} \tag{12}$$

$$d_{mn} = \sqrt{(x_s - x_d)^2 + (y_s - y_d)^2 + (SN_z - Z_k)^2} \tag{13}$$

4.7. Connectivity

The connectedness between sink node and terminal node pairs (s, t) is determined using this connectivity matrix, and the connectivity is described using (14).

$$C_q(\tau) = \begin{cases} 1 & \text{for Connected Network} \\ 0 & \text{for Unconnected Network} \end{cases} \tag{14}$$

4.8. Node Mobility

Sensor node movements in undersea scenario are independent of one another. It's an assumption in the all the mobile sensor network models; that nearly all the sensor nodes travels independently of the each other. The sensor node path is typically treated as an individual understanding of a stochastic process, such as a RW or a RWPP [37]. According to [38], at each change in time interval ($\Delta\tau$), the new node positions are assessed as a function of speed and direction using (15).

$$\begin{aligned} x_i(\tau+\Delta\tau) &= x_i(\tau)+\Delta\tau v_i(\tau)\cos\phi_i(\tau) \\ y_i(\tau+\Delta\tau) &= y_i(\tau)+\Delta\tau v_i(\tau)\sin\phi_i(\tau) \end{aligned} \tag{15}$$

4.9. *Reliability Evaluation*

The network reliability is a function of time, and it is assessed by averaging the outcomes of Q replication runs at each τ during the mission duration [39-43]. The UWSN reliability is expressed mathematically as (16).

$$R_G(\tau) = \frac{\left(\prod_{u_i \in p} R_{u_i}(\tau)\right)\sum_{q=1}^{Q} C_q(\tau)}{Q} \tag{16}$$

5. Algorithm

An algorithm has been presented and built based on the scheme described in preceding section. The technique is as follows;

- ***Step1***: Label all the set-up considerations
- ***Step2***: Randomly deploy the nodes and obtain (X_i , Y_i)
- ***Step3***: compute the mysterious depths using (7)
- ***Step4***: Obtain node condition - (9)
- ***Step5***: Channel has been modeled by (2), (3), (4), & (5)
- ***Step6***: Estimate the transmission losses using (6)
- ***Step7***: Ensure link survival using (11) and (12)
- ***Step8***: Check the connectivity and update $\tau = \tau + \Delta\tau$.
- ***Step9***: Node movements are updated using (15)
- ***Step10***: Reiterate *Step* 2 to *Step* 9 until τ reaches *tMission*
- ***Step11***: Replicate *Step* 2 to *Step* 10 for Q simulation runs
- ***Step12***: Estimate network reliability (16).

6. Simulation Parameters

The parameters considered for analyzing USWN connectivity and estimating two-terminal reliability are listed in Table 1.

Table 1. Input Parameters

Input Parameter	Specification
Network Dimension	3D
Set-up	Shallow water
Scene Depth (SD)	20 to 100m
Surface Node Depth (SND)	5 to 100m
Geographical Area (NCA)	10000 sq. m
Communication Range (TR)	100m
Number of Nodes (NS)	18
V_{max}	6m
V_{min}	3m
Level factor (θ)	1000
Shape factor (β)	1.5
Direction (ϕ)	0-180°

tMission	72 hrs
Iterations (*Q*)	10 000
Node Movement Model	RWPM
Frequency (*F*)	100Hz to 100KHz
Temperature (*T*)	16 ^{0}C to 30 ^{0}C
Salinity (*S*)	27ppt to 30 ppt
pH	8

7. Simulation Results

A network with 18 nodes and a frequency range of 100-1000Hz has been used to replicate the UASN in shallow water surroundings (*SD* of 20 to 100m) while taking acoustic channel propagation into consideration. The transmission range between all nodes is uniform within a 100m radius. The network has been set up such that it can cover an area up to 1000 × 1000 m^2. In addition to the aforementioned depths, the source node is also placed at depths ranging from 5 to 75 meters in a variety of scenario depths. Figures 3 & 4 illustrate the effect of network parameters and the sound channel parameters (frequency, temperature, and salinity) on exposure area, set-up size, and broadcast range at different *SD* (20, 50, and 100m).

The surface sink is primarily fixed at 5m depth, and is progressively increased by 5m rises to a maximum of 100m. The study unequivocally demonstrate that reliability falls down as frequency rises from 100Hz to 1000Hz whereas reliability rises as salinity and temperature fall from 30ppt to 27ppt and 30°C to 16°C, respectively. It has been found that reliability depend on frequency at any point in a mission's lifetime because the sound channel exhibits increased absorption at higher frequencies, which has an effect on node connectivity. As the effect of undersea parameters (temperature, salt content) on the acoustic signal is small compared to the surface of the earth, and the speed of sound has reached a constant level, so it can be seen that reliability improves as SD increases. Because of the pressure underwater, reliability has decreased as *SND* has risen. As a result, an ideal buoy length should be taken into account to maximise network resilience.

7.1. Impact of Underwater Parameters on UASN Reliability (SD)

Effect of Underwater parameters like salinity; temperature; scenario depth; and source node depth along with frequency on 2TR of UWSN has been simulated and studied. The variation of the salinity (see Figure 3(a)), frequency (see Figure3 (b)), and temperature (see Figure3(c)) on two terminal reliability has been studied with the variation in scenario depth (shallow water). For a constant network parameter (NS =18, TR = 100m and NCA 100 x 100 Sq. mts) with a source node deployed at 20 m, it is noticeable that as the scenario depth increase, the reliability increases.

The network reliability depends on the changes of sound speed (sound speed is a function of salt content, temperature and depth). Lower the depth (say 30m) the sound speed decreases with more randomness; it tries to stabilize at depth (say 80m) and still further the randomness in sound speed reduces beyond 100m. The value of salinity depends up on the temperature profile, therefore the variations in the salinity is high at the lower depths because of the random temperature profile at these lower depths. Hence, the salinity has higher value at the surface of the deployment scenario and the salinity is gradually decreasing with decrease in temperature at the higher depths (say from 20m to 100m). In addition, at the higher salinity level the sound propagation fluctuates more randomly at lower depths that influence the connectivity and reliability of the network.

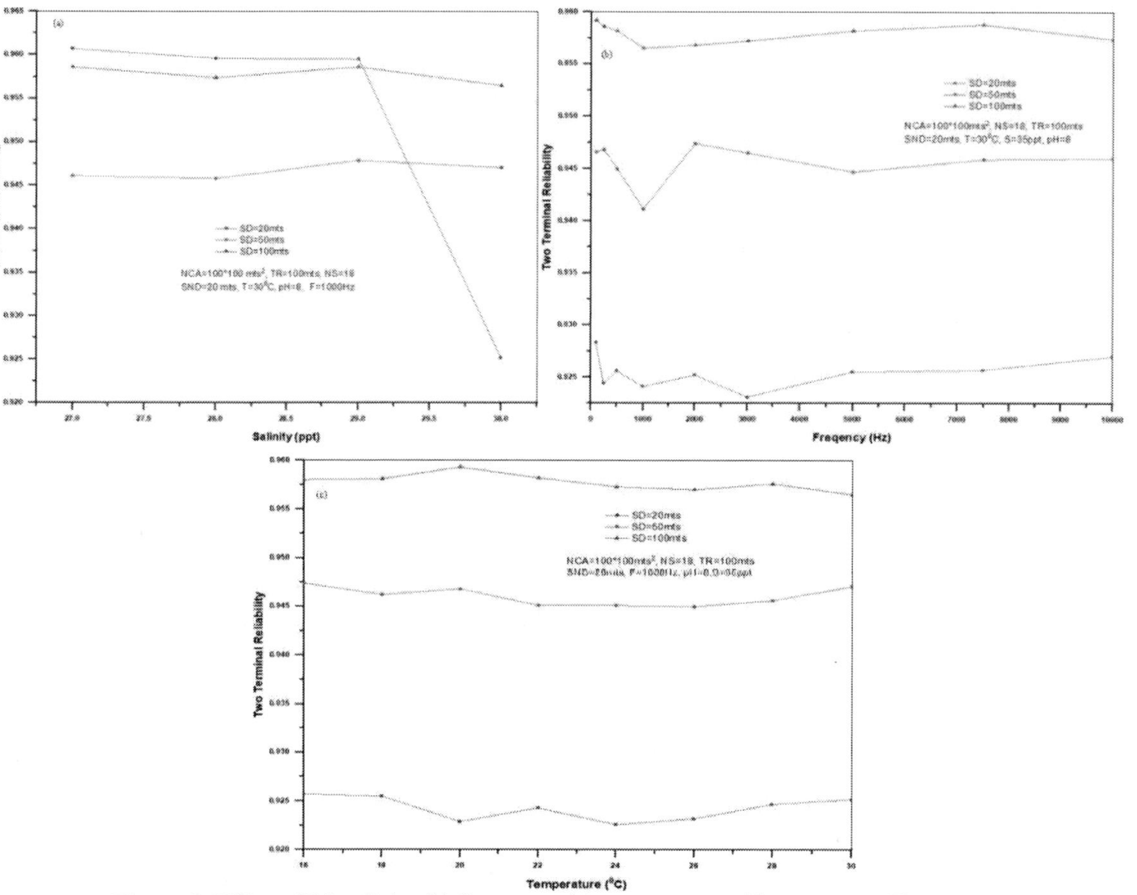

Figure.3: Effect of (a) salinity (b) frequency (c) temperature with respect to *SD* on *2TR*

7.2. Influence of Network Size on UASN Reliability(SND)

It is observed from Figures (4(a), 4(b), and 4(c)), the reliability decreases with increase in *SND* irrespective of network and underwater parameters. The decrease in reliability is due to the path loss experienced by an acoustic link among the sensor nodes in an UWSN is a function of distance. As the depth of the source node increase, the distance between the nodes alters and it revises the path loss. Greater the depth of the source buoy, the distance required to form a perfect link among the sensor nodes required is high. Hence, the path loss associated with higher distances directly influences the connectivity of the network. At this scenario, the acoustic links are formed among the sensor nodes in the multi-hop manner. Basically, at higher depths the effect of sound spread, absorption is less, but due to the deployment of source node with various buoy lengths will increase the multi-hop links among the nodes. Hence, at the lower buoy lengths the maximum reliability achieved, whereas at the higher buoy lengths the reliability achieved is minimum.

For example, it is depicted from Figure 4(a), the maximum network reliability (0.9601) is achieved at lower salinity (say 27ppt) at *SND*=5mts, whereas at the same salinity, the network reliability obtained at SND=75mts is 0.8643. The network dependability is drop down by 10%.Whenever the frequency is increases, the absorption increases and the path loss due to absorption is a frequency and distance dependent. The network reliability decreases with increase in frequency is observed (Figure 4 (b)) in accordance with decreases in *SND.* The network reliability achieves maximum (see Figure 4(c)) value at the lower temperatures (16 ^{0}C) when compared to higher temperatures (30 ^{0}C) irrespective of SND.

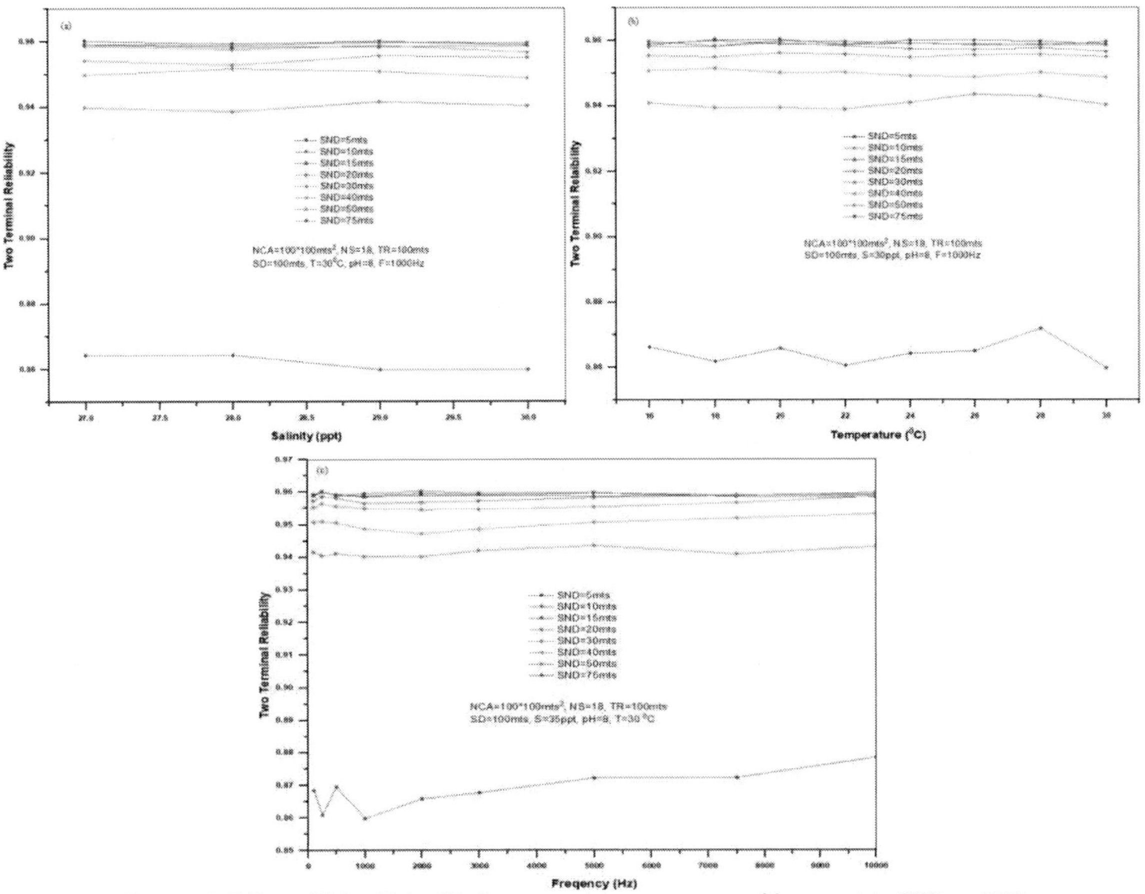

Figure. 4: Effect of (a) salinity (b) frequency (c) temperature with respect to *SND* on 2TR

7.3. Case 3: Impact of Transmission Range on UWSN Reliability

The nodes failure distribution follows the Weibull distribution, one of the best statistical distributions, as explained in 7.1 & 7.2 and the underlying assumptions. Node failure can be brought on by a number of factors, including node mobility, coverage gaps, poor transmission range, surface reflections, speed, and more. The reliability ideas state that a component, system, or network's reliability declines during period of use. The mission period of 72 hours is taken into consideration in the current study since MANETS are application-specific and short-lived networks.

The graphs (see Figures 5 (a), 5(b), & 5 (c)) demonstrate the highest reliability is accomplished at F=100Hz, S=26ppt, and T=16 0C for a NCA (100m2); TR (100m); NS (18) positioned at 100m and fixed source depth of 20m. The reliability achieved is 0.9552 at F=100Hz, whereas at F=1000Hz the achieved reliability is 0.9546 (Figure 5 (a)). Similarly, the network reliability attained maximum value (0.9586, 0.9580) at the salinity S=26ppt and temperature T=16 0C respectively (see figure 5(b) & 5(c)). However, if the designer needs to improve the obtained reliability (@100000 Sq. m) increasing either or both transmission range (500m) and the number of nodes (100) in the network by keep the values of salinity (S=26ppt), temperature (T=16 0C), and frequency (F=100Hz) is a solution.

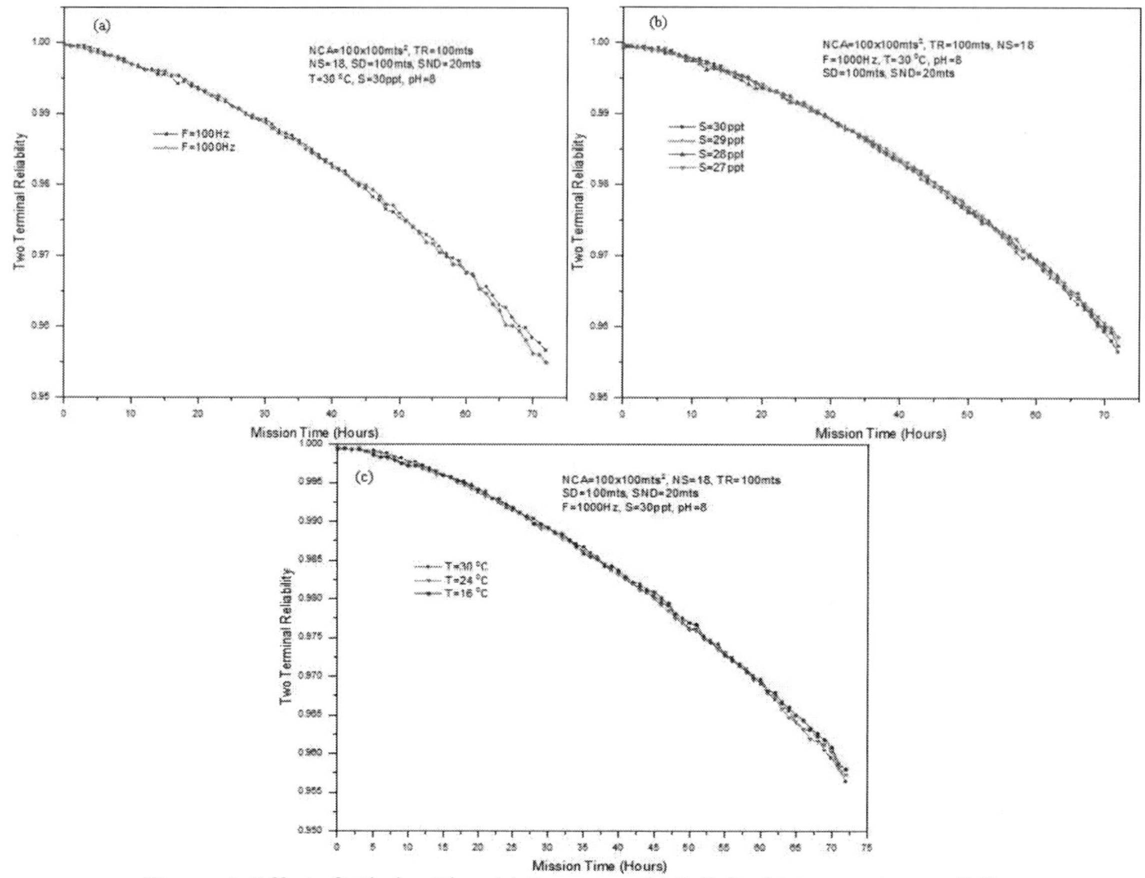

Figure. 5: Effect of Mission Time (a) Frequency (b) Salinity (c) temperature on 2TR.

8. Conclusion

The reliability analysis of underwater sensor network by considering an acoustic propagation model which illustrates the consequence of undersea constraints (salinity, temperature, depth, and pH) on network connectivity has been analyzed. A technique for evaluating the network reliability has been developed. The network reliability is a function of network parameters (network size, communication range, transmission area, mission time etc), undersea constraints (temperature, salinity, depth (pressure), pH), and also acoustic propagation parameters such as frequency, transmission losses. The reliability measures have been evaluated with different temperature (16 ^{0}C to 30 ^{0}C) and salinity (27 ppt to 30 ppt) and also varying frequency ranging from 100Hz to 100 KHz values by making the network parameters to a fixed value. The outcomes show that, the network reliability at any instant of time is a frequency dependent. The reliability is also depending on the sound speed which a function of salinity, temperature and depth.

References

[1] E. M. Sozer, M. Stojanovic and J. G. Proakis, "Underwater acoustic networks," IEEE Journal of Oceanic Engineering, vol. 25, no. 1, pp. 72-83, January 2000

[2] I.F. Akyildiz, D. Pompili, and T. Melodia, "Underwater acoustic sensor networks: research challenges", Ad Hoc Networks, vol. 3, pp. 257-279 May 2005

[3] Barbeau M, Garcia-Alfaro J, Kranakis E, and Porretta S, The Sound of communication in underwater acoustic sensor networks, 9th International Conference on Ad Hoc Networks , Sep 2017, Niagara Falls, Canada. pp.13 – 23

[4] I.F. Akyildiz, D. Pompili, and T. Melodia, "Challenges for efficient communication in

underwater acoustic sensor networks", ACM SIGBED Review - Special Issue on Embedded Sensor Networks and Wireless Computing, vol. 1, pp. 3-8, July 2004

[5] M. Stojanovic and J. Preisig, "Underwater acoustic communication channels: Propagation models and statistical characterization", IEEE Communications Magazine, vol. 47, no. 1, pp. 84-89, January 2009

[6] H. Jindal, S. Saxena and S. Singh, "Challenges and issues in underwater acoustics sensor networks: A review," International Conference on Parallel, Distributed and Grid Computing, pp. 251-255, December 2104

[7] N. N. Ismail, L. A. Hussein and S. H. S. Ariffin, "Analyzing the Performance of Acoustic Channel in Underwater Wireless Sensor Network (UWSN)," 2010 Fourth Asia International Conference on Mathematical/Analytical Modelling and Computer Simulation, pp. 550-555, May 2010

[8] Khazaie, Shahram & Chen, Xue,"Linear approximation of underwater sound speed profile: Precision analysis in direct and inverse problems", Applied Acoustics, vol.140, pp. 63-73, November 2018.

[9] M. M. Ali, Sarika Jain and Radhika Ramachandran, "Effect of Temperature and Salinity on Sound Speed in the Central Arabian Sea", The Open Ocean Engineering Journal, vol. 4, pp. 71-76 July 2011

[10] Sanjay Kumar, Shanthi Prince, Jinka Venkata Aravind & Santosh Kumar G, "Analysis on the effect of salinity in underwater wireless optical communication", Marine Georesources & Geotechnology, vol.38 no. 3, pp. 291-301, February 2020

[11] Hovem, Jens, "Underwater acoustics: Propagation, devices and systems", Journal of Electroceramics, vol. 19, no.4 pp. 339-347, December 2007

[12] Y. Kularia, S. Kohli and P. P. Bhattacharya, "Analyzing propagation delay, transmission loss and signal to noise ratio in acoustic channel for Underwater Wireless Sensor Networks," 2016 IEEE 1st International Conference on Power Electronics, Intelligent Control and Energy Systems (ICPEICES), pp. 1-5, July 2016

[13] C. V. Rao, N. Padmavathy, and S. K. Chaturvedi, "Reliability evaluation of mobile ad hoc networks: with and without interference", IEEE 7th International Advance Computing Conference, pp. 233-238, January 2017.

[14] J. Proakis, E. Sozer, J. Rice, and M. Stojanovic, "Shallow water acoustic networks". IEEE Communications Magazine, vol 39, no. 11, pp.114–119, November 2001.

[15] C C Leroy, and F Parthiot, "Depth-pressure relationships in the oceans and seas", The Journal of the Acoustic Society of America, vol. 103, no. 3, pp. 1346-1352,March 1998.

[16] N. P. Yuwono, D. Arifianto, E. Widjiati and Wirawan, "Underwater sound propagation characteristics at mini underwater test tank with varied salinity and temperature," 2014 6th International Conference on Information Technology and Electrical Engineering (ICITEE), pp. 1-5, October 2014

[17] J. Yang, L. He and C. Shuai, "The simulation of underwater acoustic propagation with the horizontal changes of sound speed profiles," 2016 IEEE/OES China Ocean Acoustics (COA), pp. 1-4, January 2016

[18] S. Misra and A. Ghosh, "The effects of variable sound speed on localization in Underwater Sensor Networks," 2011 Australasian Telecommunication Networks and Applications Conference (ATNAC), pp. 1-4, November 2011

[19] H. Shi, D. Kruger and J. V. Nickerson, "Incorporating Environmental Information into Underwater Acoustic Sensor Coverage Estimation in Estuaries," MILCOM 2007 - IEEE Military Communications Conference, pp. 1-7, October 2007

[20] S.K. Chaturvedi, N. Padmavathy, "The influence of scenario metrics on network reliability of mobile ad hoc network", International Journal of Performability Engineering, vol. 9, no. 1, pp. 61–74, January 2013.

[21] B. Venkata Sai Kumar, N. Padmavathy, "A Hybrid Link Reliability Model for Estimating Path

Reliability of Mobile Ad Hoc Network", Procedia Computer Science, vol. 171, pp. 2177-2185,2020.
[22] Jason L. Cook, Jose Emmanuel Ramirez-Marquez, "Two-terminal reliability analyses for a mobile ad hoc wireless network", Reliability Engineering & System Safety, vol. 92, no.6,pp. 821-829, June 2007
[23] N. Padmavathy, S.K. Chaturvedi, "A Systematic Approach for evaluating the reliability metrics of MANET in shadow fading environment using monte carlo simulation", International Journal of Performability Engineering, vol. 12, pp.265-282, May 2016.
[24] N. Padmavathy, S.K. Chaturvedi, "Reliability evaluation of capacitated mobile ad hoc network using log-normal shadowing propagation model", International Journal of Reliability and Safety, vol. 9, no.1, pp. 70-89, September 2015.
[25] B. Venkata Sai, N. Padmavathy, "A systematic approach for analyzing hop count and path reliability of mobile ad hoc networks", International Conference on Advances in Computing, Communications and Informatics, pp. 155-160, September 2017.
[26] N. Padmavathy, K. Anusha, "Dynamic reliability evaluation framework for mobile ad-hoc network with non-stationary node distribution", Communication and Computing Systems, CRC Press, Taylor and Francis, January 2018.
[27] N. Padmavathy, J. R. C. Teja, and S. K. Chaturvedi, "Performance evaluation of mobile ad hoc network using MonteCarlo simulation with failed nodes", 2ndInternational Conference on Electrical, Computer and Communication Technologies, pp. 1-6, February 2017.
[28] J.L. Cook, J.E. Ramirez-Marquez, "Reliability analysis of cluster-based ad-hoc networks", Reliability Engineering and System Safety, vol. 93, no.10, pp. 1512–1522, October 2008.
[29] N. Padmavathy, "An efficient distance model for the estimation of the mobile ad hoc network reliability", 4th International Conference on Information, Communication and Computing Technology, Springer, pp. 65-74, January 2019.
[30] Min H, Cho Y, Heo J, "Enhancing the Reliability of Head Nodes in Underwater Sensor Networks", Sensors, vol. 12, no. 2, pp. 1194-1210, January 2012
[31] Umar, M.A. Hasnat, M. Behzad, I. Baseer, Z.A. Khan, U. Qasim, N. Javaid, "On Enhancing Network Reliability and Throughput for Critical-range based Applications in UWSNs", Procedia Computer Science, vol. 34, pp. 196-203, December 2014
[32] Nimbalkar, D. Pompili, "Reliability in underwater inter-vehicle communications", Proceedings of the Third Workshop on Underwater Networks, WUWNET 2008, San Francisco, California, USA, pp. 19-26, September 2008.
[33] Padmavathy N and Rao C V, "Network Reliability of Underwater Sensor Network in shallow water", International Conference on System Automation, Reliability, Quality Control, Safety, Maintenance & Management, pp.45-61, December 2019.
[34] K. Akkaya, A. Newell, "Self-deployment of sensors for maximized coverage in underwater acoustic sensor networks", Computer Communications, vol. 32, pp. 1233-1244, May 2009.
[35] P. Jiang, S. Liu, J. Liu, F. We, and L. Zhang, "A depth adjustment deployment algorithm based on two-dimensional convex hull and spanning tree for underwater wireless sensor networks", Sensors, vol. 16, no.7, pp. 1-20, July 2016.
[36] K. V. Mackenzie, "Nine-term equation for sound speed in the oceans", Journal of acoustical Society of America, vol.70, no.3, pp. 807-812, September 1981.
[37] D. N. Sandeep, Vinay Kumar, "Review on Clustering, Coverage and Connectivity in Underwater Wireless Sensor Networks: A Communication Techniques Perspective", IEEE Access, 5, pp. 11176-11199, June 2107.
[38] Z. Zhou, J. Cui, and A. Bagtzoglou, "Scalable Localization with mobility prediction for underwater sensor networks", IEEE 27th Conference on Computer Communications, pp. 2198-2206, May 2008.
[39] N. Padmavathy, S. K. Chaturvedi, "Reliability evaluation of mobile ad hoc network: with and without mobility considerations", PROCEDIA Computer Science, vol. 46, pp. 1126-1139,

2015.
[40] S K Chaturvedi and N Padmavathy, "Mobile ad hoc network reliability: an imperative research challenge", Advances in Reliability and System Engineering, Springer International Publishing, Chapter 4, pp 87–119, December 2017.
[41] N. Padmavathy, S. K. Chaturvedi, "Evaluation of mobile ad hoc network reliability using propagation-based link reliability model", Reliability Engineering and System Safety, vol. 115, pp. 1-9, July 2013.
[42] Venkateswara Rao, C., Padmavathy, N, Effect of Link Reliability and Interference on Two-Terminal Reliability of Mobile Ad Hoc Network, Advances in Data Computing, Communication and Security. Lecture Notes on Data Engineering and Communications Technologies, 106, pp. 555-565 2022.
[43] N Padmavathy and Venkateswara Rao Ch, Reliability Evaluation of Underwater Sensor Network in Shallow Water Based on Propagation Model, Journal of Physics: Conference Series, vol. 1921, 012018, pp. 1-17, 2021.

PERFORMANCE COMPARISON OF CHANNEL CODING TECHNIQUES FOR OFDM SYSTEM

M. Vijaya Lakshmi [1], G.Gagana Reddy[2], AlamurSucharitha[3], N. Akshara [4], N.Vaishnavi[5]

[1.]Assistant Professor ETE Department, G.Narayanamma Institute of Technology and Science,Hyderbad, Telangana, India.
[2,3,4,5.] B.Tech Student,ETE Department, G.Narayanamma Institute of Technology and Science,Hyderbad, Telangana, India.

Abstract. Wireless communication is one of the speedily evolving and dynamic technology areas in the realm of communication. OFDM is an acronym for Orthogonal Frequency Division Multiplexing. It is a version of FDM in which orthogonal subcarriers are closely spaced to maximize bandwidth consumption. By decreasing Inter Symbol Interference (ISI) and increasing the Signal-to-Noise ratio (SNR), it has enhanced the quality of long-distance communication. It's commonly employed in digital communication systems to safeguard digital data from noise and interference while also reducing bit mistakes. Convolutional codes, Block codes, are the two fundamental forms of channel codes. Block codes are based on abstract algebra and finite field arithmetic. They are capable of detecting and correcting faults. This category includes LDPC and Reed Solomon codes. Convolution codes, alternatively are created using a distinct strong mathematical framework and are generally utilised for real-time error correction. They combine all of the data into a single codeword. Reed-Solomon codes are block-based error correcting codes used in digital communications, storage, and ADSL, among other uses. The BER performance of an OFDM system is investigated in this paper using various channel coding algorithms.

1. Introduction.

Wireless communication is a wide term that encompasses all procedures and ways of connecting and communicating between two or more devices via wireless communication technologies and devices employing a wireless signal [1]. Wireless communication works by sending electromagnetic signals through the air, physical surroundings, or atmosphere from an enabled device. The sending device can be a sender or an intermediate device capable of wireless signal propagation. When the destination or receiving intermediate device captures these signals, a wireless communication bridge between the sender and receiver device is created. As wireless communication becomes more widespread, an increasing number of devices, ranging from smart phones to medical implants, are competing for unlicensed spectrum.

The growing need for ultrahigh-data-rate services, as well as future advanced Internet of Things (IoT) and Industry 4.0 applications, provide significant technological problems. New service classes like ultra-HD video and multisensory virtual reality demand even higher spectral efficiency and the investigation of extreme frequency ranges. These radio devices may interact with each other in today's network, restricting real-world speed and capacity. Future wireless applications will necessitate high data rates. Naturally, coping with an ever-unpredictable wireless channel while communicating at high data rates is difficult.

The concept of multi-carrier transmission has recently surfaced as a way to fight wireless channel hostility and provide high data rate communications. OFDM is a type of multi-carrier transmission in which all of the subcarriers are orthogonal. OFDM offers a high user data rate transmission capacity while maintaining a fair level of complexity and precision. Because of its great bandwidth efficiency and robustness against multi-path environments, the orthogonal frequency division multiplexing (OFDM) system is a good communication approach. As digital multimedia applications become increasingly prevalent, the demand for broadband communications systems continues to rise. Over the last decade, OFDM has been the most widely used highspeed communication method [2]. Many industry executives have stated that OFDM technology is the way of the future for wireless communications.

The rapid development of mobile communications has paved the way for the history of wireless communication to follow. To correctly demodulate data, a receiver must be able to receive the entire signal. As a result, when signals are broadcast near together, they must be separated by a guard band so that the receiver may use a filter to distinguish them. With OFDM, this is not the case. Despite the fact that the sidebands of each carrier overlap, they can still be received without interference because they are orthogonal to one another. This is accomplished by setting the carrier spacing to the symbol period's reciprocal.

The data to be communicated on an OFDM signal is split amongst the signal's carriers, with each carrier carrying a portion of the payload. Each carrier's data rate is reduced as a result. Because of the decreased data rate, interference from reflections is significantly less of a concern. Over a single-channel data transmission technique, OFDM provides numerous advantages. Because the sub-channels are tightly spaced, OFDM is more resistant to electromagnetic interference and allows for more efficient use of total available bandwidth. Because there are multiple channels accessible, it is also more resistant to interference.

As a result, narrowband interference on a single sub-channel has no effect on the other channels, allowing the entire system to function. Multipath echo effects can also cause frequency specific interference fading, which can be addressed. Because the various sub-channels have a reduced data rate, guard intervals between symbols can be employed to eliminate ISI and reduce multipath errors. In this study, the BER performance of an OFDM system is examined and compared for different channel coding approaches such as Convolutional codes and Reed Solomon codes. BER performance is also implemented using linear block codes. Introduction to Wireless Communication and OFDM is the first section of the study. Following that, the OFDM system with convolution codes is examined, followed by the OFDM system with Reed Solomon Code. The last section contains all of the results and graphs, as well as a comparison of alternative methodologies.

2.OFDM SYSTEM WITH CONVOLUTION CODE

A convolutional code is an error-correcting code that creates parity signals in telecommunication. It is a widely used coding method in which the output code bits are chosen by logic operations on the current bit in a stream and a small number of prior bits rather than blocks of bits. Convolutional codes are widely used in a variety of applications, including digital video, radio, mobile communications (e.g., GSM, GPRS, EDGE, and 3G networks (until 3GPP Release 7)), and satellite communications, to enable reliable data transport. With consumer demand for more information and, as a result, rising high data bandwidth driving communications systems, error correction coding has become increasingly vital. Using an error correction approach like the Viterbi algorithm to improve the BER while retaining high

IOP Conf. Series: Materials Science and Engineering 1272 (2022) 012012 doi:10.1088/1757-899X/1272/1/012012

data reliability is one option. The Viterbi algorithm, which was created by Andrew Viterbi as an error-correction strategy for noisy digital transmission, is a useful method for increasing channel dependability through Forward Error Correction (FEC).

It is currently utilised in a wide range of digital communications systems, including LTE Physical Downlink Control Channel (PDCCH), CDMA [3] and GSM digital cellular, dial-up modems, satellite, deep-space communications, and 802.11 wireless LANs Speech recognition, keyword spotting, and computational linguistics are all applications of it. The trellis diagram is the second and more prevalent method of describing convolutional codes. The trellis diagram is a visual representation of the movement between different states as time passes. Any memory-less mapping, such as all block encoders, can be easily represented over a well defined trellis diagram, just like a convolutional encoder. The key distinction from the convolutional encoder's trellis diagram is that in this case the trellis diagram is time variable, meaning that the trellis sections change with time. The number of received bits in a data stream through a communication channel that have been altered due to noise, interference, distortion, or bit synchronization faults is referred to as the number of bit errors in digital transmission. The BER is the number of bit errors per unit time.

In a traditional serial data system, symbols are communicated one by one, with each data symbol's frequency spectrum being allowed to occupy the entire available bandwidth [4]. The data is sent out in multiple narrow-band sub channels at the same time. To begin, randomly generated data is serially transferred to the transmitter block. Then there's a serial-to-parallel conversion. One of the possible modulation systems [5] (e.g., BPSK, QPSK, QAM [6]) is used to modulate the signal for transmission. An Inverse Fast Fourier Transform (IFFT) block maps the parallel bits of data to the subcarriers.

To prevent Inter Symbol Interference, a cyclic prefix (CP) is added into the OFDM symbol and a guard interval (GI) is appended to each block of data during further signal processing (ISI). The generated OFDM symbols are translated to serial form and broadcast across a channel once the CP and GI are inserted.

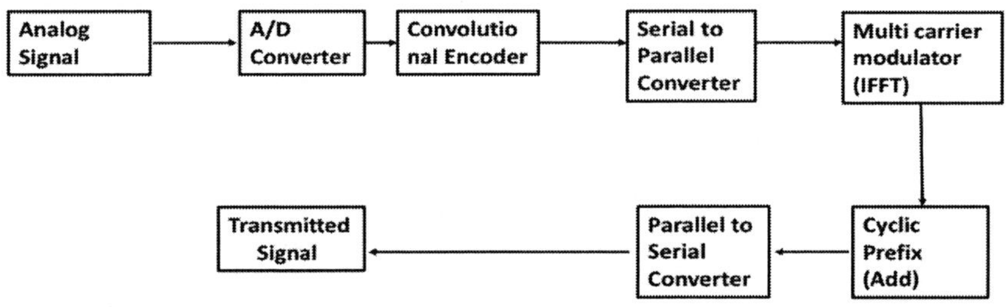

Figure 1. Transmitter of Convolutional coded OFDM System

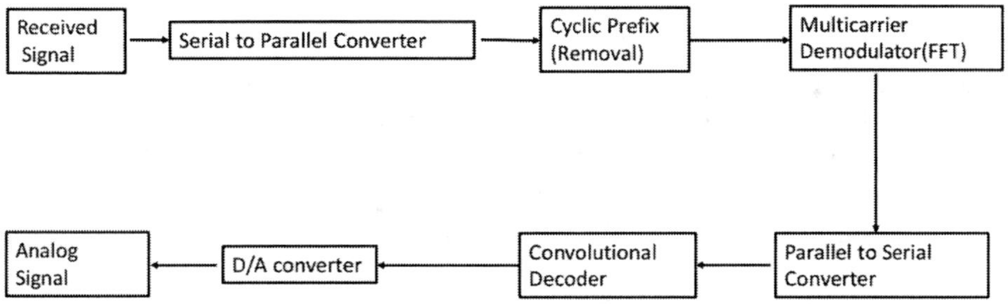

Figure 2. Receiver of Convolutional coded OFDM System

The OFDM receiver uses techniques that are the inverse of those used by the OFDM transmitter to obtain the original data. OFDM subcarrier coefficients are utilized to encode both data and random signals in a randomization strategy. Because we don't have a pre-defined set of data, a randomizer can assist us by selecting a random sample of data and performing the complete process for that data. Data modulation is the technique of transforming data into radio waves by adding information to an electrical or optical carrier signal, allowing computers to store and process analogue signals. For data modulation, the transmitter and receiver implementation are quite basic. There's no need for an analog - to - digital, and it is a cost-effective solution that can be used in the event of financial difficulties.

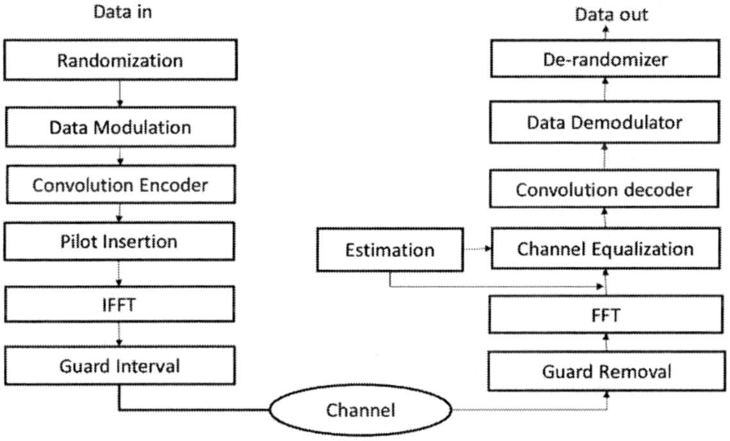

Figure 3. Algorithm of Convolutional Coded OFDM System

Data bits are fed into a K-bit shift register called the constraint length in the encoder block. As each bit enters the register from the left, the previous bits are relocated to the right, and the register's oldest bit is eliminated. A pilot signal is a reference or known signal in telecommunications that has a defined amplitude and frequency. The receiver may readily estimate the channel behaviour with the help of the pilot (attenuation, gain and delay). As a result, in order to send a signal, it must first be converted from frequency domain to time domain using the Inverse Fast Fourier Transform. The guard interval is used to reduce channel delay spread and prevent Inter Symbol (ISI) and Inter Carrier (ICI)interference.

The duration between data characters sent wirelessly is reduced when Short Guard Interval or Guard Removal is disabled. That could boost throughput in the right radio setting. At the receiver Fast Fourier Transform is used to send them back to the frequency domain. Channel Estimation, which plays a key part in an OFDM system, is used to increase the capacity by enhancing system performance interms of BER.

Moving forward, the Equalizer block aids in the Equalization process, which is the reversal of signal distortion caused by transmission across a channel. Equalizers are used to render the frequency process, and they have the advantage of being extremely simple to construct for OFDM. In contrast to the Convolutional encoder, the decoder estimates the data stream based on the received code bit sequence and encoder state diagram knowledge. A demodulator is used to extract the information content from the modulated carrier wave, and then a De-Randomizer is used to reduce congestion and other losses.

3.OFDM SYSTEM WITH REED SOLOMON CODE

In the proposed system, Reed Solomon codes are used for encoding and decoding data at the source and the receiver end. The message signal is divided into blocks via the Reed Solomon algorithm, with some blocks containing actual data and others containing parity bits. These data blocks (encoded data) are transmitted through the communication channel. Reed Solomon code finds and corrects faults even if the channel is unstable and the message is garbled.

Irving S. Reed and Gustave Solomon created the Reed Solomon code. The Reed-Solomon code is a non-binary BCH code subtype. Reed- Solomon codes' encoder varies from binary encoders in that it works with multiple bits rather than single bits. Reed-Solomon codes aid in the recovery of damaged communications transmitted over a network. The Reed-Solomon encoder receives data and adds some parity bits to our original data bits before sending it over the noisy network. The Reed-Solomon decoder, on the other hand, detects and recovers corrupted messages.

3.1. Reed-Solomon code parameters

- 2m-1symbols represent block length (n).
- The message size in Reed-Solomon codes is (n-2t), where t is the number of errors rectified.
- The size of the parity check is supplied by (n-k) or 2t symbols.
- The minimum distance (a) is equal to (2t+1).
- The message's size is k bits.

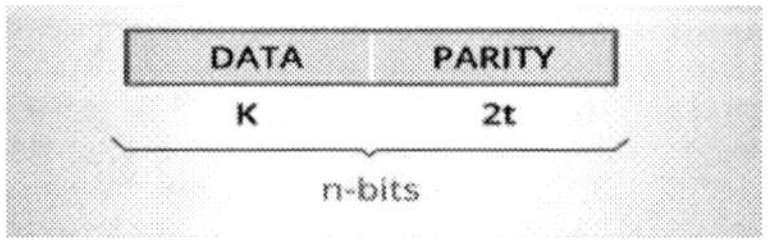

Figure 4–Representation of N-bit Solomon Codes

3.2. Encoding

The method of encoding in Reed Solomon code has the following steps -

The message is represented as a polynomial p(x), and then multiplied with g(x).

Mapping of the message vector $[x_1,x_2,x_3.....x_k]$ to a polynomial of degree less than k such that

$p_x(a_i) = x_i$ for all i = 1,...k

Interpolation methods are used for evaluating polynomial like Lagrange Interpolation.

Using this polynomial, the other points $\alpha_{k+1}....\alpha_n$, are evaluated. Then Calculated encoded message is s(x) = p(x) * g(x). The sender sends this encoded message along with the g(x).

3.3. Decoding

Errors and Erasures can be corrected using Reed-Solomon algebraic procedures. A decoder can be used to correct *t* errors or up to *2t* erasures. It is utilized in storage media such as CDs and DVDs, wireless or mobile communication for data transfer, satellite communication, digital TV BAR codes and QR codes and it has the best redundancy efficiency. Reed-Solomon codes allow for the adjustment of block length and symbol size, as well as a wide range of code speeds and efficient decoding procedures.

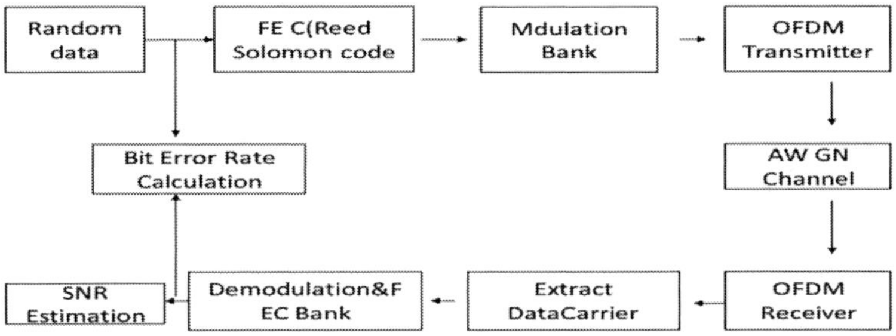

Figure 5. Reed Solomon coded OFDM system

FEC is a method of obtaining error control in data transmission in which the source sends redundant data and the receiver recognizes only the portion of the data that contains no apparent errors, and modulation is applied to this data to improve system performance and convert data into radio waves by adding information.

This data is then transferred by an OFDM transmitter, which operates digital data across multiple carriers. The AWGN channels add white Gaussian noise to the data, which is received by the OFDM receiver[8]. This allows digitized data to be carried by multiple carriers, and then the Data-Carrier is extracted to perform demodulation and FEC Bank in order to recover the original data from the modulated Carrier Signal. The BER is determined using both random and original data.

4. Results.

OFDM with Convolution and RS Codes is simulated with the help of MATLAB version 13b.The BER performance of OFDM system is improved with convolutional codes and it is further improved by Reed-Soloman codes. In The OFDM system initial input data is modulated and applied to the input of IFFT for sub carrier mapping of data and it is passed to the cyclic prefix to convert linear convoluted channel into circularly convoluted channel. The output of Cyclic prefix is applied to the power amplifier and it is transmitted to the output. At the receiver side cyclic prefix is removed and data is demodulated and decoded to get the perfect error recovery. Figure 6 shows the plot for SNR vs BER for an original OFDM system with PSK modulation technique. As SNR increases the BER performance is improved in OFDM system with PSK modulation. Bit Error Rate (BER) indicates the ratio of number of bits corrupted to the total number of bits transmitted to the output

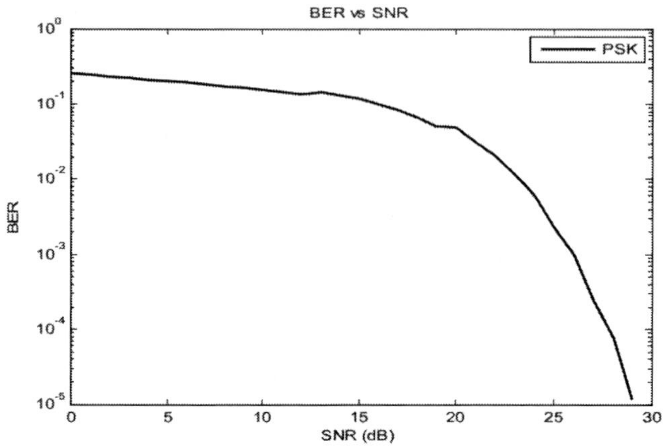

Figure 6. BER Vs SNR plot for OFDM system

Table. 1 BER for the OFDM system

SNR(dB)	OFDM(BER)
2	0.35
4	0.3
6	0.2
8	0.15
12	0.025
14	0.003
16	0.0002

Convolutional Coded OFDM System: Convolutional codes are added to the OFDM system for improving the BER performance.The initial data is convolutional coded and it is applied to the input of modulator to increse the efficiency and toreduce the band width of the system.The output of the modulator is apllied to the IFFT block and cyclic prefix is added to reduce the Intersymbol, Intercarrier interferences and then it is transmitted to the output.At the receiving end reverse operations are performed to recover the data.

Figure 7 shows the plot of SNR vs BER for an original OFDM system with convolutional coding technique. With Signal to Noise Ratio on the X-axis and Bit Error Rate on the Y-axis, the graphs for Convolutional Coded OFDM system illustrate Bit Error Rate for various parameters. The SNR ratio and the BER are found to be inversely related to one another. The higher the Signal-to-Noise Ratio the lower the Bit Error Rate and the performance is improved with convolutional codes.

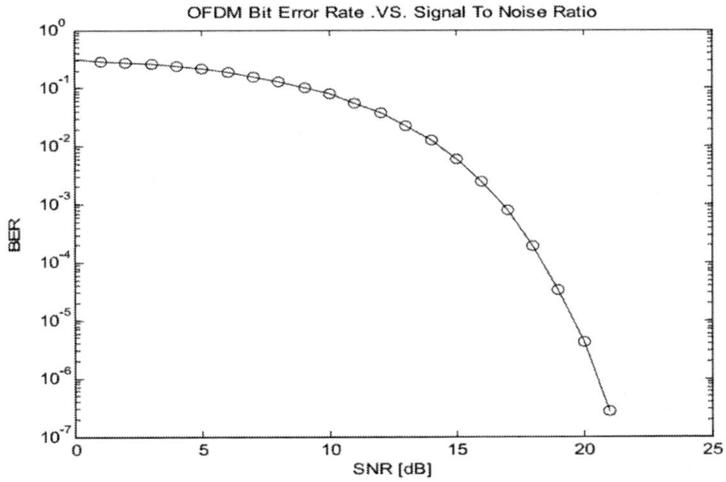

Figure7. SNR vs BER plot for Convolutional Coded OFDM system

Table.2. BER for Convolutional coded OFDM system

SNR(dB)	OFDM with Convolutional codes(BER)
5	0.15
10	0.07
15	0.005
20	0.00007

Reed-Solomon-Coded OFDM system:
These codes are added to the OFDM system for improving the BER performance.The initial data is Reed-Solomon coded and it is applied to the input of different modulators to improve the spectral the efficiency and to reduce the band width of the system.The output of the modulator is apllied to the IFFT block and cyclic prefix is added to reduce the Intersymbol, Intercarrier interferences and then it is transmitted to the output.At the receiving end cyclic prefix removal,demodulation and decoding operations are performed to recover the data.
Figure 8 shows BER performance of Reed Solomon Coded OFDM system for different modulation techniques The P_e is reduced and BER performance of OFDM system is improved as SNR increased from5dB to 20 dB and QPSK system out performances when compared to the other modulation techniques.

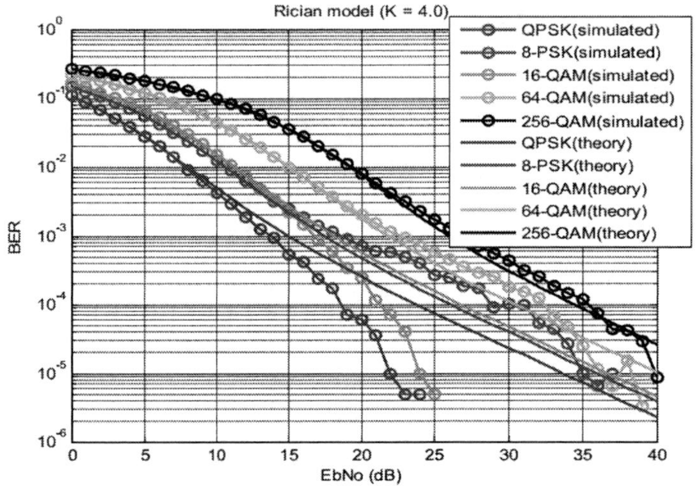

Figure 8.BER for Reed Solomon Coded OFDM system for different modulation techniques.

Table.3 BER for Reed-Solomon code OFDM System.

SNR (dB)	QPSK (BER)	BPSK (BER)	16-QAM (BER)	64-QAM (BER)	256_QAM (BER)
5	0.02	0.04	0.08	0.1	0.12
10	0.004	0.01	0.01	0.05	0.1
15	0.001	0.009	0.003	0.001	0.03
20	0.0009	0.0005	0.0004	0.0095	0.001
25	0.00007	0.000095	0.0009	0.0005	0.0009
30	0.000001	0.00003	0.00005	0.00001	0.0007
35	0.0000003	0.000008	0.000001	0.000004	0.00001
40	0.00000007	0.0000009	0.0000005	0.0000001	0.000008

Table.4 Parameters Comparison

SNR (dB)	Uncoded OFDM (BER)	Convolutional Codes (BER)	RS Codes (BER)
0	0.4	0.2	0.1
5	0.3	0.15	0.08
10	0.15	0.07	0.01
15	0.005	0.005	0.003
20	0.0002	0.00007	0.0095

Table 4 shows the PER performance of OFDM system, Convoluional codes and Reed Solomon codes. When comparing alternative coding systems, it is clear that RS code has a lower bit error rate, a more efficient use of redundancy, a wider range of code rates, and efficient decoding approaches.

5. Conclusion

The BER performance of an OFDM-based wireless communication system for various channel coding schemes is shown in this paper. The impact of digital modulations under convolutional coding and other fading channels is highlighted by a variety of system performance outcomes. Reed-Solomon codes are complex to implement, but because they can handle large bursts of mistakes, their performance is substantially better and consistent. Reed-Solomon codes are recommended for systems that demand very high performance, and convolutional codes are preferred for systems that do not allow intricate structure. The BER performance of an RS coded OFDM system is improved when compared to convolutional codes. BER performance can be improved further by implementing Discrete Wavelet Transform (DWT) OFDM system for wireless application. It can also be done by modifying the modulation methods and taking into account different fading channels.

6. References

[1] Candice King, 2014 "Fundamentals of Wireless Communication",67th Annual Conference for Protective Relay Engineers.

[2] Y.J.Liu, 2019 "Introduction to OFDM Receiver Design and Simulation", ARTECH.

[3] Heung Gyoon Ryu, Tran Phuong Hoa, Nguyen Thanh Hieu, JinJianxue, NOVEMBER 2004, "BER Analysis of Clipping Process in the Forward Link of the OFDM-FDMA Communication System", IEEE Transactions on Consumer Electronics, Vol. 50, No. 4.

[4] S. Das, Elisabeth De Carvalho, Ramjee Prasad, 2009, "Performance Analysis of OFDM Systems with Adaptive Sub Carrier Bandwidth" - (IJCSIS) International Journal of Computer Science and Information Security, Vol. 5,Issue No.1.

[5] N. H. Maruf Bhuyan, Jia Uddin, Monirul Hoque, Md. Ahasan Habib,Oct-Dec-2012. "Performance Analysis of Wi-MAX:Modulation Scheme Versus Bit Error Rate", International Journal of Electronics & Communication Technology (IJECT)-Vol 3, Issue 4, Version 2.

[6] Rune Svensson, December 2007,"An Introduction to Adaptive QAM Modulation Schemes for Known and Predicted Channels" - Proceedings of the IEEE on Adaptive modulation and transmission in wireless system | Vol. 95, No. 12.

[7] Dulal Haque, Shaikh Enayet Ullah and Razu Ahmed,2008: "Performance Evaluation of a Wireless Orthogonal Frequency Division Multiplexing System under Various Concatenated FEC Channel Coding Schemes", International Journal of Electrical and Power Engineering, vol.2, no.6, pp.389-392, Pakistan.

[8] Nita J. Borkar, Dattatray. S. Barman, 2012, "BER Performance of OFDM System with Adaptive Modulation", Complex Systems (ICCS), 2012 International Conference on, 5-6 Nov.

[9] Rishad Ahmed Shafik, Md. Shahriar Rahman, AHM Razibul Islam,2006 "On the Extended Relationships among EVM, BER and SNR as Performance Metrics", International Conference on Electrical and Computer Engineering.

[10] TeodorIliev, Izabella Lokshina, DimitarRadevGeorgi Hristov, 2008, "Analysis and evaluation of Reed-Solomon codes in Digital Video Broadcasting systems", Wireless Telecommunication Symposium.

Generation of Pulse Sequence Using EMG Signals for Application in Transfemoral Prosthesis

R Dhanush Babu, Mahesh Veezhinathan*, Dhanalakshmi Munirathnam and V Aishwarya

Department of Biomedical Engineering, Sri Sivasubramaniya Nadar College of Engineering, Kalavakkam, India - 6030110

Email: maheshv@ssn.edu.in

Abstract: The percentage of people having a lower leg amputation is high, and the incidence of unemployment among these amputees is likewise rising. Hence, it requires the intervention of an innovative solution to serve the function of a lost limb. Electromyogram (EMG) signals is a result of the potential generated by muscles during contraction. In this work, an attempt has been made to extract EMG signals from four set of muscle groups and the acquired signals were pre-processed and transformed to pulses to extract the contraction phase of the signal. Furthermore, the processed signals were subject to feature extraction process where in the Mean Absolute Value (MAV), Integrated EMG Feature (IEMG) and various statistical parameters associated with the signal such as the mean, median, standard deviation, variance, kurtosis, skewness was calculated in order to serve as an input to drive the stepper motor of a transfemoral prosthesis. To promote real time acquisition and control, a transfemoral socket with an ischial containment has been designed.

Keywords: Electromyography, Transfemoral Prosthesis, Integrated EMG

1. Introduction

According to the World Health Organization (WHO) there are nearly 30 million amputees with limb loss and about 80% of that population do not have access to an active and affordable prosthesis [1]. There are about 2,30,000 amputees living in India who are in need of a low-cost and effective prosthesis to assist them in walking and other activities [2]. Electromyography is a measure of the electrical activity of the muscle. An action potential is developed in the muscle fibers as a result of depolarization and repolarization [3]. Muscle activation resulting in generation of EMG signals is achieved by a sequence of events from cortical excitation to motor unit activation and excitation-coupling [4]. An Electromyograph detects the potential generated during muscle activation and records it as an Electromyogram (EMG). EMG is a bio signal that is very useful in the diagnosis of disorders related to muscles and also in assessing various parameters associated with muscle like muscle fatigue, muscle spasms etc. [4,5]. EMG signals are vital in the control of exoskeletons, prosthetic hands and legs, and wheelchairs, as well as in amputee rehabilitation [6,7,8,9]. Various assistive and rehabilitation devices were designed and developed using EMG signals. Mingxing et al., constructed an EMG-controlled knee exoskeleton to aid in the rehabilitation of stroke patients by allowing them to play a training game with a small-time delay of 64 ms [6]. Sudarsan et al., created a myoelectric prosthetic arm that was operated by the amputee's own EMG signals after appropriate processing [7].

Similarly, in the work conducted by Hayder et al., EMG signals were utilized to control wheelchair motions such as forward, backward, left and right turn, and stop situations, as well as one muscle position commanding five conditions and the control performance of such systems can be greatly improved by the use of wearable electrode systems [8,9]. Furthermore, to aid the amputees Amanpreet et al., designed and developed an EMG based Human-Machine Interface (HMI) which performed specific wheelchair motion control by using distribution of muscle power and frequency [10]. It is recommended to reduce the number of EMG recording channels in order to increase the number of commands for effective control of devices [11]. EMG signals also play a vital role in the development of Physiological Devices (PDs).

In this work, the surface EMG signals were acquired using the BIOPAC MP45 acquisition device and software system from a healthy individual with a particular focus on the hamstring (Bicep Femoris and Semitendinosus) and quadriceps (Sartorius and Rectus Femoris) muscles. The signals were acquired from the hamstring and quadriceps muscles [12,13] and the acquired EMG signals were processed using the MATLAB software version 2019. The EMG signals obtained from the residual limb were useful and significant to drive the stepper motor of the prosthetic limb thereby achieving mobility in the desired direction. A wearable Ischial containment socket embedded with surface electrodes to record the EMG signals was designed using Autodesk Fusion 360 software. The EMG signals obtained from the residual limb were useful and significant to drive the stepper motor of the prosthetic limb thereby achieving mobility in the desired direction [14].

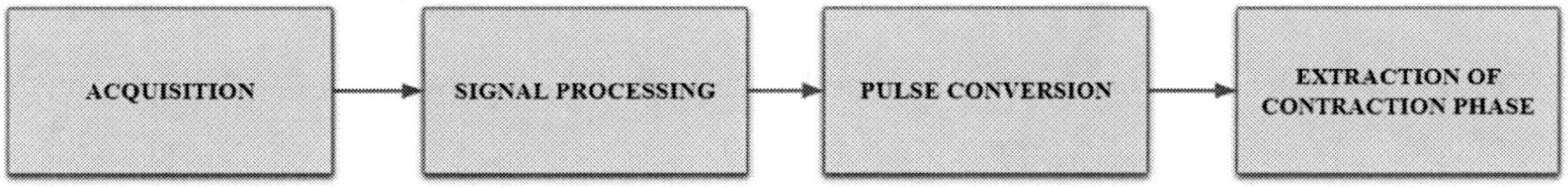

Figure 1. Block diagram of the proposed work

2. Methodology

2.1. EMG signal acquisition

The overall block diagram of the proposed work is shown in Figure 1. The EMG signals were acquired using surface electrodes placed on appropriate muscle locations. In this work, Kendall Covidien 200 series electrodes were used since it has a gel radius of 5 mm and easily sticks to the surface of the skin. The BIOPAC MP45 device and software was since it is flexible and easy to acquire the signals with minimal noise. BIOPAC MP45 is a two channel biomedical data acquisition device with an input voltage range of ±200 μV - 2V, an A/D sampling resolution of 16 bit and a signal to noise ratio of 89 dB. The sampling ratewas set to be 250 Hz and a Notch filter was selected in order to remove the power line interference of 50 Hz. The participant's skin was abraded with spirit to reduce the impedance of the skin and to remove any dead skin cells that might affect the recording of EMG signal. The electrodes were positioned on the respective muscle locations and the subject was informed about the recording setup and was asked to stand in an erect position. The hamstring and the quadriceps muscles is activated producing an action potential during flexion motion and the EMG signals are generated, hence the subject was asked to exhibit flexion movement for 30 secs followed by a 10 secs baseline period [12,13]. The EMG amplitude is influenced by muscle activation [15]. During the Flexion motion of the knee both the agonist and antagonist muscles arebeing activated leading to generation of EMG signals on the anterior and posterior regions of the thigh [16].Hence the signal was acquired during the flexion phase. The subject was monitored carefully throughout the procedure and visual feedback of the signal recording was given.
In this work, the signals were obtained from four different muscles namely the sartorius, rectus femoris, biceps femoris and semitendinosus muscles which constitute the anterior and posterior

IOP Conf. Series: Materials Science and Engineering 1272 (2022) 012013 doi:10.1088/1757-899X/1272/1/012013

segment of the thigh respectively. The electrode location and placement are shown in Figure 2(a) and Fig 2(b). The EMG signals acquired from the anterior and posterior compartment of the thigh were stored as a text file and plotted in MATLAB as shown in Figure 3(a) and 3(b) respectively.

2(a)

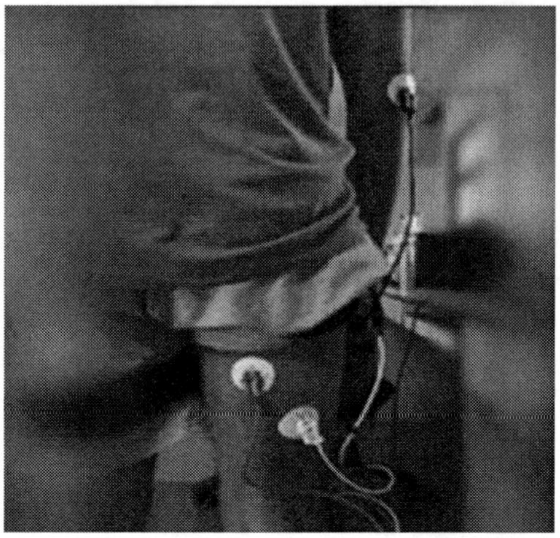

2(b)

Figure 2(a). Electrode placement on the anterior segment of thigh. **2(b).** Electrode placement on the posterior segment of the thigh.

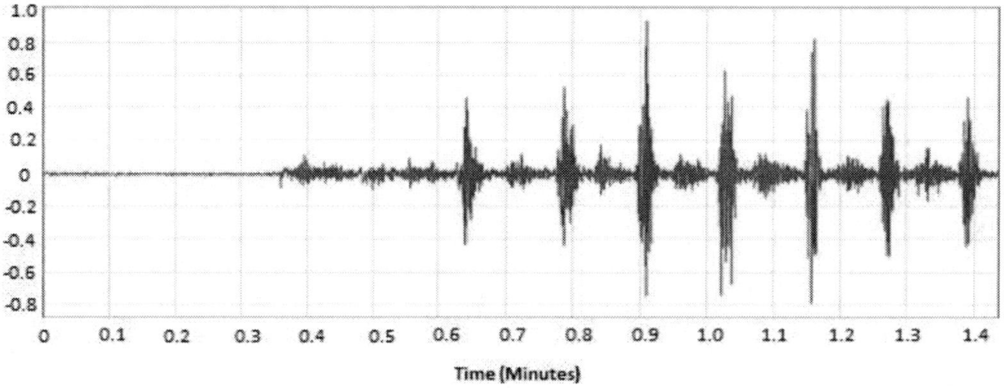

3(a)

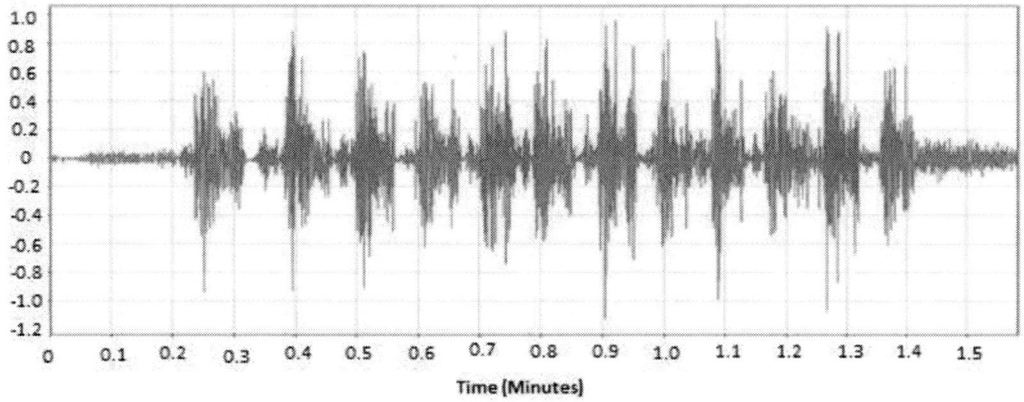

3(b)

Figure 3(a). EMG signals acquired from the anterior segment of the thigh.

3(b). EMG signals from the posterior segment of the thigh plotted using MATLAB

2.2. EMG signal processing

The obtained raw EMG signal must be processed in order to extract information from it [17]. To remove the artifacts and noises present, the Raw EMG signal was processed by filtering, detrending, rectifying and smoothening to extract the unintended features [11]. The EMG signal possessed a lot of noise and hence was filtered with a suitable bandpass filter. In this work, a bandpass Butterworth filter with a cutoff frequency of 50-150 Hz was chosen since Butterworth filters are efficient when studies are related to processing EMG signals obtained from the lower limb [18,19]. An IIR Butterworth bandpass filter of order 78 was used and its magnitude and phase response is shown in Figure (4). Lower cutoff frequency of the bandpass filter removes the baseline drifts that occur due to movement, perspiration etc, and higher cutoff frequency removes higher frequency noises and prevents aliasing [19]. The filtered signal was rectified in order to remove the negative components and to get the absolute value. Furthermore, the signal was smoothened using the moving mean method with a window duration of 450 ms. The processed signal is shown in Figure 5(a) and 5(b) respectively.

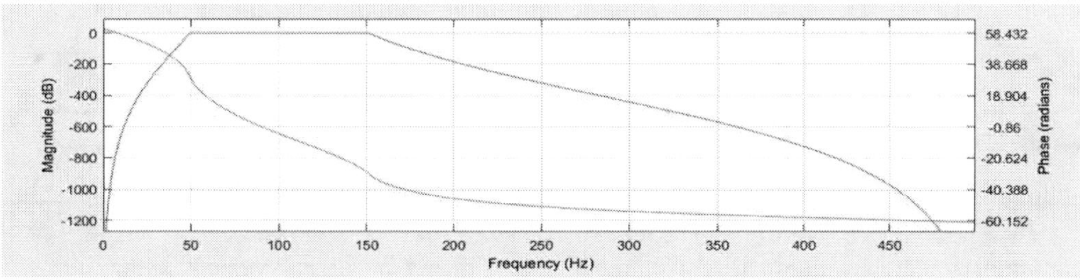

Figure 4. Magnitude and phase response of the designed filter

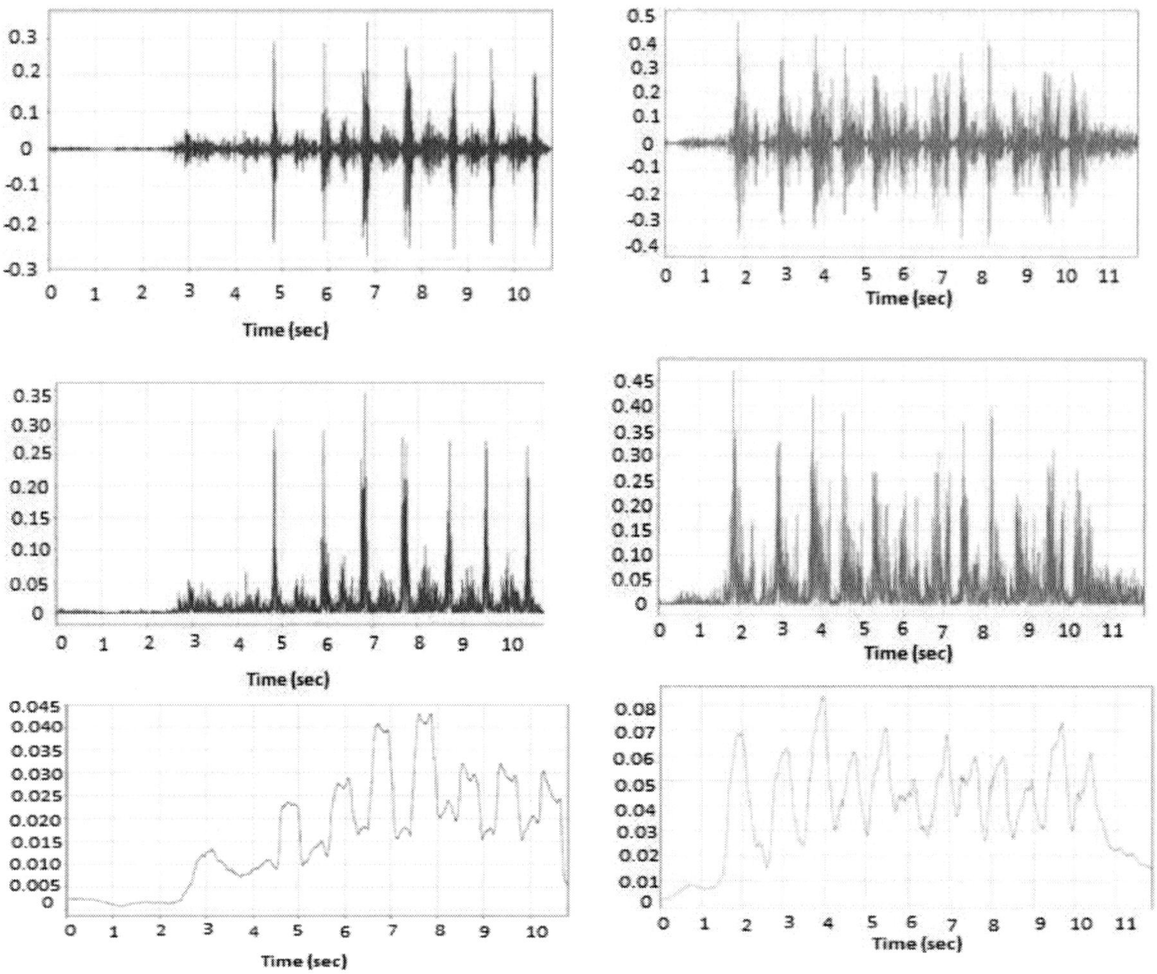

5(a) 5(b)

Figure 5(a). Filtered, rectified and smoothened signal obtained from anterior segment of the thigh. **5(b).** Filtered, rectified and smoothened signal obtained from the posterior compartment of the thigh

Table 1. Statistical parameters of the acquired EMG Signal

Muscles Position	**Mean**	**Median**	**Standard Deviation**	**Variance**	**Kurtosis**	**Skewness**
Anterior	-0.004901	-0.005085	0.064304	0.004135	40.582344	0.445286
Posterior	-0.005502	-0.00672	0.134531	0.018099	11.835418	0.174744

2.3. EMG feature extraction

Statistical parameters and features were extracted from the acquired bio signal in order to strengthen the implementation aspect of the current work. The statistical parameters that were obtained can be seen in Table 1. Furthermore, two features namely the Mean Absolute Value (MAV) and the Integrated EMG Feature (IEMG) were obtained from the raw EMG signal in LabVIEW software

IOP Conf. Series: Materials Science and Engineering 1272 (2022) 012013 doi:10.1088/1757-899X/1272/1/012013

version 2019. The LabVIEW block diagram is shown in Figure 7. The mean absolute value is used to identify and assess muscular contraction levels, whereas the IEMG is used to determine the muscle's pre-activation index [20]. The Mean Average Value was obtained using the Mean Average Rectifier block for both the anterior and posterior segment EMG signals. The output can be seen in Figure 6(a) and 6(b) respectively. The IEMG and MAV were found to be very efficient in constructing a bio signal powered prosthesis [14]. The IEMG of the signal was derived by rectifying the raw EMG signal followed by subsequent integration. The integration width was set to 142 and the obtained waveform was a sawtooth waveform which can be seen in Figure 6(c) and 6(d) respectively.

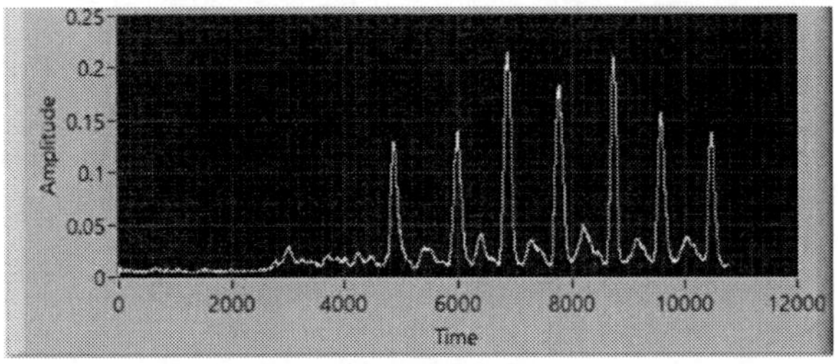

6(a)

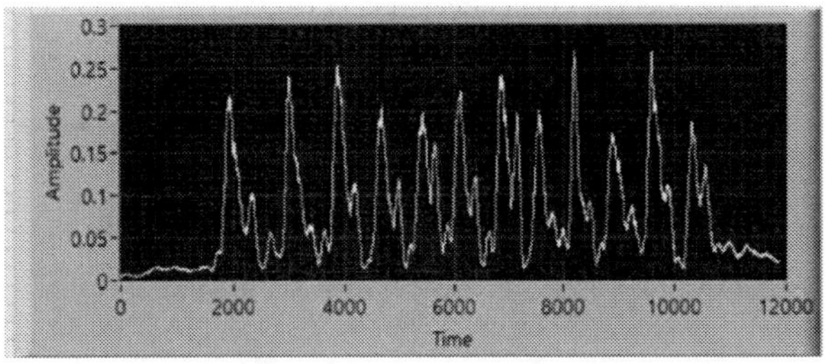

6(b)

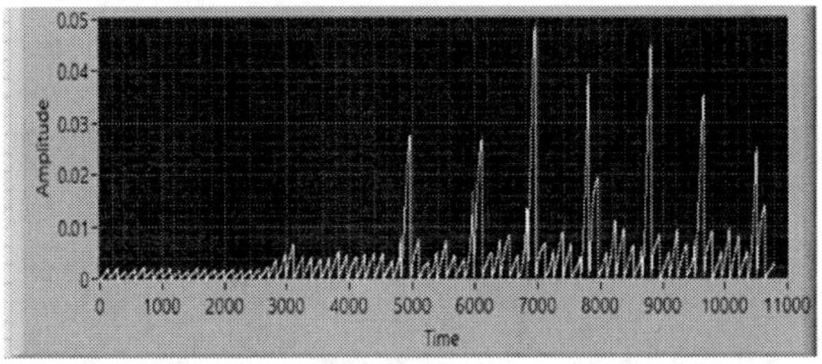

6(c)

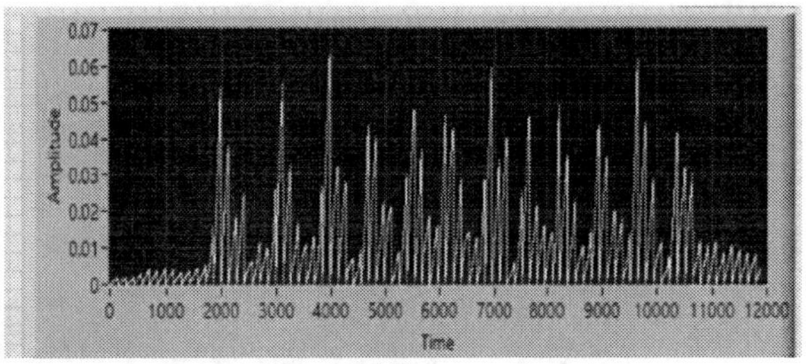

6(d)

Figure 6. Mean Average value plot for **6(a)** anterior segment, **6(b)** for posterior segment EMG signals. Sawtooth waveform of **6(c)** anterior segment, **6(d)** posterior segment EMG signals.

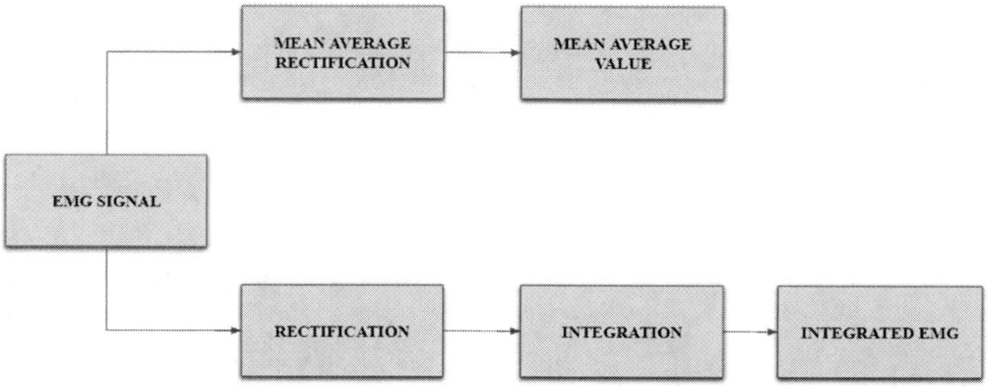

Figure 7. Overall block diagram of EMG signal processing and feature extraction

2.4. Pulse conversion

The processed EMG signal was converted to a pulse in order to trigger and drive a stepper motor. The signal was initially visualized in the LabVIEW software and positive comparators were used with a threshold of 0.4 in order to generate the pulse output. The output of the comparators is shown in Fig 8(a) & 8(b).

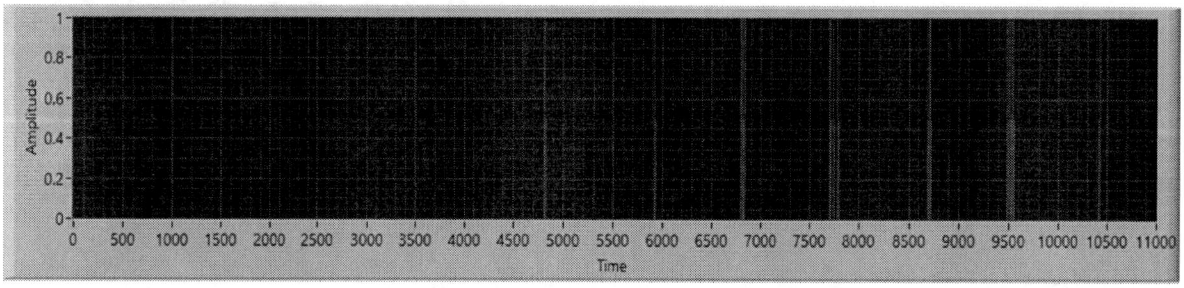

8(a)

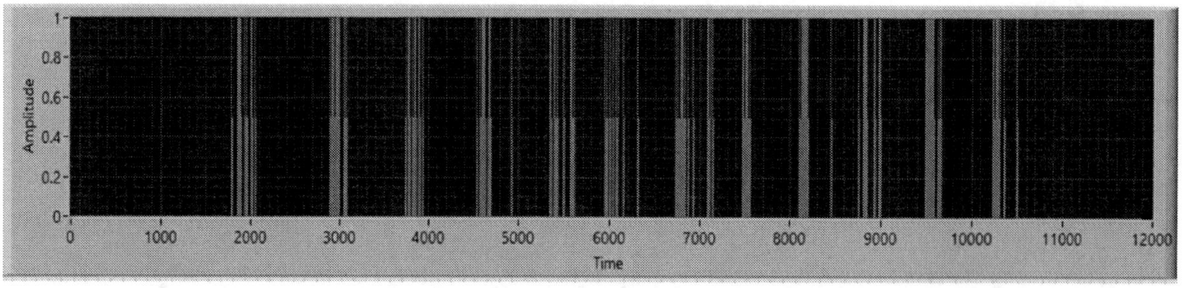

8(b)

Figure 8. Pulse plot of the EMG signal from **8(a).** the anterior segment, **8(b).** the posterior segment.

2.5. Socket Design

To assist the transfemoral amputees, an ischial containment socket was designed using Autodesk Fusion360 software. An Ischial containment socket design is usually preferred over Quadrilateral socket design because of its superiority in terms of comfort and energy efficiency [21,22]. The comparison of features of the two types of prosthesis is given in Table 2.

Table 2. Comparison of properties of quadrilateral and ischial containment socket [21,22]

Quadrilateral socket design	**Ischial containment socket designs**
It utilizes ischial weight bearing concept.	It utilizes weight bearing through narrow mediolateral ischium,ischial tuberosity and the ramus inside the socket.
Does not have a bony lock system.	By slanting the posterior wall and having a higher lateral wall, a locking or wedging action is created to stabilize the femur. To limit motion and promote stability, a three-point pressure mechanism is employed to lock the femur into adduction.
Femur stabilization is poor when the gluteus maximus activates. Quadrilateral socket moves laterally, as ischium has no effect of stopping weight bearing shift.	By bearing pressure on the skeletal parts, the narrow mediolateraldesign increases stability and reduces motion loss via soft tissue.

Surface EMG Electrodes are placed inside the designed socket to record the EMG signals from the hamstring and quadriceps muscles. Two pairs of electrodes are placed in the anterior and posteriorregions of the socket design. The 3D model of the socket design is shown in Fig 9(a) and 9(b) respectively. The designed 3D model can be used for additive manufacturing (3D printing) [15].

9(a) 9(b)

Figure 9(a). Top view of the designed prosthesis. **9(b).** Side view of the designed prosthesis

3. Results and Discussion

The raw EMG signals from the anterior and posterior regions of the thigh were obtained, and the signals were further processed to get the desired characteristics. An IIR bandpass filter of order 78 with a cut-off frequency of 50-150 Hz was designed. Graphical plots of raw, filtered, detrended, rectified and smoothenedsignals were plotted using MATLAB software. A moving mean filter with a window duration of 450 ms was adapted to smoothen the signal. To get more information about the data acquired from thigh muscles (anterior / posterior), a statistical analysis was performed and the following parameters were acquired such as mean (-0.004901 / -0.005502), median (-0.005085 / -0.00672), standard deviation (0.064304 / 0.134531), variance (0.004135 / 0.018099), kurtosis (40.582344 / 11.835418) and skewness (0.445286 / 0.174744). The various features of the EMG signal were determined. The integrated EMG (IEMG) feature, which helps identify and assess muscular contraction levels, and the mean absolute value (MAV) feature, which offers information about the muscle's pre-activation index, was determined. The IEMG and MAV were found to be very efficient in constructing a bio signal powered prosthesis. The contraction phase of the EMG signal was converted to a pulse for the purpose of driving a stepper motor. The threshold voltage wasset to 0.4 during pulse transformation. A 3D model of an Ischial containment socket design was designed using Autodesk Fusion 360 software in order to assist the transfemoral amputees.

4. Conclusion

In this study, the contraction phase of the EMG signal was identified and converted to a pulse in order to drive a stepper motor and a 3D model of the Quadrilateral socket was designed to help the transfemoral amputees. The EMG signal acquisition and its processing has been performed followed by the statistical study of the data. More information about the acquired raw data was obtained by determining the statistical parameters. The EMG signals were acquired from a healthy volunteer using surface EMG electrodes and the processing was carried out in terms of filtering, detrending, rectifying and smoothening. Valuable EMGsignal features such as Integrated EMG feature and Mean absolute value was extracted. It was found that the MAV and the IEMG features are efficient in the construction of a bio signal powered prosthesis. The study also found that the contraction phase of the signal that was converted to pulse was used to drive a stepper motor. A comparative analysis between the Ischial containment socket and quadrilateral socket design was performed. It was found from the comparison, that the Ischial containment socket design is usually preferred over Quadrilateral socket design because of its superiority in terms of comfort and energy efficiency it provides. Hence, a 3D model of the Ischial containment socket design with a provision for electrode placement was designed to record the EMG signals from the hamstring and quadriceps muscles to promote real time acquisition. It was also found that the proposed design can be 3D printed using additive manufacturing. This Ischial containment socket design model was proposed in order to help the transfemoral amputees wherein the surface electrodes which are placed within the socket acquire the EMGsignals and those acquired signals are processed and converted into pulse. This pulse signal is then used in the control of a stepper motor, thereby helping the amputees control their own movement with their own bio signal with the help of a comfortable prosthesis.

Conflict of interest

No conflict of interest was declared by the authors.

References

[1] Hamner, S R., Narayan, V G and Donaldson, K M, 2013. *"Designing for scale: Development of the remotion knee for global emerging markets"*. Annals of biomedical engineering, **41**(9), pp.1851-1859.

[2] Narang, Y S, Arelekatti, V M and Winter, A.G., 2015 "*The effects of prosthesis inertial properties on prosthetic knee moment and hip energetics required to achieve able-bodied kinematics*." IEEE Transactions on Neural Systems and Rehabilitation Engineering, **24**(7), pp.754-763.

[3] Nazmi N, Abdul Rahman M A, Yamamoto S I, Ahmad S A, Zamzuri H and Mazlan S A, 2016" *A review of classification techniques of EMG signals during isotonic and isometric contractions*." Sensors, **16**(8), p.1304.

[4] Yousif H A, Zakaria A, Rahim N A, Salleh A F B, Mahmood M, Alfarhan K A, Kamarudin L M, Mamduh S M, Hasan A M and Hussain M K, November 2019 "*Assessment of muscles fatigue based on surface EMG signals using machine learning and statistical approaches: a review*." In IOP *Conf.*, Series: Materials Science and Engineering (Vol. **705**, No. 1, p. 012010).

[5] Sikder F, Sarkar D, Schwartz O. and Thomas C K, December 2017 "*Method for concurrent processing of EMG signals from multiple muscles for identification of spasms*." In 2017 IEEE Signal Processing in Medicine and Biology *Symp.*, (SPMB) (pp. 1-6). IEEE.

[6] Lyu M, Chen, W H, Ding X, Wang J, Pei Z and Zhang B, 2019 "*Development of an EMG-controlled knee exoskeleton to assist home rehabilitation in a game context*." Frontiers in neurorobotics, **13**, p.67.

[7] Sudarsan S Student Member I.E.E.E. and Sekaran, E.C., 2012" *Design and development of EMG controlledprosthetics limb*." Procedia engineering, **38**, pp.3547-3551.

[8] Hayder A Azeez, Norasmadi Abdul Rahim, Muhammad Juhairi Aziz bin safar, Ammar Zakaria. 2016 "*An Intelligent EMG Controlled Wheelchair – A Review*" - International Acadmey of Engineering and Medical Research.

[9] Zhao S, Liu J, Gong, Z, Lei Y, OuYang X, Chan C C and Ruan S, 2020. "*Wearable physiological monitoring system based on electrocardiography and electromyography for upper limb rehabilitation training*." Sensors, **20**(17), p.4861

[10] Kaur A, 2021 "*Wheelchair control for disabled patients using EMG/EOG based human machine interface:a review*." Journal of Medical Engineering & Technology, **45**(1), pp.61-74.

[11] Nazmi N, Rahman, MAA., Mazlan SA, Zamzuri H. and Mizukawa M, "*Electromyography (EMG) based signal analysis for physiological device application in lower limb rehabilitation*." In 2015 2nd *Int., Conf.*, on Biomedical Engineering (ICoBE) (pp. 1-6). IEEE, March 2015.

[12] Ha KH, Varol HA and Goldfarb M, 2010," *Volitional control of a prosthetic knee using surface electromyography*". IEEE Transactions on Biomedical Engineering, **58**(1), pp.144-151.

[13] C Michael Schuch, C P O *Transfemoral Amputation: Prosthetic Management.*

[14] Celia Pradeep, Shalu George K - October 2015– "*Control of Stepper Motor Using Surface EMG*" -IJAREEIE

[15] Amrutsagar L, Parit G, Ghyar R. and Bhallamudi R, 2020" *Parametric Design and Hybrid Fabrication of Above-Knee Prosthesis*." Indian journal of orthopaedics, **54**(3).

[16] Jarić S, Ropret R, Kukolj M and Ilić D B, 1995 "*Role of agonist and antagonist muscle strength in performance of rapid movements*." European journal of applied physiology and occupational physiology, **71**(5), pp.464-468.

[17] Chowdhury R H, Reaz M B, Ali M B M, Bakar A A, Chellappan K and Chang T G, 2013" *Surface electromyography signal processing and classification techniques*." Sensors, **13**(9), pp.12431-12466.

[18] Nazmi N, Abdul Rahman M A, Yamamoto S I, Ahmad S A, Zamzuri H and Mazlan S A, 2016 "*A review of classification techniques of EMG signals during isotonic and isometric*

contractions." Sensors, **16**(8), p.1304.
[19] William Rose, 2019 - *Electromyogram analysis*- Mathematics and Signal Processing for Biomechanics.
[20] Christopher Spiewak, MdRasedul Islam, Md Assad-Uz Zaman and Mohammad Habibur Rahman, 2018"*A Comprehensive Study on EMG Feature Extraction and Classifiers*" - - Mechanical Engineering Department, University of Wisconsin-Milwaukee, USA.
[21] Panel SusanL KappCPO, *Transfemoral Socket Design and Suspension Options* - Author links open overlay
[22] C Wehe, *An Overview of Transfemoral Socket Designs*, Bradley University of North Dakota

LoRa-based IoT applications on campus: experimental demonstrations and performance evaluation

Vladilen Kolesnikov[1], Marcelo Neves[1], Bhargavi Goswami[1]
[1] Queensland University of Technology, 4001, QLD, Australia

Abstract. Lora technology is widely used in solving many IoT problems because of its long-range capability and low-cost. However, the model which works in the 2.4GHz frequency and without regional limitations is a novelty and its full capabilities are unknown. In this research the device is test for its transmission capabilities on a university campus because this is the conditions with high 2.4GHz interference from Wi-Fi, Bluetooth, etc. The authors also try to find the packet delivery rate and define at what distance LoRa end-node starts to lose connection with its gateway. Along the way receiving signal strength indicator is registered in order to understand the device power level. All of it is done by placing the gateway in the middle of the campus so that it equally reaches the furthermost points, connecting it with a LoRa-based end-node and testing this connection by placing the end-node device in different spots within the area and analysing the connection. LoRa 2.4GHz has shown lower effectiveness in terms of distance in comparison with its sub-GHz counterparts. Receiving signal strength indicator also demonstrated a lower power level in comparison with the mentioned device specifications. However, the LoRa technology still provides bigger distance than Wi-Fi or Bluetooth and is cheaper than other protocols used in IoT. Higher results can be achieved by using a better antenna and by eliminating physical obstacles.

Keywords: LoRa; Sensors Networks; IoT; Internet of Things; LPWAN

1. Introduction

Internet of things (IoT) is a growing field and sensor network solution is one of its parts that has been attracting more attention every day. The Ericson Mobility Report stated in their findings that the number of IoT devices would grow from 7 billion in 2017 to 20 billion in 2023 [1]. In the past many sensors were wired and it drastically limited its area of use, wireless technologies started to change the situation but many aspects had to be taken into account such performance capabilities and deployment costs. New solutions arise and the need for the most "perfect" one is needed. LoRa is a new low-power wide-area networks (LPWANs) device that helps to solve many problems related to IoT. According to the authors of the official documentation [2], it's built using spread-spectrum modulation technique derived from the existing Chirp Spread Spectrum (CSS) technology which makes it possible to establish long distance communication by reducing data rate. The modulation process is thoroughly described mathematically in [3]. LoRa is also much cheaper to implement than its counterparts [4] and doesn't require any repeaters along the distance, its power consumption is also less: that is why it has grown in demand.

In this paper, the performance of 2.4GHz LoRa will be analysed since these industrial, scientific and medical (ISM) unlicensed bands are the most commonly used in the world by other technologies such as Wi-Fi, Bluetooth, etc. Initially LoRa used to be available only with reginal restrictions called Regional Parameters [5] which means that every device has different specifications for different regions, for instance, the bands of 863 MHz – 870 MHz are available for Europe, 915-928 MHz for Australia, 470-510 MHz for China, 902-928 MHz for the United States, etc. In this case the Effective Isotropic Radiated Power (EIRP) which represents power of antenna in single direction is also different:

the maximum allowed IERP for Australia and the USA is +30 dBm, +16 dBm for Europe and +12.15 dBm for China. Even the duty cycle can be different: less than 1% for China and no limitations for Australia, yet no dwell time limitation for the first and 400ms of dwell time for the latter. All these limitations make the production more complicated as well as the number of possible transmissions is quite limited for LoRa sub-GHz devices. Likewise, the use of different frequencies makes it impossible to use the same device in such applications as ship logistics, however 2.4 GHz LoRa can solve these issues. The device allows higher data rates since it can function in higher bandwidth, and yet longer distances than other technologies in the same frequency such as Wi-Fi or Bluetooth [6], that is why the application of this technology extends further: it can be used for warehouse management as indoor scenario or construction site monitoring and livestock racking as outdoor scenario. Along with the existing use, LoRa can be further applied in future development [7].

Because of its novelty, the new device is not completely studied thus its performance capabilities are not well understood. As it has been seen before, Lora 2.4GHz doesn't only provide its users with new features but also raises many questions of what it can do and how. In this study the authors experimentally tested the connection between Lora gateway and its end-node on a university campus because these are the conditions of high 2.4GHz interference. Various sensors are known to be widely used in modern cities: for smart traffic lights, winds speed, temperature and many other applications, however these are the places where many people live and work at the same time which means high usage of Wi-Fi, Bluetooth and other devices, most of the Wi-Fi traffic still goes in the 2.4GHz frequency: all of this means that LoRa 2.4GHz has to find its place in the same frequency. University campus simulates similar conditions that is why the results obtained from the experiment can be referenced while applying Lora IoT solutions in a modern busy city centre. The researchers will also try to answer the question of how to improve the quality of connection with fewer packets lost.

Therefore, this paper is organized as follows: Section I contains Lora background information and introduction to the topic, while Section II briefly outlines previous studies about LoRa range and performance analysis in related works. Section III introduces method used in the experiment. The results obtained from it along with following discussion and conclusion are presented in Section IV.

2. Related works

Many researchers have been trying to answer the questions mentioned above: for example, the authors of [6] did a mathematical calculation using Electronic Communication Committee (ECC-33) model and came to the conclusion that the maximum range of LoRa 2.4GHz in an urban environment was 867 meters. Another experimental study conducted in Melbourne CBD described in [8] says that the maximum possible range in an urban environment was 600 meters, however it was done in the 902–928MHz bands.

Since 2.4GHz frequency is widely used by many technologies and the most numerous in a busy city centre is supposedly Wi-Fi, its interference with LoRa has to be taken into account. The conference article [9] shows that due to its modulation, LoRa has a high interference tolerance and under some conditions still can operate up to 1.7km, however the paper doesn't provide information on the experimental environment and the maximum SF used is 10. In [10] the researchers created a MATLAB-based LoRa PHY simulator which emitted baseband signal. They decided to test scenarios both with co-channel and in-band coexistence scenarios. They found out that in the first situation the resistance to the interference could be very high as long as the bandwidth was kept as low as possible, 125KHz in their test, and SF kept above 9 with SF of 12 showing the best results. In the second case they discovered that different Wi-Fi technologies has different level of interference. For instance, IEEE 802.11b was less aggressive than IEEE 802.11n. OFDM-based Wi-Fi signal interference was practically on a constant level. Similar findings regarding Wi-Fi interference are described in [11], the research was conducted by Semitech Corporation in their laboratory. They also prove high Wi-Fi interference immunity and say that it can be maximized increasing spreading factor.

Regarding packet delivery rate, authors of [12] tested LoRa in a tree farm and the closest SF to ours was 11, the minimum PDR at 200-meter distance was 82.76%, yet the environment had different type of physical obstacles and the device worked in the US 915 MHz ISM bands. The lowest average RSSI was -95.04 dBm. In a similar environment to the one of this paper, at Graz University of Technology, LoRa 868 MHz was tested by the authors of [13]. In one of the experiments, they placed the two devices indoor without clear line of sight at the distance of 115 meters and tried to find the best settings in order to achieve the highest reliability in terms of packet delivery rate. As the result, the utmost average PDR was achieved with the highest SF of 12 and lowest BW of 250 MHz. The researchers in [14] did a performance evaluation of LoRa 915 MHz on the campus of National Chiao Tung University located in Hsinchu, Taiwan. The conclusion stated that in the indoor scenario when end-nodes were placed inside the buildings while gateway was still outside, packet loss achieved 60-90% with the distance of 1200m which is not reliable for IoT applications. In the outdoor scenario, the packet loss at 1360m was at the average of 16% some of which may have been provoked by existing physical obstacles some of them by 4G base stations that operate at 900 MHz in Taiwan. Another interesting finding was that packet delivery loss increased by 20% in rainy weather.

Other researchers in [15] tested LoRa 433 MHz and 868 MHz with 3 dBi and 2.2 dBi antenna gains respectively in the urban area of Dortmund, Germany. It was noted that the worst-case scenario for achieving stable connection was when the end-node was positioned on the ground floor of a building in the central district: 2% PDR for 868 MHz device and 0% PDR for 433 MHz. The researchers of [16] tested LoRa 433 MHz with 6 dBi antenna gain on a lake at the average distance of 3 km between the nodes and the gateway. They found out that the packet loss rate could be as little as 0.26% in the area with few physical obstacles and as big as 63.26% with more obstacles even though the distance was shorter which meant that LoRa was capable to operate on relatively big distances as long as there were no physical obstacles.

Performance of LoRa 2.4GHz in urban environment is not completely studied since this is quite a new piece of technology. Based on the literature findings, readers can discover the approximate maximum communication range for urban busy environment or realize that physical obstacles are the main problem for establishing stable connection between LoRa devices. However, it must be mentioned that most of the experiments mentioned above were conducted in sub 1 GHz bands which automatically excludes 2.4 GHz interference from multiple Wi-Fis, Bluetooth and other devices: all of them can decrease LoRa 2.4 GHz propagation. Sub-GHz LoRa also had its own drawbacks such as 4G stations in Taiwan which may not be the case for LoRa 2.4GHz. Physical obstacles are said to deteriorate the connection as well as weather conditions. Likewise, real-life scenarios can differ from laboratory measurement setups. Packet loss issue also must be carefully addressed because it can have terrific consequences for such application as health monitoring systems.

3. Method

The experiment of this study was carried out in the city if Brisbane, QLD, Australia at the Queensland University of Technology, campus at Gardens Point. It was a sunny Sunday day which meant that the weather and 2.4Ghz interference from other devices was supposedly at its minimum in comparison with weekdays. Although the test started in the afternoon that is the campus wasn't completely empty. The HiQ building was chosen for the gateway because it was situated right in the centre of the area and had similar distance to all of the extreme points of the studied area. The distance from the furthermost point on the campus, Gardens Point medical Centre, was approximately 200 meters (Fig. 1).

LoRa gateway was placed on the top 7th floor right in the middle because the access to the roof was limited (Fig. 2). It has to be noted that this is the researchers' centre floor, the other floors below are dedicated to library and study spaces that is why there still were people working using their devices in the same frequency. The gateway operated in three receiving channels, the default settings were used for this study and the channels were set as follows: Rx1 channel was set at 2403 GHz, Rx2 was set at 2479 GHz and Rx3 was set at 2425 GHz. Bandwidth for all of the channels was 812kHz, spreading factor was set to 12, antenna had height of 3cm and its gain was 13 dBi, coding rate was 4/8LI, packet payload size was 30 bytes (Fig. 3).

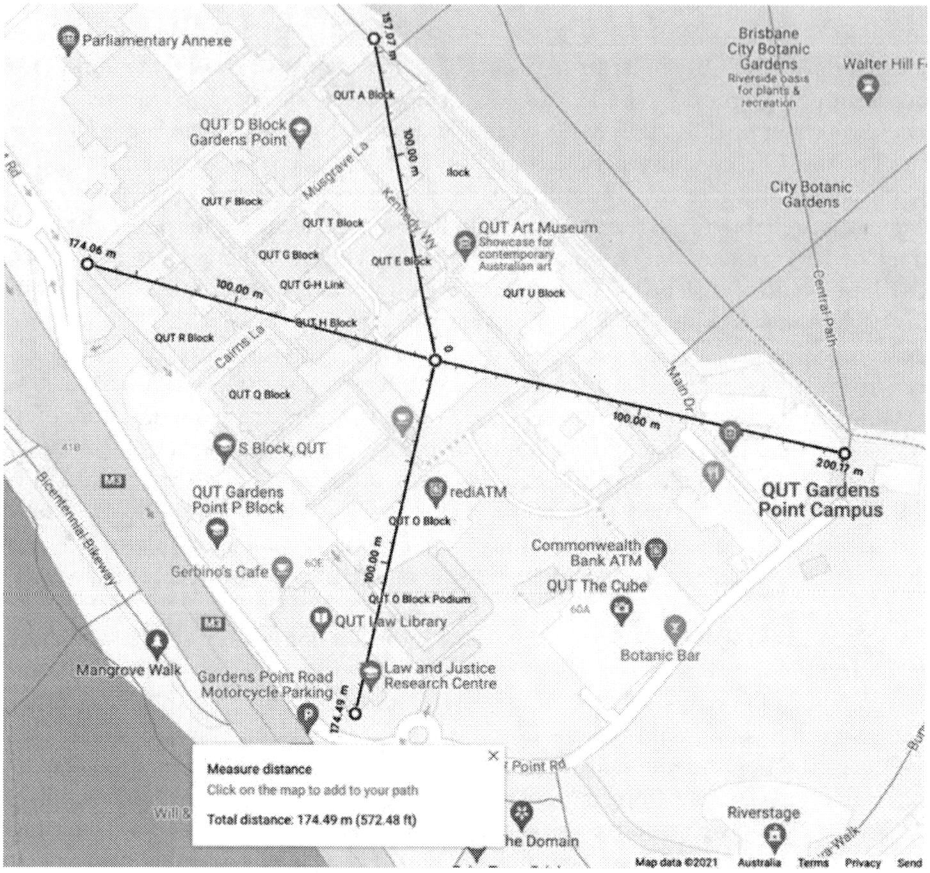

Figure 1. Gateway position and its distance to all furthermost points.

Figure 2. Gateway position.

The SX1280 LoRa 2.4GHz end-node was controlled by STM32 Nucleo-L073RZ board (Fig. 4) and it operated with the same settings, however, it should be mentioned that the sending channel of the end node was dynamic, which meant it sent packets in all of the three above mentioned channels in random order. All the codes and instructions for running the devices are available on github [17].

Figure 3. LoRa 2.4GHz gateway with RaspberryPi.

Figure 4. SX1280 end-node.

For instance, it could send 5 packets consequently in the 2425 GHz frequency even though the microcontroller was restarted all of the 5 times to send those packets, another time it could do the same but in 2479 GHz frequency. Yet, this was exactly the type of connection that had to be taken into account while drawing the results: LoRa didn't use connection like TCP which meant the packets weren't resent in case of no delivery. However, the packets were sent in all of the three channels, that is why one out of the three packets delivered could be considered as established connection.

While gateway was statically situated at the same place during all the experiment, end-node was placed at different positions within the campus. At each of the points there were sent approximately 20 to 40 packets. However, when the so-called "blind spots" were found with 0% or close to this PDR level, the fixed number of 20 packets was sent to make sure that the connection was completely lost. The researcher with the end-node using Putty terminal for macOS would emit a packet and over the phone tell the other researcher with the gateway what channel it was sent from (Fig. 5 (a)). It should be mentioned that the end-node logs didn't tell which gateway channels corresponded to which frequencies that is why the researcher with the end-node had to write them down, e.g., channel 0 had the frequency of 2403, etc. Further, the node researcher would use the channel number over the phone instead of saying the whole frequency because that would take more time., e.g. a packet was sent over channel 1 and a packet within channel 1 was expected to arrive. In order to emit a packet, it was simply needed to press the black button of the microcontroller and after the restart LoRa would immediately send a packet on a random channel. This was the fastest way to conduct the experiment because by the program LoRa would emit packets after a fixed period of time and the periods would be increased with each following emission which meant that sending 20-40 packets could take half an hour, while it could be done in 5-10 minutes by pressing the button. On the other end, the researcher would pay attention to the logs of the mentioned channel and wait until the logs were updated. Most of the times it would take a second for the packet to appear in the logs. The received packet would come with all the relevant data in a JSON format. Information such as RSSI, payload size, coding rate and other was available, however for this experiment only RSSI was needed (Fig. 5 (b)). The researcher with the gateway would tell these details over the phone and the other researcher would write them down for each packet. If the packet wasn't delivered it was also mentioned and noted down. Taken into account the limitations of this study, only uplink connection was tested. In order to test RSSI at no distance, end-node antenna was kept next to the gateway antenna at the 0cm range.

The methodology of this study is focused on a more empirical approach in comparison with those mentioned in the related works section for LoRa 2.4GHz: the biggest part of them was made in the laboratory settings. Yet, similar statistical method was used in the other papers for LoRa with regional restrictions. "Real-world" settings were considered to be more valuable for future use of the technology and LoRa 2.4GHz findings in this scenario are quite limited.

IOP Conf. Series: Materials Science and Engineering 1272 (2022) 012014 doi:10.1088/1757-899X/1272/1/012014

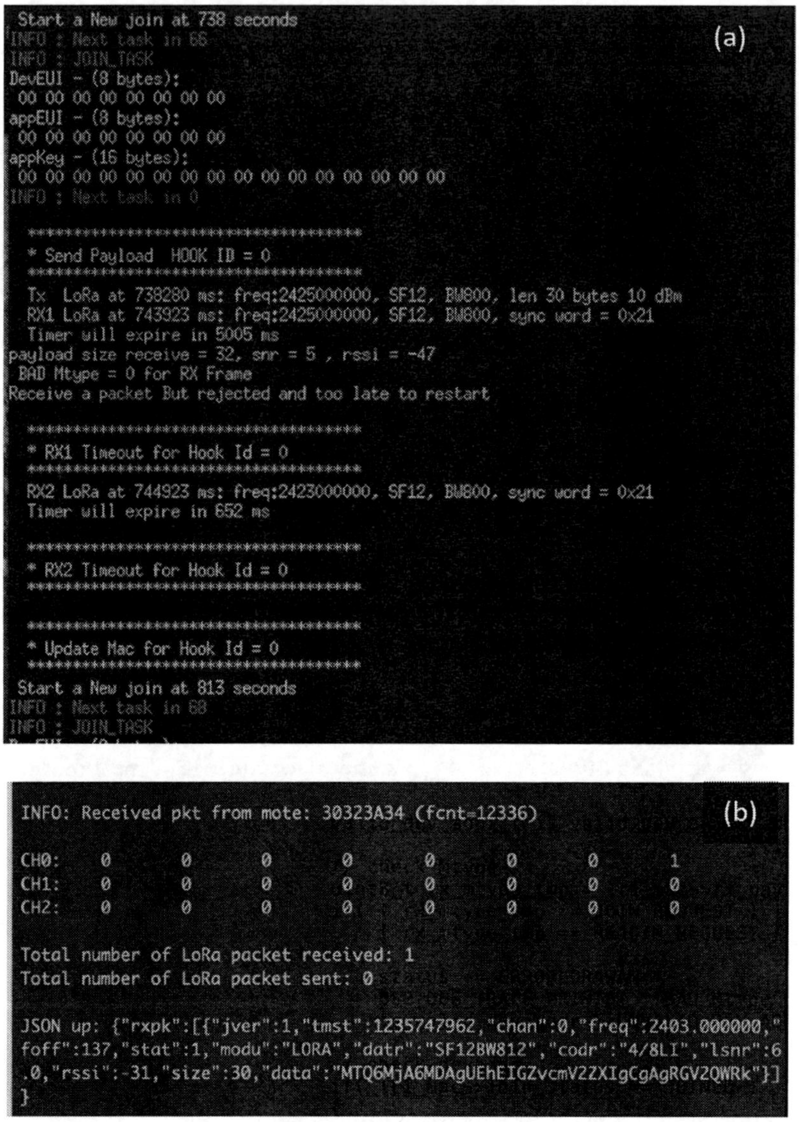

Figure 5. (a) LoRa gateway logs in macOS terminal; (b) End-node logs in Putty.

4. Results

Having analysed the collected data, the authors of this study have been able to draw a map with the points tested and packet delivery rate from those points (Fig. 6). A peculiar trend was noticed: when the connection was stable, the gateway received 100% of the packets losing 0-1 packet only. These points are marked with the light green colour. However, as soon as the packet loss started to take place, PDR sharply dropped to approximately 50%. Since the this is still more than a half, this percentage was allocated to a separate category and was marked yellow. At some positions the PDR dropped lower than that, and even though the device had no TCP protocol so the packets lost weren't retransmitted and were considered lost, the end-node sent the same packet in three different channels, and it was noticed that the packets could be received consequently on one of the channels while completely lost on the other two which meant the message was still delivered. That is why the cases when one third of the packets were successfully delivered were considered as existent communication even though it could be improved with better hardware by applying a bigger antenna: the gateway antenna height was approximately 3cm, this factor also influences the delivery rate according to the studies in [12]. This category is marked orange. The next category was when the PDR dropped below 35% but some of the packets were still received, yet this was rated as poor communication because it was unreliable for delivering any kind of messages. It is marked light red. The last category is marked dark red, at these points the gateway was able to receive at most one packet: those were considered as "blind spots". Most of the times it happened when the devices were separated by many buildings.

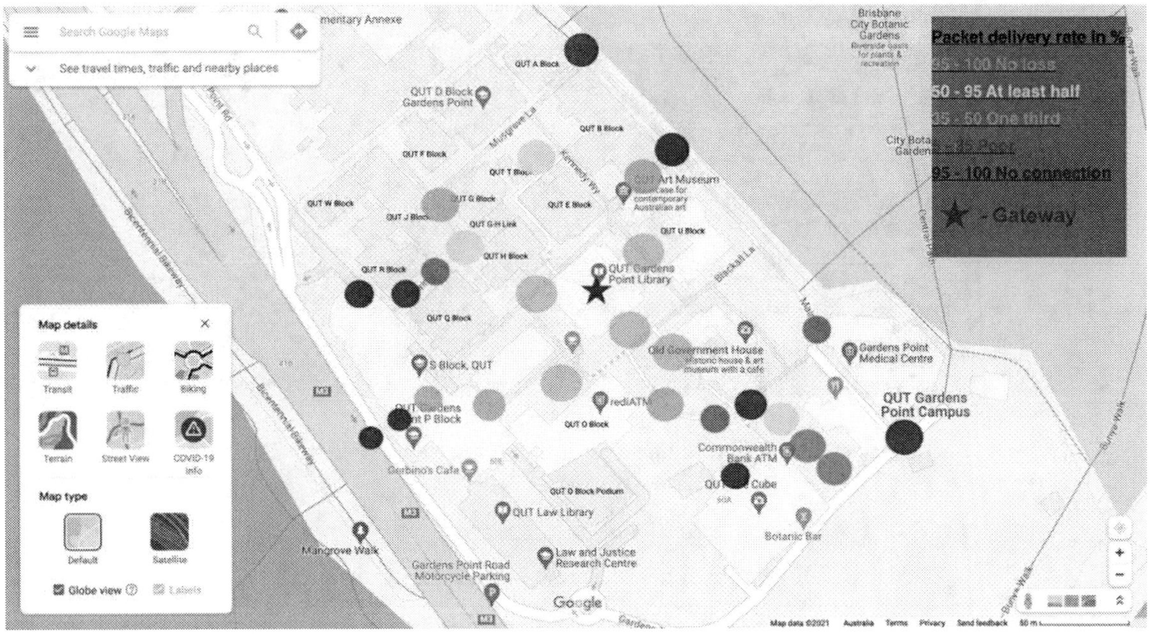

Figure 6. Packet delivery rate map.

Based on the map, a line graph was composed which demonstrates correlation of the average distance and average PDR categories (Fig. 7). The graph allows the readers to understand the average distance at which PDR starts to drop. While the distance was up to 50m the PDR was 100%, up to 80m the rate still stayed above 50%. One third of the packets was still delivered at the distance of up to 125m, however the connection became unreliable further than that, and blind spots appeared after the 179m point. Yet, these results only revealed average numbers: the closest blind spot was found at the distance of approximately 100m behind Old Government House Museum while 100% PDR was also possible at 60m.

Having evaluated performance metrics of LoRa 2.4GHz, the researchers have been able to analyse the received signal strength indicator which helps to understand the sensitivity of the gateway. The following table shows the average correlation of distance to PDR and its RSSI (table 1). While placing the two devices next to each other RSSI stayed in the range from -3 dBm to -18 dBm. As soon as the node was take outside but still had perfect PDR without losses the average RSSI was about -77 dBm. Then it started dropping with the distance and the lowest RSSI we registered was -102 dBm. In comparison we can refer to another study in [18], the authors of the experiment stated that the highest RSSI they achieved was -78 dBm while the lowest was -125 dBm. However, the research was done in Greece using SX1272 LoRa with regional restrictions working in sub-GHz.

Based on the analysis, with the settings and the environment of the experiment the maximum distance can be established around 179m (Fig. 8). The PDR at this point was 12% which belonged to the category of poor communication where less than 35% of packets were delivered, yet it was the furthermost position where the gateway still received messages.

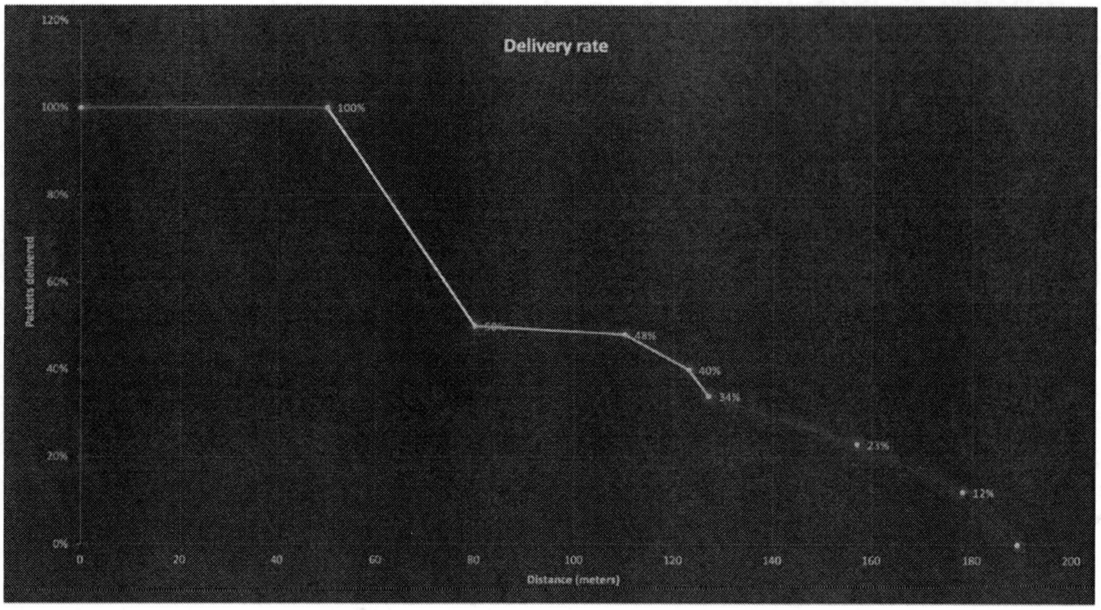

Figure 7. RSSI in relation to distance and its PDR with SF of 12.

Table 1. Packet delivery rate in relation to distance.

Distance	Delivery Rate	RSSI	SF
0m	100%	-10.4	12
50m	100%	-77.5	12
80m	50%	-94.65	12
110m	48%	-94.5	12
123m	40%	-96.73	12
127m	34%	-96.6	12
157m	23%	-97.7	12
178m	12%	-99.1	12
189m	0%	0	12

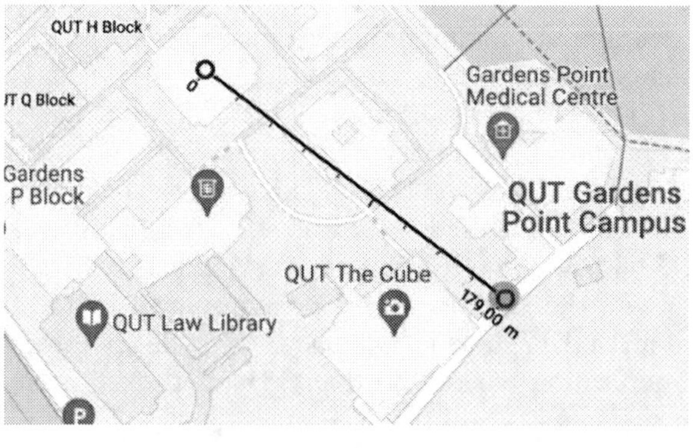

Figure 8. Maximum distance according to the main settings of the experiment.

Discussion

Based on the literature review, it can be understood that physical obstacles are the main issue for LoRa connectivity, thus, to prove the hypothesis there was taken a decision to position the gateway in a more open space. One of the tallest points with free access was on the last floor on the staircase of the Y block at the QUT Gardens Point campus. This experiment isn't described in the method section because initially it wasn't planned to take place. All of the research questions were expected to be answered in the first experiment; however, the results were not satisfying: 100% PDR only was possible to achieve at the distance of up to 50 meters. Merely for the sake of proving the hypothesis another experiment was agreed to take place.

This time the only obstacle at the staircase was a glass wall which was much less of a barrier compared to the previous experiment. To prove the point with less points tested, the node was positioned at a longer distance than the maximum achieved in the past in order to if the PDR would be higher in this case. It should be mentioned that the experiment took place on Tuesday that is why 2.4GHz interference was even bigger. As the result, the first point was approximately 220m away from the gateway, the line of sight wasn't clear but had less obstacles due to the height of the gateway position: PDR was at 48% (Fig. 9). Likewise, in order to reinforce the result shorter distances were tested. When the end-node was positioned at120m and 160m the PDR was 84% and 52% respectively. This time it was also possible to reach PDR in the range 90% and 60%, which wasn't achieved in the previous experiment: PDR would start to drop dramatically from 100% to approximately 50% as soon as packet loss took place. This time the average RSSI for the shortest distance was -75.4 dBm, -93.43 dBm for the middle position and -92.24 dBm for the furthermost point.

In the consequence of the new findings, it became possible to answer the question of how to improve the quality of connection and have a better PDR. While 2.4GHz interference and weather conditions may play its role, the main challenge for LoRa is physical obstacles, avoiding them the communication can be established at much bigger distances.

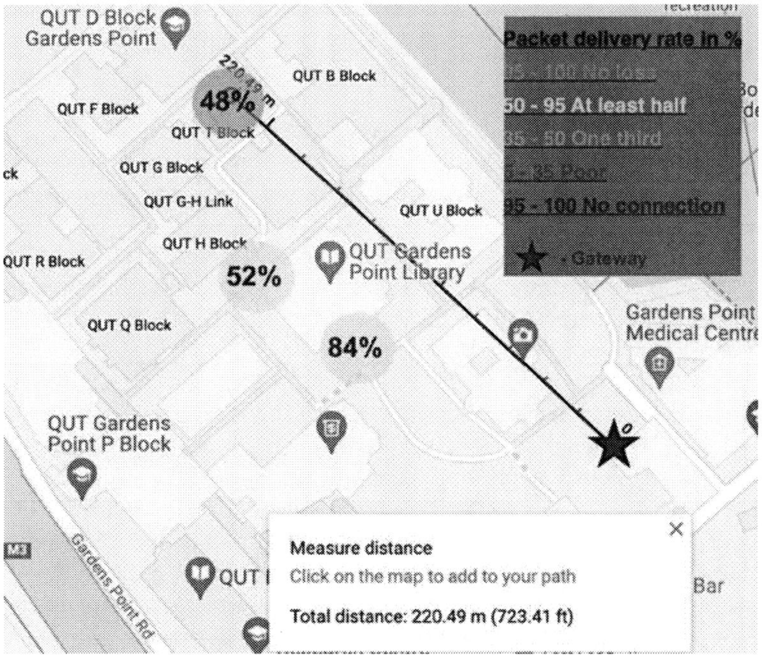

Figure 9. Test with less physical obstacles.

One of the limitations of these experiments to be mentioned is the settings. The devices functioned with the default SP of 12 and 812kHz BW. According to the literature review above, a lower BW could improve the connection, however it could also increase latency [19]. This performance may be unsuitable for some applications such as healthcare sensors while it may be enough for others such as weather stations. In this research no particular use case of the technology was meant to be studied that is why the default settings were used.

Conclusion

Within this experiment the researchers have been able to evaluate performance of LoRa SX1280 2.4GHz uplink communication. The maximum achievable distance on the QUT Gardens Point Campus was established at 176m. It was also noticed that PDR drops to nearly 50% as soon as the connection starts to deteriorate and keeps going down. Yet, the behaviour of packet delivery loss is different when there are fewer physical obstacles: 84% PDR in the second experiment. Therefore, it has become possible to prove the point that LoRa is able to reach bigger distances as long as physical obstacles are minimized even though interference from other devices is bigger than before.

For future studies it is important to evaluate the performance with bigger antenna as it is shown in the experiment in [14]. This 2.4GHz LoRa development gateway is meant for testing purposes [20] and had an antenna of only 3cm which would need to be much bigger in a real-life scenario.

Likewise, such details as use-case scenario have to be taken into account for future evaluation because based on the final purpose, LoRa settings can be adjusted. If the payload is relatively small but the distance has to be big, smallest BW and biggest SF can be used. If the network design requires low latency in trade of distance, BW can be increased.

Another concern for future studies is coexistence of LoRa 2.4GHz with other technologies in the same frequency. Based on the results from the first experiment, it could be seen that the transmission with one third of PDR had a specific behaviour: almost all the packets were delivered on one channel while completely lost on the other two. Thus, a hypothesis could be built that some of the channels are more susceptible to interference than the others. Before applying LoRa 2.4GHz on a university campus or busy city centre, each channel performance evaluation should be studied.

Last finding is that Lora sensitivity has turned out to be less that the one mentioned in [2]: -102 dBm against -129 dBm (at SF12 BW 812.5 kHz).

Acknowledgments

Thanks to Dr. Glen Tian and Dr. Yanming Feng for approving the research and providing us with the hardware.

References

[1] P. Cerwall, P. Jonsson, R. Möller, S. Bävertoft, S. Carson and I. Godor, "On the pulse of the networked society—Hg. v. Ericsson.," Ericsson Mobility Report. Ericsson, Stockholm, Sweden., 2015.

[2] S. Corporation, "Wireless RF LoRa 2.4GHz" [Online]. Available: https://www.semtech.com/products/wireless-rf/lora-24ghz/sx1280zxxxxgw1.

[3] L. Vangelista, "Frequency Shift Chirp Modulation: The LoRa Modulation" IEEE Signal Processing Letters, pp. 1818 - 1821, 2017.

[4] A. H. Jebril, A. Sali, A. Ismail and M. F. A. Rasid, "Overcoming Limitations of LoRa Physical Layer in Image Transmission" Sensors, pp. 18(10), 3257, 2018.

[5] TTN, "Regional Parameters" [Online]. Available: https://www.thethingsnetwork.org/docs/lorawan/regional-parameters/.

[6] N. B. Thomas Janssen, M. Aernouts, R. Berkvens and M. Weyn, "LoRa 2.4 GHz Communication Link and Range" Sensors (Basel, Switzerland) published by Multidisciplinary Digital Publishing Institute (MDPI). https://www.mdpi.com/1424-8220/20/16/4366, 2020.

[7] [R. M. Sandoval, A.-J. Garcia-Sanchez and J. Garcia-Haro, "Performance optimization of LoRa nodes for the future smart city/industry" EURASIP Journal on Wireless Communications and Networking volume, p. Article number: 200, 2019.

[8] P. J. Radcliffe, K. G. Chavez, P. Beckett, J. Spangaro and C. Jakob, "Usability of LoRaWAN Technology in a Central Business District" in 2017 IEEE 85th Vehicular Technology Conference (VTC Spring), Sydney, NSW, 2017.

[9] A. F. Rander, K. D. Ballal, M. N. Petersen and S. Ruepp, "Ranging Capabilities of LoRa 2.4 GHz" in 2020 IEEE 6th World Forum on Internet of Things (WF-IoT), New Orleans, LA, 2020.

[10] L. Polak and J. Milos, "Performance analysis of LoRa in the 2.4 GHz ISM band: coexistence issues with Wi-Fi" Telecommunication Systems. Springer Link. https://link.springer.com/article/10.1007/s11235-020-00658-w#citeas, p. 299–309, 2020.
[11] S. Corporation, "Application Note: Wi-Fi immunity of LoRa at 2.4 GHz" June 2017. [Online]. Available: https://semtech.my.salesforce.com/sfc/p/#E0000000JelG/a/44000000MDcO/Ll4bon.4HPwcyXv9fegcfcgbpvLYd7Lx_aZLMzYNLIQ.
[12] Y. C. J. C. H. S. D. J. S. K. S. K. A. S. A. R. Daeun Yim, "An experimental LoRa performance evaluation in tree farm" in an experimental LoRa performance evaluation in tree farm, Seoul, 2018.
[13] M. Cattani, C. A. Boano and K. Römer, " Experimental Evaluation of the Reliability of LoRa Long-Range Low-Power Wireless Communication" QoS in Wireless Sensor/Actuator Networks and Systems, p. 6(2):7, Journal of Sensor and Actuator Networks.
[14] N. C. T. U. T. Shie-Yuan Wang Department of Computer Science, Y.-R. Chen, T.-Y. Chen, C.-H. Chang, Y.-H. Cheng, C.-C. Hsu and Y.-B. Lin, "Performance of LoRa-Based IoT Applications on Campus" in IEEE 86th Vehicular Technology Conference (VTC-Fall), Toronto, ON, Canada, 2017.
[15] P. Jörke, S. Böcker, F. Liedmann and C. Wietfeld, "Urban channel models for smart city IoT-networks based on empirical measurements of LoRa-links at 433 and 868 MHz" in IEEE 28th Annual International Symposium on Personal, Indoor, and Mobile Radio Communications (PIMRC), Montreal, QC, Canada, 2017.
[16] L. Lingling, J. Ren and Q. Zhu, "On the application of LoRa LPWAN technology in Sailing Monitoring System" in 13th Annual Conference on Wireless On-demand Network Systems and Services (WONS), Jackson, WY, USA, 2017.
[17] V. Kolesnikov, "GitHub LoRa Code" [Online]. Available: https://github.com/vladilenbunnnny/lora2.4_gateway_end-node_code.
[18] G. Kazdaridis, S. Keranidis, P. Symeonidis, P. Tzimotoudis, I. Zographopoulos, P. Skrimponis and T. Korakis, "Evaluation of LoRa Performance in a City-wide Testbed: Experimentation Insights and Findings" in WiNTECH '19: Proceedings of the 13th International Workshop on Wireless Network Testbeds, Experimental Evaluation & Characterization, 2019.
[19] J. Petajajarvi, K. Mikhaylov, A. Roivainen, T. Hanninen and M. Pettissalo, 14th International Conference on ITS Telecommunications (ITST), Copenhagen, Denmark, 2015.
[20] Semtech, "How to Perform Ranging Tests with SX1280 Development Kit" 2017. [Online]. Available: https://semtech.my.salesforce.com/sfc/p/#E0000000JelG/a/44000000MDcY/ZsmAVCVenZkc0lUrr3RuxWSfdFxY2Tj_msk4N9DAhBo.

Supervised Learning System for Pest Disease Identification and Recommendations using Wireless Sensor Network in Agriculture Domain

A C Amune[1], H M Pande[2], V P Musale[3]

[1] Vishwakarma Institute of Technology, Pune, India
amruta.amune@gmail.com

[2, 3] Dr. Vishwanath Karad MIT World Peace University, Pune, India
[2] himangi.pande@mitwpu.edu.in
[3] vinayak.musale@gmail.com

Abstract

Designing wireless sensor networks (WSNs) that will sustain for longer time have been a challenging issue. For achieving this, efficient utilization of available energy should be anticipated from energy-constrained sensors running autonomously for long periods. In the field of agricultural research, farming related activities, assistance to farmers using advanced IoT and sensor network technologies is a need of time. Initially this paper briefed about smartphone based application systems that forewarns the farmer about pests/diseases that can occur on the crop. The designed system assists farmers to foresee the crop disease using machine learning algorithm. The disease can be predicted based on the environmental parameters as soil moisture, temperature and humidity obtained from sensor node and other parameters like rain fall, evaporation rate and sun shine hour obtained from open weather map application programming interfaces. A prototype is developed using ultra low powered micro-controller with interconnected sensor's module. While comparing with the Arduino micro-controller based system, it was observed that the developed prototype system utilized less energy with effectively forewarned the farmers about crop diseases using Internet of Things (IoT). Thus it would assist the farmers proactively to reduce the losses due to crop diseases.

Keywords: Energy; Networks; Routing; Sensor; Wireless.

1. Introduction

Nowadays, due to recent technological advances, the sensor networks become most popular in various fields like industrial, agriculture, healthcare, defense, etc. Agriculture sector in India has been diminishing day by day which affects the crop production capacity. The utilization of advanced technologies is growing rapidly in agriculture domain to monitor the field. The cotton crop is one of the important cash crops in India that is cultivated on large scale that affects the Indian economy in so many ways. The crop yield had been diminishing due to pests' attacks and many farmers predicted it out of their previous experiences or expert assistance. There is the possibility of an inaccurate prediction of diseases, substantiating the need for latest technology.

Understanding of agricultural issues, farmers need to have relevant knowledge of the happenings at the surroundings for making fruitful decisions. In agriculture domain through the development of an information management system, farmers get assisted with the help of telephonic communication and helped with multimedia which is easily accessible. The various climatic conditions lead to pest attacks on the crop, leading to tremendous amount of economic losses to the farmers, which mostly is the outcome of unawareness regarding diseases and their preventions among farmers.

So there is a vital need to solve the problem in the domain and to design automatic system to spread awareness of the pest attacks on the crops among the farmers. It will help to restore crop's health and put it back on higher growth. This paper proposes an e-Agriculture cloud based android application framework consisting of soil health monitoring and considering environmental parameters that are taken from open weather map web APIs. The different modules of implemented system help farmers to make gainful decisions with needful information required throughout the entire farming cycle.

The proposed system spreads awareness of the disease attacks on the crops. Also it will forewarn farmers about the disease that can occur on their crops which will be detected by data collected from the various sensors. The proposed system got trained using machine learning algorithm to predict disease attacks and message farmer about the same.

2. Related Work

In one of the research, authors briefed about the current key issues in agriculture field to prevent crops infected due to pests. It also tells about the traditional methods like manual investigation, geographical information system etc. To overcome the disadvantages of these methods this paper proposes an IOT system using wireless sensor networks which are having high density, wide range, and strong dynamic characteristics, so it has become an effective method for collection and transmission of agricultural information [1].

The paper [2] highlighted a revolutionary technology that represents the future of computing and communications. Modern technologies can control the cost, maintenance and monitoring performance. Satellite and aerial imagery play a vital role in modern agriculture. Also the authors surveyed some typical applications of agriculture sensor monitoring network technologies using cloud computing as the backbone. The simple agriculture model was being addressed with wireless networks. This survey is used to understand the different technologies and to build sustainable smart agriculture.

In the research paper [3], authors provided the information about network architecture in agriculture along with the pest monitoring system. GPS technology is used to get the location of tagged objects and acquire the relevant information. For the transmission of information, wireless sensor networks (WSN) and GPRS/GSM are the mainly used network technology in the field of agriculture. The information platform of disease and insect pests monitor based system integrated all kinds of information such as the agricultural management knowledge, purchase, warehousing, delivery and retail. It also realizes the information interchange between different phases.

The authors in their research described the way that communication technologies and intelligent context-service system provides autonomous decision without human interactions. They used LoRa WAN technology and compared it with ZigBee network. Also the paper presented the design of smart communication system manager used as low cost irrigation controller which can get controlled remotely by mobile phone. There exist many additional sensors and devices which could add new and useful features to the agro-system if they could only be integrated with it. They aimed to make more efficient multimedia platform to determine the most efficient decision to manage and monitor their agricultural fields, especially for fertilizing tasks [4].

The work presented by authors in [5] is the result of a preliminary analysis to realize an integrated microwave system for weed seed bank depletion. An overview of the main sub-systems was given along with the prototype realization. The paper presented a general preliminary overview of a system connected to communication infrastructure in order to improve agriculture. The core of the system was the weed seed bank depletion by means of microwave based heating solution. The microwave technique was supported by a vision system, which would be used both in visible and infrared bands according to different crops in order to maximize its efficiency and a GPS based localization system. All the parameters to program and control the proper working of the systems, as well as measured data, was available with a specific application, making the whole system perfectly integrated in the world of Internet of Thing (IoT). Some technology solutions are addressed and some possible choices to realize each section are reported. The description of the whole system is reported as well.

3. Design and Implementation

In agriculture, each kind of diseases and pests is considered to be harmful to plants and has a abundant adverse effect on agriculture. Therefore, the IoT based system is designed to help and reduce the frequent use of insecticides and fungicides. Also to predict when the pests appear in order to lower the appearance of pests. The forewarn and monitoring system of agricultural disease is divided into four different categories such as sensor node for acquisition of monitoring information, intelligent data processing modules as a weather station, prediction modules, user application, web-server and wireless communication technology. Weather stations gather and forward real time weather information. The designed system uses local weather forecast data for cumulative temperature in next 48 hours.

The humidity, temperature and soil moisture data are taken from sensors attached to sensor node and other required rest of the environmental parameters like rainfall, evaporation rate, sunlight and wind speed which are responsible for causing diseases in crop fields are taken from open weather map web APIs [7]. The various machine learning techniques are applied to data sensed from environment through sensors. The results obtained by this algorithm are useful for farmers to take decision about disease conditions in advance for further implantation in the crop monitoring system. The following figure 1 shows the implementation framework of the designed system.

This system consists of ESP8266 NodeMcU module which is the main part of the system used for interfacing purposes. Three different types of sensors are used for measuring soil conditions and levels of water. These three sensors are namely RHI-112A humidity sensor, YL-69 Soil Moisture sensor and Davis 6470 stainless steel Soil temperature sensor probe. All these sensors are interfaced with NodeMcU. The Support Vector Machine based regression system is implemented for identification and classification of pest diseases.

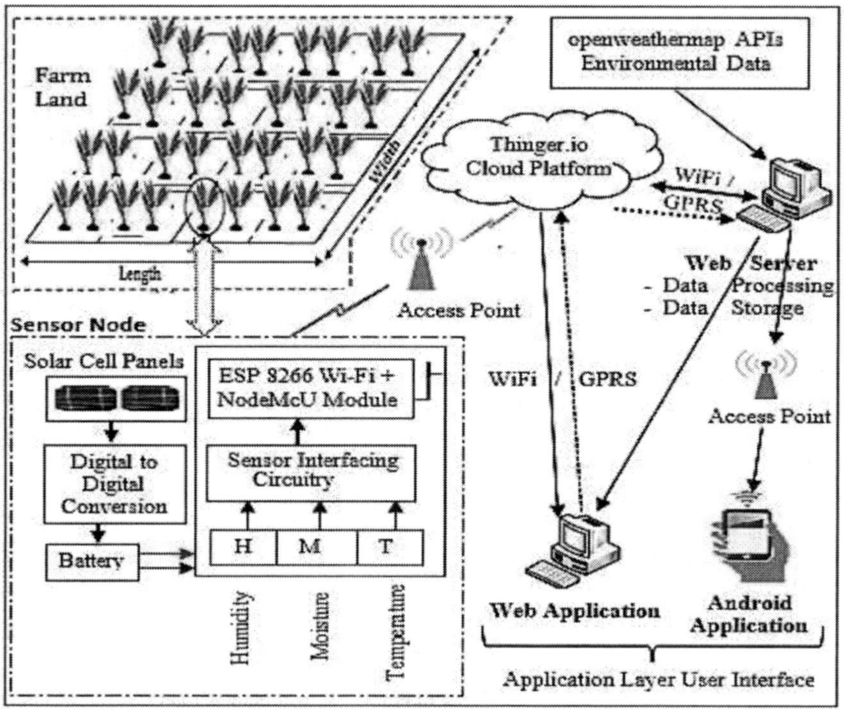

Figure 1. Implementation framework of IoT enabled Smart Agro System

Support Vector Machine (SVM) is a supervised machine learning algorithm that is used for both classification and regression challenges. However, it is mostly used in classification problems. In this algorithm, each data item is plotted as a point in n-dimensional space (where n is a number of features that are considered) with the value of each feature being the value of a particular coordinate. Then, the classification gets performed by finding the hyper plane that differentiates the two classes very well as shown in following figure 2.

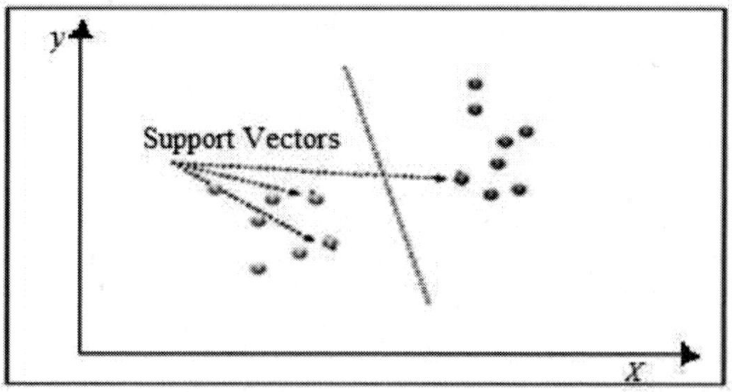

Figure 2. SVM Classification

Sensor nodes are laid on targeted farmland areas to get real-time data and collected environment signals are treated as a source of information. The data collected is stored in a database. The sensor technology is used for data acquisition and is mainly focuses on to capture part of the existing data. Also it can be used to correlate and synchronize, analyze these data elements and finally, carry out a responsive activity without user intervention. After processing, a disease that can occur will be identified or predicted, if disease occurred then a farmer gets informed with message in his local language informing the same and guided for corrective actions. This system is a low cost effective system that will boost the production of crops in agriculture and will lead to fewer losses and a happy farmer.

The farmers who are users of system should be able to register themselves by installing a "Fore Warner" web application and request. The sensor Node hardware module will get installed on farmer's farm.

The Administrator will manage all backend functionalities through a website. The farmer is able to do the following functions:

- Make new registration with device Id
- Login into system with valid username and password
- Add crop details
- Monitor the field and view result

The Administrator can do the following different functions:

- Get detailed entries of registered farmers
- Manage database records
- Control all network functions

The multi-target regression algorithm is used for implementation of this system which will fit parameters like Maximum Temperature, Minimum Temperature, Maximum Humidity, Minimum Humidity, Rainfall, Evaporation rate and Sun Shine Hours (SSH) and predict the pest value of the disease if a value greater than zero occurs then a logic called predicting logic will be used which will then predict if the disease actually occurs using the threshold value for disease. This logic will give accurate results for the pest prediction.

- **Multi-target Prediction**

For a feature vector x, predict accurately a vector of response y using function $h(x)$:

$$x=(x_1, x_2, x_3, \ldots, x_p) \xrightarrow{h(x)} y=(y_1, y_2, \ldots, y_m) \quad (1)$$

The main challenge is to appropriately model target dependencies between targets.

- **Marginal and conditional dependencies**

$$P(Y) \neq \prod_{i=1}^{m} P(Y_i) \qquad P(Y \mid x) \neq \prod_{i=1}^{m} P(Y_i \mid x) \quad (2)$$

marginal (in) dependence ≠ conditional (in) dependence

The following table 1 shows a sample data set which is used to train the prediction module. The dataset consists of various parameters like pest value, maximum temperature, minimum temperature, relative humidity, rainfall, wind speed, sunshine hour, evaporation, disease, and city. This dataset is used for training the SVM classifier model. This data set is taken from Central Research Institute for Dryland Agriculture (CRIDA) website [8] which is an ISO 9001 certified.

Table 1. Sample Dataset for Disease Prediction Module

Pest value	Max Temp. (°C)	Min Temp. (°C)	Relative Humidity 1 (%)	Relative Humidity 2 (%)	Rain Fall (mm)	Wind Speed (kmph)	Sunshine Hour (hrs)	Evapo-ration (mm)	Disease
0	26.1	9.6	66.6	27.9	0	3.5	4.3	3.3	No disease
0	28.6	10.9	75.4	33	0	2	4.6	3.4	No disease
0	30.8	15.3	69.4	26.7	0	3.7	5.1	3.9	No disease
0	29.6	14.2	75.7	27.7	0	4	6.1	4.3	No disease
0	28.5	11.4	68.3	26.1	0	3.7	6.8	4.3	No disease
0	29.7	17.3	60.6	25.3	19	4.5	6.1	3.9	No disease
0	31.7	16.5	74	36	0	3.5	6.9	4.6	No disease
0	34.3	18.9	68.7	31.3	4	4.1	7.4	5	No disease
0	35.1	18.2	53.1	22.6	0	5	7.5	6.2	No disease
0	36	18.8	79.3	44	0	3.7	7.7	6.5	No disease
0	35.4	19	62	31.1	0	4.8	6.2	6.5	No disease
0	38.8	20.4	62.6	24.9	0	3.8	7.6	8.3	No disease
0	39.2	22.9	56.4	21	0	4.4	7.5	8.9	No disease
0.02	27.5	23.3	77.9	49.9	198	1.1	5.7	2.3	American BollWorm
0.32	29.6	23.5	85.3	68.9	56	2.2	5.7	3.1	American BollWorm
0.18	28.3	23.1	89	75.9	54	1.1	5.7	3.2	American BollWorm
1.52	32.5	22.9	88.9	83.3	0	3.4	4.8	4.9	American BollWorm
1.5	34.6	23.1	92.3	79.6	7	1.3	6.3	5.4	American BollWorm

4. Observations

The efficient communication framework for wireless sensor networks helps to transfer data accurately in IoT applications. The new approach based on the idea of sharing a dynamic cryptic credentials ensures security to sensor based applications without increasing communication cost for wireless networks with minimal energy consumption.

The system monitors the changes in the environmental conditions like temperature, humidity, rainfall, wind speed, sun-shine-hour etc. The pest/diseases which affect the crop mostly requires given environmental conditions to be in specific range for a pest to attack the crop. The humidity, moisture and temperature values are collected through sensors i.e. the real time data to detect the possibility of pest attack. The detection of crop disease along with disease name and its occurrence is getting identified by monitoring farm field conditions based on the parameters mentioned in table 1. Different supervised learning algorithms can be used to do this kind of classification like Decision Tree (DT), Artificial Neural Network (ANN), Bayesian Network (BN), k- Nearest Neighbors (kNN) and Support Vector Machines (SVM) algorithms. Metrics points that shows efficiency SVM over all others can be given as below;

Accuracy (A): precision of the classifier in predicting a value.

Tolerance to noise (TN): ability of the classifier to handle erroneous data in the training process.

Speed of learning (SL): speed of learning of the classifier.

Speed of classification (SC): speed of the algorithm in classifying a new instance.

These metrics can have values between 1 and 4, with 4 being Excellent, 3 Good, 2 Average, and 1 Poor. The results obtained are presented in figure 3 as shown below.

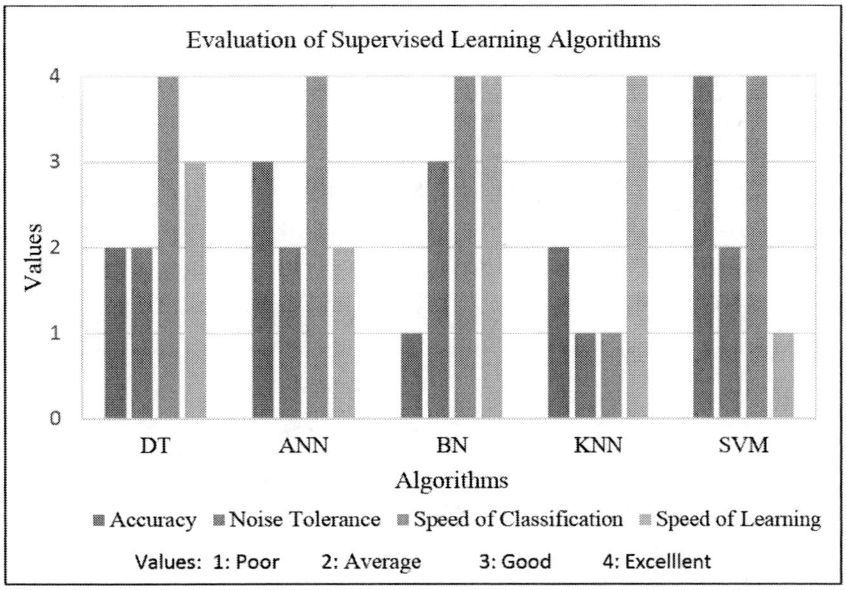

Figure 3. Evaluation of Supervised Learning Algorithms

This IOT based system by considering soil temperature, moisture, humidity, and other abiotic stresses along with SVM classification is also evaluated for different kinds of crops and for probable diseases that can be caused.

The following figure 4 shows the relationship among temperature, humidity, moisture and other environmental Parameters for the prediction of American Bollworm (Larva) Disease related to cotton crop. The machine learning algorithm predicts the appropriate disease based on the training datasets and the input parameters provided to it.

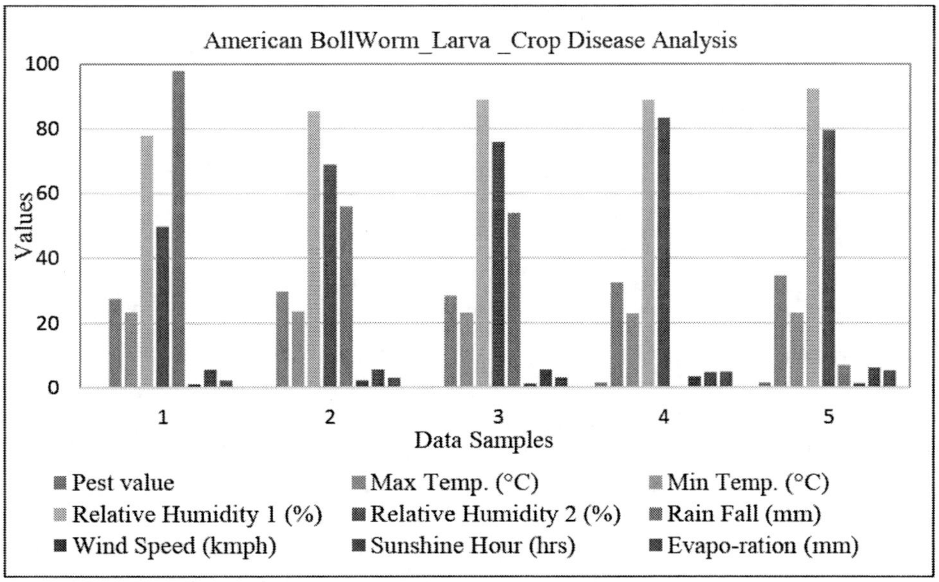

Figure 4. A relationship between temperature, humidity, moisture and other environmental parameters for prediction of American Bollworm (Larva) Disease

Support Vector Machine (SVM) method proposed by Vladimir Vapnik for classification and regression problems can also use to predict the severity of the fungus called rice blast, which attacks rice crops. Used training data were collected between 2000 and 2004 [14]. However, they compared SVM with other approaches. These approaches as Conventional Multiple Regression (REG), Generalized Regression Neural Network (GRNN) and Back

Propagation Neural Network (BPNN). Mentioned algorithms are applied to obtain results to calculate the absolute percentage error (APE). It is observed that SVM is the most accurate in predicting the severity of rice blast with %APE equals to 36.69. Similarly other algorithms presents results as GRNN with %APE equals to 46.37, REG with %APE = 65.44 and BPNN with %APE equals to 52.27.

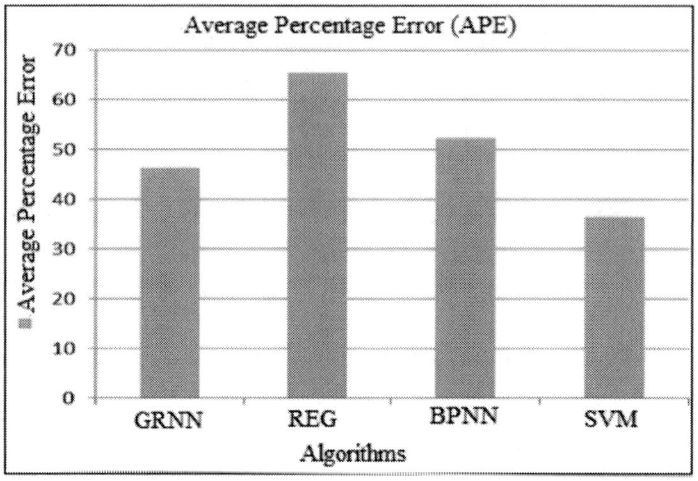

Figure 5. Comparative Analysis of Average Percentage Error (APE) of different algorithms for predicting the severity of the fungus Pericularia oryzae

5. Conclusion

The system is going to perform prediction on real-time data sensed by sensors which will reduce hectic image processing task and can be used to predict pests and diseases that may affect their crop by using machine learning algorithm. Also according to temperature and humidity when the infection rates are high, the system alerts the end user (farmer) in their native language. In addition to this, beforehand it also notifies the control timing of the harm caused by the cumulative temperature. It will reduce the burden of farmers by preventing them from manual monitoring of the field.

References

[1] Yang Jing, Zhan Yuzhi, Du Dan, Wang Xiao, Yang Ping, Chen Lingfang, Sun Yuefang, Li Zetaol, "An Early Warning System of Diseases and Pests for Blueberry Based on WSN", Chinese Control Conference 2017.

[2] Mahammad Shareef, Mekala, Dr. P. Viswanathan, "A Survey: Smart Agriculture IOT with cloud computing", International Conference on Microelectronic Devices, Circuits and Systems (ICMDCS) 2017.

[3] Yun Shi, Zhen Wang, Xianfeng Wang, Shanwen Zhang, "Internet of Things Application to Monitoring Plant Disease and Insect Pests", International Conference on Applied Science and Engineering Innovation, 2015.

[4] Carlos Cambra, Sandra Sendra, Jaime Lloret, Laura Garcia, "An IOT Service-Oriented System for Agriculture Monitoring", IEEE 2017.

[5] M. Allegretti, L. Gilli, L. Quaglia, "A microwave system connected to an IOT infrastructure for weed seed bank depletion", IEEE2017.

[6] Jinpeng W, Yibo C. Jean-Pierre C., "An integrated Survey in Plant Disease Detection for precision agriculture using Image Processing and wireless multimedia sensor network", (ICACEEE 2014) France, 2014.

[7] https://openweathermap.org/

[8] http://www.crida.in/

[9] S. Azfar, A. Nadeem, A. B. Alkhodre, K. Ahsan, N. Mehmood, T. Alghmdi, Y. Alsaawy, "Monitoring, Detection and Control Techniques of Agriculture Pests and Diseases using Wireless Sensor Network: A Review", International Journal of Advanced Computer Science and Applications, Vol. 9, No. 12, 2018.

[10] Heamin Lee, Aekyung Moon, Kiyeong Moon, Youngjae Lee, "Disease and Pest Prediction IOT System in Orchard: A Preliminary Study", International Conference on Ubiquitous and Future Networks 2017.

[11] T. N. Qureshi, T. Shah and N. Javaid, "EESAA: Energy Efficient Sleep Awake Aware Intelligent Sensor Network Routing Protocol", Proceedings of 15th International Multi Topic Conference (INMIC), Islamabad, PK, 2012.

[12] Nikolaos A. Pantazis, Stefanos A. Nikolidakis and Dimitrios D. Vergados, "Energy-Efficient Routing Protocols in Wireless Sensor Networks: A Survey", IEEE Communications Surveys & Tutorial, vol. 3, pp. 1-41, 2012.

[13] P. Minet, "Energy Efficient Routing in Ad Hoc and Sensor Wireless Networks: Architectures: Algorithms and Protocols", Bentham Science 2009.

[14] V. Musale, D. Chaudhari, "Efficient and Secure Keying Mechanism for Communication in Sensor Networks", Lecture Notes on Data Engineering and Communications Technologies book series (LNDECT, volume 26), pp. 989–995, 2018.

Pipe inspection robots: a review

*Binil John[1] and Shafeek M[2]

[1]Department of Electronics and Communications, College of Engineering Trivandrum, Kerala, India ,[2] Department of Mechanical Engineering, College of Engineering Trivandrum, Kerala, India, * biniljohn0198@gmail.com

Abstract. Pipelines are one of the safe and most economical ways of transporting fluid matters in industries and households. These pipelines get deteriorated due to aging, internal and external corrosion, and so on. This may cause fluid leakages and may contaminate the environment. So, pipelines are inspected regularly to maintain proper functioning and cost-effectiveness. In-pipe inspection robots (IPIR) that move inside pipelines are widely used to carry out inspection operations. IPIR along with visual and other r non-destructive tests (NDT) are proven to be useful compared to manual inspection. Wheel-driven robots, track-driven robots, wall-press robots, pipe inspection gauge (PIG) robots, and many other bio-inspired designs are available. The design parameters and specializations of each type of robot will be presented in this article. Because each robot is created with different factors in mind, such as pipeline diameter, mobility, adaptability, working conditions, and so on, these robots are unique and have different constraints. These topics will be covered in depth in this article, and will ultimately aid in the selection of certain robots based on specific criteria. New design strategies in the pipe inspection industry that have yet to be explored will also be covered in depth.

1.Introduction

Pipelines are one of the most commonly used economical way of transporting gas and liquids like oil, water, sewage etc. Pipelines are constantly aging and sustaining damages; thus, it is important to periodically inspect to prevent or cure the damages. Faulty pipelines can cause environmental damage and financial loss to the owner. The pipelines usually have long and complicated structures and mostly they are buried underground, thus it is difficult to inspect the pipelines manually. Pipe inspection robots are the most economical and reliable method of inspecting the pipelines. The increase in pipelines is also pointing to the fact that using the robots are more suitable. Robots are designed considering various factors like maneuverability, stability, adaptability to various pipeline diameters.

Pipeline inspection robots reduces the human intervention, can work in hazards environments and can perform inspection in parts which are inaccessible to the humans. Robots made the inspection work much simpler, safe and economical. A large number of research work is published in this field, now also it is a concerned field of research. Many different mechanisms were introduced into this field, we will go over the advantages and disadvantages of each design in detail in this paper. Each robot is created by taking into account a variety of factors such as pipe diameter, size, shape, and maneuverability.

2. Pipe Inspection Robot: Classification

Out-pipe inspection robots and in-pipe inspection robots are the two primary types of pipe inspection robots. The out-pipe inspection robot (OPIR) moves over the pipeline's outer surface, gripping onto the pipe with some mechanism. While the in-pipe inspection robot (IPIR) goes through the inside of the pipe lines and performs the essential inspection work, they are extremely useful in checking underground and underwater pipelines.

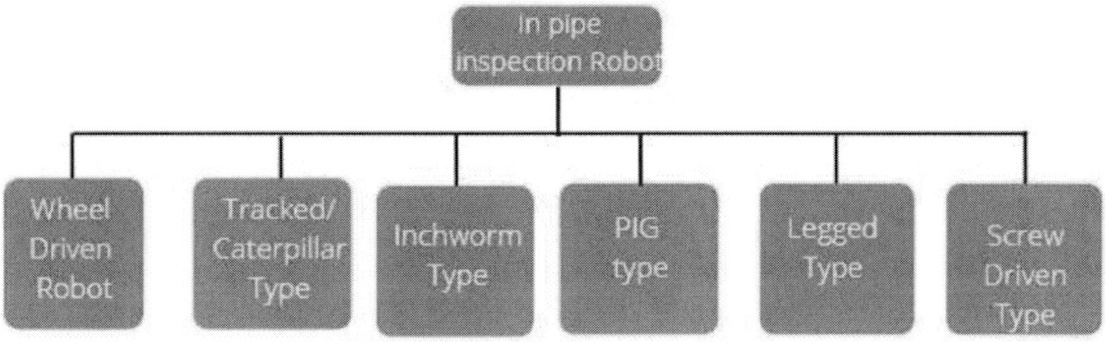

Figure.1. Classification of In-pipe Inspection Robots

2.1 Wheel Driven Pipe Inspection Robot

Wheel-driven pipe inspection robots are one of the most researched types of inspection robots, with a variety of wheel-based designs available. As indicated in Figure 1, there are two types of robots: simple structure robots and wall-press robots. Wheels powered by actuators/gearmotors help the robot move. The majority of wheeled robots use a variety of devices to maintain contact with the surface. Simple structured wheel driven robots look like regular wheeled robots and are mostly used for horizontal pipe inspection. It doesn't have any additional links to keep in contact with the inner surface of the vertical pipelines since it doesn't have any.

Gunatilake et al. [1] have used 6 wheeled simple structured sewage inspection robots (fig.2.a) where the robot is attached with a stereo camera vision combined with infrared laser profiling unit and have successfully tested in buried water pipes in Sydney, Australia Another form of inspection robot is the wall press type, which has wheel modules positioned at an angle and is always in touch with the inner surface of pipe lines. The number of wheel modules on this sort of pipe inspection robot can range from two to six (as shown in figure .3).

IOP Conf. Series: Materials Science and Engineering 1272 (2022) 012016 doi:10.1088/1757-899X/1272/1/012016

Kwon et al. [3] presented a wheel-driven robot with two 180-degree-apart wheel modules. Only these two components are used to control the steering and driving modes. Each wheel contains two motors, one for driving and the other for steering. a flat wheel-type robot that can drive and steer with only two-wheel chains When compared to a three-wheel train, one of the key advantages of this robot is its manoeuvrability. More space for sensor equipment was also noted as a plus. It contains three CMOS cameras: one in the front and two on each side (figure 2.b).

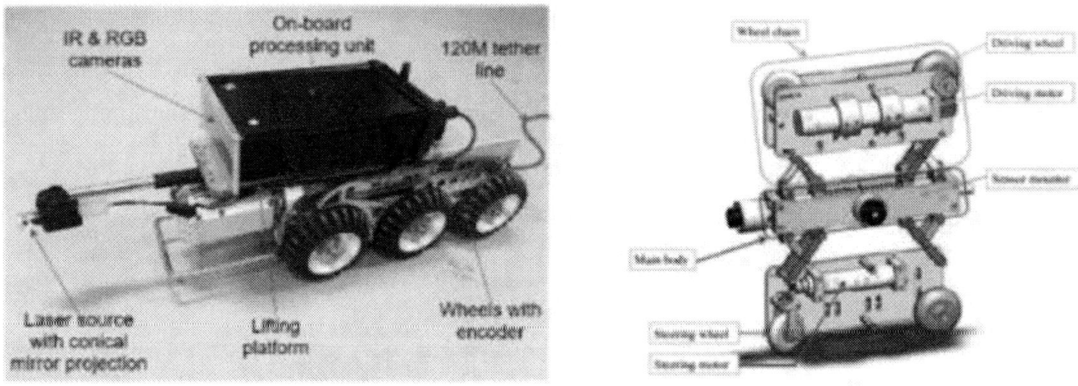

Fig.2. (a) Simple structured wheel driven pipe inspection robot [1] ,(b)Wall press wheel driven pipe inspection robot with wheel modules arranged at 180 degrees with each other.

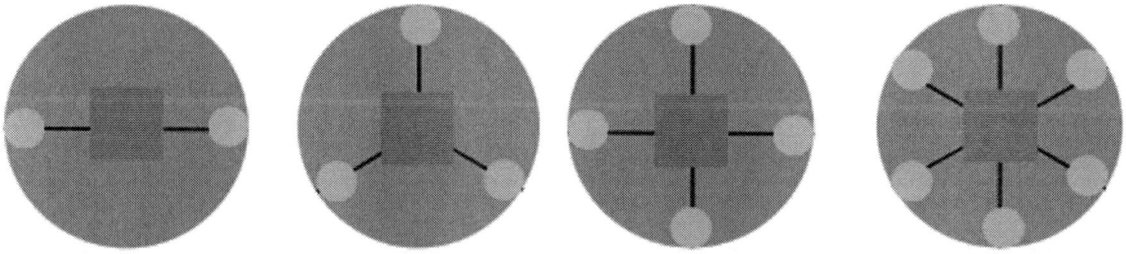

Fig.3 Different wheel module arrangements.

The MRINSPECT (Multifunctional Robotic crawler for INpipe inSPECTion) inspection robots are a series of wheel driven walpress type pipe inspection robots which has introduced a series of advancements in the wallpress type robots [2] [3] [4] [5]. S. G Roh et al [2] proposed wheel driven wall pess type robots MRINSPECT III and IV for the inspection of urban gas pipelines. It can pass throughhorizontal as well as vertical pipe lines and it can travel along the reducers elbows. It's three dimensional steering capability provides outstanding mobilityu along the pipelines. The MRINSPECT III is made up of two portions connected by a steering device called a Double Active Universal Joint. Each segment has three wheel modules that are 120 degrees apart and are positioned circumferentially around the body. Each wheel segment has a pantograph mechanism with a sliding base. This wheel module can adjust to tiny differences in pipe diameters. The rear wheel portion receives the driving force, while the front wheel segment is only used to guide the motion.

For urban gas pipes with a nominal diameter of 4 in internal diameter, the MRINSPECT IV [5] design is proposed. Body frame, driving module, and charge-coupled device (CCD) assembly are the three primary components. The three driving modules are circumferentially linked at 120 degrees apart, and the wheel modules' expansion and retraction are regulated by a spring. Mathematical expressions are used to discuss strategies for navigating through elbows and branches.

The MRINSPECT VI [6] is a wall press type inspection robot with wheels that can pass through pipelines with an inner diameter of 150 mm. A multi-axial differential gear mechanism is used, which is powered by a single motor. It can adjust to different diameters; nevertheless, the key improvement in this model over the previous model is the ability to mechanically modulate the velocities of the active wheels without exerting any control effort. Three active and three passive wheels are included. When it comes to MRINSPECT VII [7], the design has been upgraded into a robot that can pass through pipe lines with sizes ranging from 150 to 200 mm. They converted the wheel expansion and retraction into a radial movement, which improved the robot's stability. It is equipped with a rescue mechanism that allows the robot to be retrieved in any situation. A 2-2D differential gear system is used by the robot to mechanically control the velocity of each active wheel individually. A separate BLDC motor controls the expansion and retraction of the wheels.

2.2 Tracked/ Caterpillar Type Pipe Inspection Robot

For mobility in the pipe inlets, the robots employ tracks instead of wheels. Tracks increase the amount of surface area in contact with the pipe inlet, increasing traction and stability. More body weight and heavy equipment can be supported by tracked robots. Slow speed and mobility are also shortcomings of the robots. There are two types of tracked robots: simple structured and wall-press type.

Zhao et al [8] discuss a long-distance pipeline inspection robot mechanism. The robot features three crawlers and electric putters, each with its own speed and radius control. They confirmed that the robot locomotion in a huge pipe is stable. The robot is designed to inspect pipelines with interior diameters ranging from 950 to 1200 mm. Many sensors are built into the robot. Pressure sensors are mounted to the crawlers to evaluate the traction and encoder sensors are utilized to acquire the robot's actual speed.

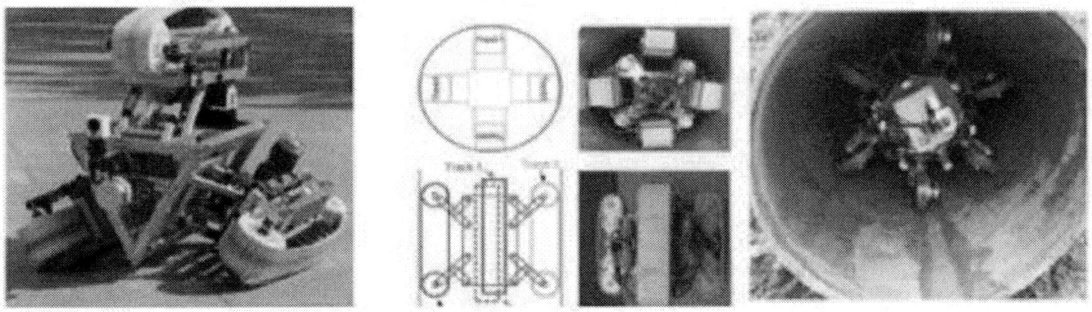

Fig.4. **(a)** Large in-pipe Robot prototype with three-wheel train, **(b)** FAMPER: A Fully Autonomous Mobile Robot for Pipeline Exploration, [6], **(c)** Oil pipeline inspection robot [7]

Kim et al [9] created FAMPER, a completely autonomous robot with four-wheel modules that are 90 degrees apart. The Wheel modules can adjust to different internal pipe diameters, and the track can tilt to overcome obstacles. To achieve maximum traction, the robot's diameter can range from 127 mm to 157 mm. The robot can travel through horizontal and vertical pipes, as well as U and T junctions.

Dae-Won Kim and colleagues [11] The active pipe inspection robot proposed, in this work employs a DC motor to regulate the wall-pressing force. It was designed for long-distance inspection of seawater pipes, such as horizontal and slope linear pipelines, with enormous variable diameters ranging from 600mm to 800mm. Its mechanical architecture is divided into two modules: drive and control. Three pantograph type links spaced at 120° with three caterpillar track wheels make up the driving module. This design allows for pipe diameter adaption as well as adjustment of the wall-pressing force.

Abidin et al. [12] developed the D 300 wall press pipe inspection robot for checking 300 mm inside diameter sewage pipe lines (fig 5). The robot's design is based on a wall-pressed caterpillar that may be used for both horizontal and vertical pipelines, as well as elbows. For travelling in the pipeline's corner, this robot is equipped with ultrasonic sensors, a high-performance microprocessor board, and a powerful geared motor. The IPR-D300 is made up of three separate drive modules that can change the speed depending on the pipeline characteristics.

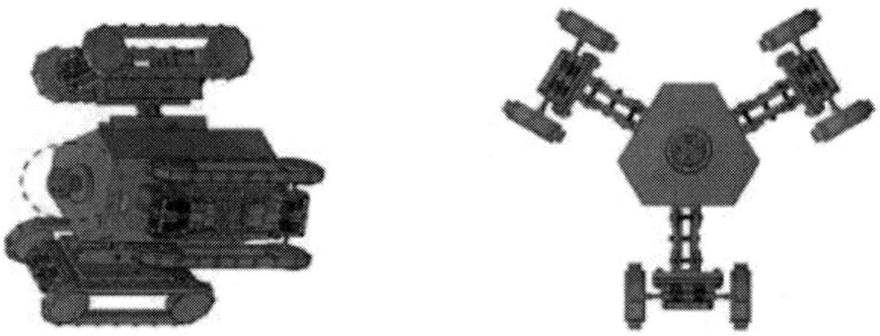

Fig. 5. Conceptual design of IPR-D300

2.3 Inchworm Type Pipe Inspection Robot

IPIR is a bio inspired robot, and the worm-type robot is one of them. Their movement resembles that of an earthworm. Repetitive contraction and expansion actions propel the robot forward. These have slower driving speeds and generate a lot of friction force while moving.

Chablat et al [13] have developed a pipe inspection robot that mimics the movement of a caterpillar inside pipe lines. The mechanism is divided into three sections: one for motion and two for legs attached to the pipe's inner part. Three legs with slot follower mechanisms assist the robot in mimicking caterpillar motion. The static method has been discussed, and the relationship between the contact force between the pipe and the legs has been determined using the Wrench equation and the coulombs law of friction. However, the study made no mention of robot joint mobility.

Hayashi et al. [14] developed a robot that employs fluidic flexible actuators that are Extension type flexible pneumatic actuators (EFPAs). The new ring-shaped tube given at the ends nodes has increased the holding mechanism (it can create 160 N clamp force). To reduce volume change when expanding, the enhanced holding mechanism is built similarly to an automotive tyre wheel. Pipe inspection robots with dual pipe holding mechanisms have been developed for pipe diameters of 75 and

50 mm. The robot uses three EFPA units (fig. 6.a) to achieve steering in six directions. Electric leakage and short circuit are no longer issues with this new design. The outcomes of this robot's experiments demonstrate that it can manipulate through complex formations without getting stuck.

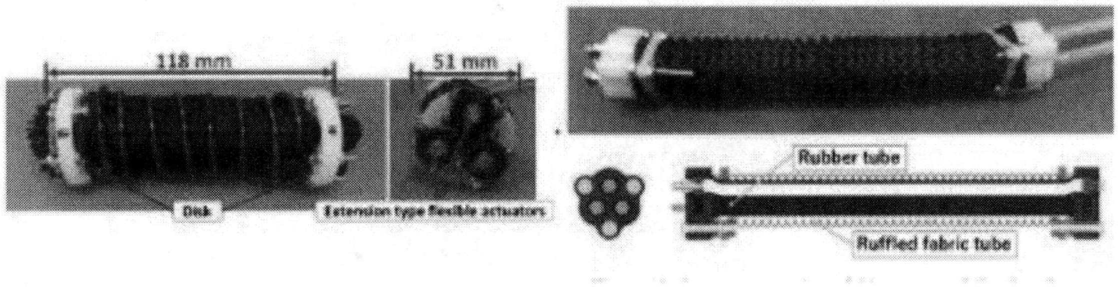

Fig.6. (a) Sliding/bending mechanism using three FFPAs, (b) Inner construction of thinner type sliding/bending mechanism [8].

Takagi et al. [15] offer a prototype design for a mobile inchworm robot for use in tight locations that incorporates DN (Double network hydrogel) gel on the inchworm robot's foot to improve friction qualities. Electric current is applied to the DN gel. Controlling the friction properties is possible. By moving the gel arear one by one, the robot was able to create its repeating peristaltic motion. Two DC gear motors and ball screws are used as guides for the gel pad on the robot. The prototype took 40 seconds to complete a single step with a stroke distance of 28.4 mm. Although the DN gel activation requires a conducting surface (cathode rail). The prototype shows less velocity characteristics and flexibility which is aimed to overcome in the next advancements.

2.4 Pipe Inspection Gauge (PIG) Robot

Pipe Inspection Gauge (PIG) Type robots are a popular type of pipeline inspection robot. It was used in oil and gas industries mainly. The movement of the robot is controlled using the fluid velocity and other means. The one disadvantage of this type of model is that it is designed and suitable for only one particular diameter. The robot consists of a number of sensors attached in order to check for the cracks and other anomalies. Since these robots are wireless a battery module is provided in the robot, and the inspection results are checked only after receiving the PIG robot. The PIG robot will be entering and leaving the pipes through the set points. Navigation of the PIG robot is difficult hence it is not used in complex structures.

Nguyen et al [9], In this paper they have done the dynamic design and analysis of PIG type robot for 90 degree bent pipes. They have considered the 90-degree bent as 3 sections and the dynamic model is derived by using Lagrange equation. They have solved both the downstream and upstream considerations, and the efficiency of the derived models were tested using simulations and found to be effective.

Mishra et al [10], proposed a PIG pipeline inspection robot which is able to move autonomously for cross country crude oil pipe lines. Inside the robot there is a power source to power the sensors inside. And no power is given to the motion of the inspection robot, it is important to control the speed of the robot in side the pipeline to avoid damage to the equipment. They have used a kinematic model to estimate the robot position inside the pipeline.

Fig.7. Pipe Inspection Gauge (PIG) Robot

2.5 Legged Type Pipe Inspection Robot

The legged type robot is moving through the inner surface of the pipe lines with the help of legs. These legs will be having high degree of freedom and it can move through the pipeline avoiding the obstacles. Thus, it is more commonly used in sewage pipelines, and it is able to provide better steering characteristics. Even though each leg of the legged robot will be requiring more than one actuator and in turn the overall system is complex. The path planning, path replanning, calibrations etc. are requires good system controller and equipments. Thus, the inspection using legged robots are costly, but it is helpful in certain specific cases

Savin et al. [11], they have studied a state observer model for walking pipe inspection robot. The state observer model with Riccati equation is giving improved characteristics to the controller. They have used a number of sensors to obtain good results from the robot. Their test results are showing good results compared with the previous models

Savin et al [12], designed a six-legged robot or the pipe inspection operation, with each leg having three actuated joints. They have proposed a geometric algorithm to find the position and orientation of the robot at each procedure. The main aim was to reduce the leg extension in each step and to reduce the extensive calculations

2.6 Screw-Drive Type Pipe Inspection Robot

Screw-Driven type pipe inspection robot is also a type of wheeled robot but with better steering characteristics. The robot will be having a certain number of wheels with adjustable links similar to the wheeled type. The main difference is that the forward movement of the robot is achieved by the helical movement of the wheels, hence the robot has both translational and rotational movement at any given time. The robot movement is of comparatively low speed but it offers good power characteristics and weight reduction, but reversing the motion of the robot is a difficult operation.

Nishimura et al [13], have designed a screw driven type pipe inspection robot where it is designed as two segments connected using a universal joint, the first segment is having a wheel connected 120 degrees apart and they move in a helical path inside the pipeline. This first segment is responsible for the steering of the robot in T sections and other joints. The robot is designed with only two actuators and it have three locomotion modes screw driving, steering and rolling modes. They have made the prototype and found out that it is giving good performance characteristics

Pradhan et al [14], Have designed a screw driven type pipe inspection robot (fig.8.b). The robot is able to travel through pipelines of diameter 127 to 152 mm. The robot is segmented into 3 namely rotor, stator and control unit, it uses the wall press mechanism hence it can move through horizontal as well as vertical pipelines. The rotor module consists of three-wheel modules arranged in 120 degrees apart and with a helix angle of 15 degrees and they will move in a helical path and the robot will move along the pipe line. And the stator wheels provide the stability to the robot when the rotor is rotating. They have conducted testings and verified the efficiency of the robot in various pipelines.

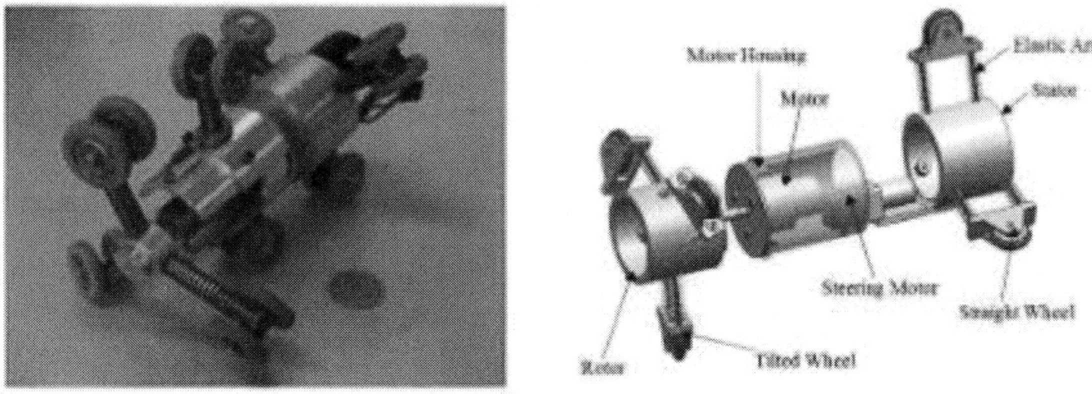

Fig.8. (a) Screw driven type IPIR [13], (b)Robot CAD model of the Screw Driven Type IPIR [14]

3. Results and Discussion

Different types of Pipe Inspection robots are discussed with the main advanced papers in each section. The comparative study of the robots is given in the table 1. It is evident that wall pressing type IPIR robots are the dominant class of robots. The robots can be chosen according to the respective needs. When the inspections are only in horizontal pipes wall pressing robots are not required and using them in these types of applications is a waste of time and effort. Hence in these cases simple structured robots are good, and the complexity in the design, fabrication and operating of these robots are simple. But when the horizontal pipes to be inspected are sewage pipes it is usually difficult to operate these wheeled robots in these types of cases tracked or legged type robots can be used. Even though they are complex in nature the efficiency of inspection will compensate for this complexity

Pipe lines with constant section diameter and characters like large lengthy oil and gas pipelines can be inspected with PIG robots, but before inspecting a pipeline with PIG robots they usually follow a certain number of procedures like cleaning the pipeline, checking for any internal bulging that may cause damage to the inspection robot. These procedures make the inspection using PIG robot a lengthy procedure.

Inchworm type robots are most suited for pipelines with small diameters since it can efficiently move inside small pipelines, but the movement is very slow. Even though inch worms are the best solution for small dia pipelines. Tracked/ Caterpillar type robots are suitable for medium and large sized

IOP Conf. Series: Materials Science and Engineering 1272 (2022) 012016 doi:10.1088/1757-899X/1272/1/012016

pipelines and it can carry heavy equipments and have a good traction characteristic. These properties make it suitable for inspections which involve heavy equipments

	Wheel Driven Type	Tracked/ Caterpillar Type	Inchworm Type	PIG type	Legged Type	Screw Driven Type
Manoeuvrability	√√√	√	√	×	√√√	√√
Adaptability	√√√	√√√	√	×	√√	√√√
Horizontal Movement	√√√	√√√	√√√	√√√	√√√	√√√
Vertical Movement	√√	√√	√	×	×	√√
Stability	√√	√√√	√	√	√√	√
Drive Speed	√√	√	√	√√	√	√

Table .1. IPIR comparative Study

4. New Design Considerations

Pipeline Inspection robots can take other design concepts also, so many bio inspired models can be used in this field. One of the main disadvantages we observed in these models are most of the robots are tested in an ideal situation which is not the actual case. Most of the pipelines which are undergoing the inspection procedure will be having a wet or semiwet environment. These environments demand a more practical solutions, one of the main models which can be adapted for this particular purpose is amphibious designs.

Amphibious robots are important classification of robots which are bio inspired, the robots can move through land, water and irregular surface. Even though the design will be complex the application fields we will be able to address by a single design will be high. And by considering Indian conditions, even now we are depending moistly on manual operations in sewage and other inspections. So, developing these more practical reliable designs will be a great milestone in the pipeline inspection field and it will help in reducing the risk involved in this field.

Tensegrity robot models are also a good choice for inspection robots in large pipelines where the movement will not be affected by the obstacles in the pipelines. And the actuators required will also be less, even though the speed that can be attained will be less. We are expecting more research work on these topics in the coming years and it will be great to see these robots helping the mankind to develop.

5. Conclusion

Different pipeline inspection robots are discussed in the paper, and a comparative study result are given. There are wide range of pipeline inspection robots namely wheeled type, legged type, Tracked/Caterpillar type, Inchworm type and so on. Each of these robots are designed for a particular application with certain pipe diameter and complexities. Hence the robots are able to process in a particular et of pipelines only. According to the requirements the robots can be chosen

The paper also mentions new research areas that can be explored in pipe inspection robot. Amphibious robots are a type of robots that can locomote through a number of surfaces and designing pipeline robots for this Pipeline inspection will be a large step towards making the robot closer to the practical application and expecting new works in this field The paper is able to give a proper overview of the pipe inspection robots and current advancements in the field.

References

[1] Gunatilake, A., Piyathilaka, L., Tran, A., Vishwanathan, V.K., Thiyagarajan, K. and Kodagoda, S., 2020. Stereo vision combined with laser profiling for mapping of pipeline internal defects. *IEEE Sensors Journal*, *21*(10), pp.11926-11934.

[2] Kwon, Y.S., Lee, B., Whang, I.C., Kim, W.K. and Yi, B.J., 2011, May. A flat pipeline inspection robot with two wheel chains. In *2011 IEEE International Conference on Robotics and Automation* (pp. 5141-5146). IEEE.

[3] Roh, S.G., Ryew, S., Yang, J.H. and Choi, H.R., 2001, May. Actively steerable in-pipe inspection robots for underground urban gas pipelines. In *Proceedings 2001 ICRA. IEEE International Conference on Robotics and Automation (Cat. No. 01CH37164)* (Vol. 1, pp. 761-766). IEEE.

[4] Roh, S.G. and Choi, H.R., 2005. Differential-drive in-pipe robot for moving inside urban gas pipelines. *IEEE transactions on robotics*, *21*(1), pp.1-17.

[5] Kim, H.M., Suh, J.S., Choi, Y.S., Trong, T.D., Moon, H., Koo, J., Ryew, S. and Choi, H.R., 2013, November. An in-pipe robot with multi-axial differential gear mechanism. In *2013 IEEE/RSJ international conference on intelligent robots and systems* (pp. 252-257). IEEE.

[6] Kim, H.M., Choi, Y.S., Lee, Y.G. and Choi, H.R., 2016. Novel mechanism for in-pipe robot based on a multiaxial differential gear mechanism. *IEEE/ASME Transactions on Mechatronics*, *22*(1), pp.227-235.

[7] Zhao, W., Zhang, L. and Kim, J., 2020. Design and analysis of independently adjustable large in-pipe robot for long-distance pipeline. *Applied Sciences*, *10*(10), p.3637.

[8] Kim, J.H., Sharma, G. and Iyengar, S.S., 2010, March. FAMPER: A fully autonomous mobile robot for pipeline exploration. In *2010 IEEE International Conference on Industrial Technology* (pp. 517-523). IEEE.

[9] Li, H., Li, R., Zhang, J. and Zhang, P., 2020. Development of a pipeline inspection robot for the standard oil pipeline of China national petroleum corporation. *Applied Sciences*, *10*(8), p.2853.

[10] Kim, D.W., Park, C.H., Kim, H.K. and Kim, S.B., 2009, August. Force adjustment of an active pipe inspection robot. In *2009 ICCAS-SICE* (pp. 3792-3797). IEEE.

[11] Abidin, A.S.Z., Chie, S.C., Zaini, M.H., Pauzi, M.F.A.M., Sadini, M.M., Mohamaddan, S., Jamali, A., Muslimen, R., Ashari, M.F. and Jamaludin, M.S., 2017. Development of in-pipe robot D300: Cornering mechanism. In *MATEC Web of Conferences* (Vol. 87, p. 02029). EDP Sciences.

[12] Chablat, D., Venkateswaran, S. and Boyer, F., 2018. Mechanical design optimization of a piping inspection robot. *Procedia Cirp*, *70*, pp.307-312.

[13] Hayashi, K., Akagi, T., Dohta, S., Kobayashi, W., Shinohara, T., Kusunose, K. and Aliff, M., 2020. Improvement of pipe holding mechanism and inchworm type flexible pipe inspection robot. *International Journal of Mechanical Engineering and Robotics Research*, *9*(6), pp.894-899.

[14] Takagi, M., Yoshida, K., Hoshino, H., Tadakuma, R., Suzuri, Y. and Furukawa, H., 2019. Sliding Walk With Friction Control of Double-Network Gel on Feet of Inchworm Robot. *Frontiers in Mechanical Engineering*, p.44.

[15] Nguyen, T.T., Kim, D.K., Rho, Y.W. and Kim, S.B., 2001, July. Dynamic modeling and its analysis for PIG flow through curved section in natural gas pipeline. In *Proceedings 2001 IEEE International Symposium on Computational Intelligence in Robotics and Automation (Cat. No. 01EX515)* (pp. 492-497). IEEE.

[16] Mishra, D., Agrawal, K.K., Abbas, A., Srivastava, R. and Yadav, R.S., 2019. PIG [Pipe Inspection Gauge]: An artificial dustman for cross country pipelines. *Procedia Computer Science*, *152*, pp.333-340.

[17] Savin, S., Jatsun, S. and Vorochaeva, L., 2018. State observer design for a walking in-pipe robot. In *MATEC Web of Conferences* (Vol. 161, p. 03012). EDP Sciences.

[18] Savin, S., Jatsun, S. and Vorochaeva, L., 2017. Trajectory generation for a walking in-pipe robot moving through spatially curved pipes. In *MATEC Web of Conferences* (Vol. 113, p. 02016). EDP Sciences.

[19] Nishimura, T., Kakogawa, A. and Ma, S., 2012, August. Pathway selection mechanism of a screw drive in-pipe robot in T-branches. In *2012 IEEE international conference on automation science and engineering (CASE)* (pp. 612-617). IEEE.

[20] Nayak, A. and Pradhan, S.K., 2014. Design of a new in-pipe inspection robot. *Procedia Engineering*, *97*, pp.2081-2091.

[21] Al-Masri, W.M., Abdel-Hafez, M.F. and Jaradat, M.A., 2018. Inertial navigation system of pipeline inspection gauge. *IEEE Transactions on Control Systems Technology*, *28*(2), pp.609-616.

[22] Kwon, Y.S. and Yi, B.J., 2012. Design and motion planning of a two-module collaborative indoor pipeline inspection robot. *IEEE Transactions on Robotics*, *28*(3), pp.681-696.

[23] Xu, F. and Wang, H., 2021. Soft robotics: Morphology and morphology-inspired motion strategy. *IEEE/CAA Journal of Automatica Sinica*, *8*(9), pp.1500-1522.

[24] Tan, G.B., Zhang, S.M. and Zhu, X.X., 2011. Design of the speed regulating pig with butterfly bypass-valve. In *Advanced Materials Research* (Vol. 201, pp. 429-432). Trans Tech Publications Ltd.

[25] Savin, S., 2018, November. RRT-based Motion Planning for In-pipe Walking Robots. In *2018 Dynamics of Systems, Mechanisms and Machines (Dynamics)* (pp. 1-6). IEEE.

Magnification of energy transmission ratio using miniature cycloidal gear box for humanoids

Umesh Chavan[1], Atharv Joshi[2], Yash Kolambe[2], Harsh Gwalani[2], Harish Chaudhari[2], Aniket Khalate[2], Pravin Hujare[3]

[1, 2] Department of Mechanical Engineering, Vishwakarma Institute of Technology Pune, Maharashtra, India

[3] Department of Mechanical Engineering, Vishwakarma Institute of Information Technology,
Pune, Maharashtra, India

Abstract. Design challenges arise in applications like humanoids where high torque is required to move and self-balance the bot. Present research focuses on development of a cycloidal gearbox for humanoid shoulder joints. Light weight gear box assembly was designed with mass of 360g. Structural design and analysis of cycloidal gearbox is carried out. A 3D printed prototype is developed to validate motion analysis. It was actuated by NEMA 17 step motor (0.45 N-m) which is controlled by A4988 motor driver module on the Arduino platform. The speed reduction ratio is achieved 20:1 in a very small space of 60 x 60 x 20mm. Motor torque 0.45Nm is magnified up to 8.82Nm. Result shows proposed cycloidal gearbox satisfies requirement of humanoid shoulder joints such as high torque, light weight and compact assembly. Additional features noticed are back drivable, vibration and noise free transmission which are highly desirable for humanoid shoulder joints.

Keywords: Torque, Cycloidal Gearbox, Humanoid, Motion study, 3D printing

1. Introduction

Robotics is a multi-disciplinary domain where mechanical, electronics, and computer science work together to get the best possible solution. In robotic arms, high torque is required at the output for holding larger loads. To increase the torque of the motor, it is essential to use a gearbox which can efficiently increase the torque in a very small amount of space. There are various gearboxes available which are very compact and can also transmit power at a very high reduction ratio (in the range of 100 and above). Forinstance, García et al. [1] proposed a review on the performance of harmonic and cycloid drives. along with emerging transmission technologies such as REFLEX torque amplifier, Archimedes drive, Nugear, Bilateral drive, Galaxie drive etc. used in modern robotics. Jonathon W. Sensinger and James H. Lipsey [2] also compared cycloidal and harmonic drives Study done by Biser Borislavov et al. [3] Study shows designing of a cycloidal reducer. Mirko Blagojevic et al. [4] presented a new concept of the two-stage cycloidal speed reducer development. Logan C. Farrell et al. [5] vibrations caused due to gearbox results in noise and affects its overall life. If a single plate is used, a counter balance must be added to avoid substantial vibration. Mathematical model and design procedures to design the epicycloid planet gear of cycloid drives was discussed by Ta-Shi Lai [6].Wei Bo et al. [7] carried out load distribution model and presented equation for design. Following researchers presented some of the potential applications of cycloidal geometry based power transmission benefits and issues. Abhi J.

Rawal [8] developed cycloidal based drive joint for wearable exoskeletons. Hiroshi Matsuki et al. [9] presented bilateral drive gear that is highly back drivable reduction gearbox for robotic actuators. Benedetto Allotta et al. [10] redesigned the cycloidal drive for novel applications in machines. Li, X. et al. [11] and Wan-Sung Lin et al. [12], presented design of a two-stage cycloidal gear reducer with tooth modifications. Li Yawei and Wu Yuanzhe [13] developed RV reducer i.e. one type of cycloid reducer. [14] In this paper, methodology was developed to draw cycloidal gear in minutes and also reduces the possibility of human error and the time to draw each gear which is the main fact why the wider use of cycloidal gears is limited. Stress analysis of cycloidal speed reducer was done by using Finite Element Method (FEM) for the case of single meshing which is the most critical case. Experimental analysis was carried using the strain gauges method [15]. Kinematic analysis of cycloidal speed reducer was performed using the softwares – Autodesk Inventor and Solidworks [16]. In the research work [17] modelled in a dynamic simulation environment, on the rotating parts of cycloidal reducer. The study [18] gave an insight about the internal load sharing of the rotating parts and their capability of carrying the shock loads of cycloidal gears. Present work focuses on the design of a cycloid power transmission unit (gearbox) that possess characteristics of a high reduction ratio in a very compact assembly, much less weight and operating reliably with very less vibrations for humanoid shoulder joints. Computer codes are developed to get the cycloidal profile. A 3D printed prototype is produced to perform tests and validate motion analysis. In order to reduce the vibrations, two cycloidal discs are used which have a phase difference of 180^0. It was actuated by NEMA 17 step motor (0.45 N-m) which is controlled by A4988 motor driver module on the Arduino platform.

2. Creative Design of Cycloidal Gearbox

2.1 Conceptual design

Conceptual design of cycloidal gearbox (CG) shown in Fig. 1. It consists of an eccentric camshaft (input shaft), cycloidal gears, annular cycloidal gear (output gear), flange, fasteners (pins) and motor. Exploded view shown in Fig. 1 elaborates power transmission elements involved in CG. Fig. 2 shows assembly of proposed gearbox with motor drive. An eccentric camshaft transmits motor power to annular gear through cycloidal gears. Which drives a cycloidal gears. In present work as a case study, torque requirement of humanoid shoulder joint is considered to design CG. Requirement is to magnify the motor torque by 20 times. This implies when eccentric cam shaft rotates in 360^0 annular cycloidal gear must rotate 18^0, so that after 20 rotations of the input shaft, the output shaft completes one rotation. Thus cycloidal gear box is designed for the transmission ratio of 20:1

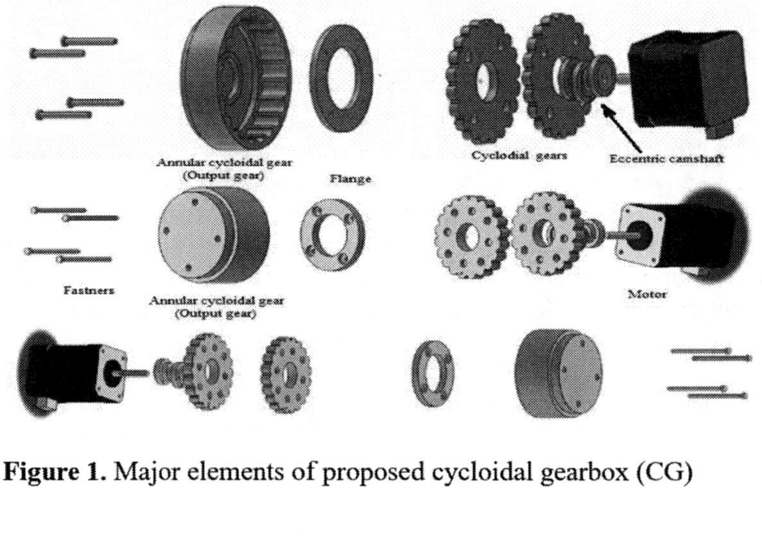

Figure 1. Major elements of proposed cycloidal gearbox (CG)

IOP Conf. Series: Materials Science and Engineering 1272 (2022) 012017 doi:10.1088/1757-899X/1272/1/012017

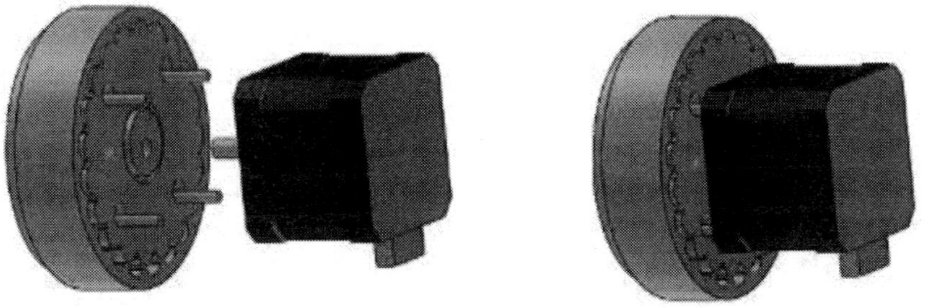

Figure 2. Assembled view of proposed cycloidal gearbox (CG)

2.2 Generation of cycloidal profile

The rotor has a very unique cycloidal shape. There are standard equations available to design the profile of the rotor [13]. The parametric curve equations were chosen because of the two equations needed to define the rotor as shown in Fig. 3. Once the rotor radius (R), eccentricity (E), roller radius (R_r), and the number of rollers (N) is set, the given equations can then be used to generate cycloidal rotor profile coordinates Cx and Cy (X and Y respectively)

$$X = R\cos(\theta) - R_r\cos(\theta - \psi) - E\cos(N\theta) \quad (1)$$

$$Y = -R\sin(\theta) + R_r\sin(\theta - \psi) + E\sin(N\theta) \quad (2)$$

$$\psi = -\tan^{-1}\left[\frac{\sin((1-N)\theta)}{\left(\frac{R}{EN}\right) - \cos((1-N)\theta)}\right] \quad (3)$$

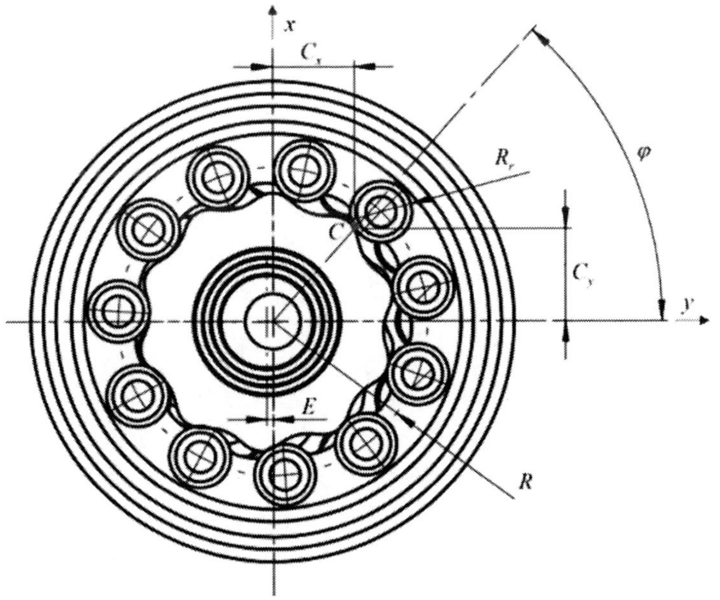

Figure 3. Cycloidal gear profile generation parameters

Contact angle between the cycloidal rotor teeth and the rollers ψ. Equation can be utilized to compute the contact angle using different θ values ranging from 0 to 360^0. The precision of the rotor profile depends on the step value of the angle θ. In other words, the meshing between rotor teeth and rollers will be precise, resulting in negligible backlash.

IOP Conf. Series: Materials Science and Engineering 1272 (2022) 012017 doi:10.1088/1757-899X/1272/1/012017

Table 1. Cycloidal profile parameters

Parameters	Values
No. of teeth on gear (N)	20
No. of roller pins (n)	21
Rotor radius (R)	30 mm
Roller pin radius(Rr)	3 mm
Eccentricity (E)	1.5 mm

MATLAB code was written to generate the cycloidal profile of gear tooth as shown in Fig. 4. Profile of cycloidal gear is a critical parameter to its design, so developed a software program in MATLAB. It takes the number of lobes and pins as input and uses a constraint that the number of pins are equal to number of lobes plus one. Rotor radius as per packaging was 30mm, eccentricity was again a critical design parameter which was 1.5mm. Using the appropriate equations other parameters were calculated housing circle diameter and transmission ratio. The shape of the rotor is very unique and standard equations are available to design the profile of the rotor. Mainly, two different types of equations are available and can be used, explicit equations and parametric equations. So, the parametric equations have been used here in the code and the angle is varied from -180 degrees to 0 degrees for the lower half and from 0 to 180 degrees for the upper half. This was done to maintain continuity and obtain a smooth profile. The results were also plotted to get a cycloidal profile as can be seen in Figure 4. The main aim behind writing the code was to see the effect that various parameters have on the profile of the rotor, the size of the rotor, the eccentricity, the transmission ratio, etc. without making CAD models for each iteration.

```
Code:
% For Transmission Ratio
n = 20; %Number of Lobes
N = 21; %Number of Pins (Rollers)
% N = n + 1 is the condition

TR = (n/(N - n)); %Transmission Ratio

R = 30; %Rotor Radius

% Housing Circle Circumference
HCC = 2 * R * pi ();

Rr = (HCC/ (4 * N)); %Roller Radius

E = 1.5; %Shaft Eccentricity

for t = linspace (-180, 0, 362)
  X = (R*cosd(t))-(Rr*cosd(t+deg2rad(atan(sind((1-N)*t)/((R/(E*N))-cosd((1-N)*t))))))-(E*cosd(N*t));
%   plot(X, Y, 'r*');
%   hold on;
  Y = (-R*sind(t))+(Rr*sind(t+deg2rad(atan(sind((1-N)*t)/((R/(E*N))-cosd((1-N)*t))))))+(E*sind(N*t));
  plot(X, Y, 'r.');
  hold on;
end

for t = linspace (0, 180, 362)
  X = (R*cosd(t))-(Rr*cosd(t+deg2rad(atan(sind((1-N)*t)/((R/(E*N))-cosd((1-N)*t))))))-(E*cosd(N*t));
%   plot(X, Y, 'b*');
```

```
%   hold on;
  Y = (-R*sind(t))+(Rr*sind(t+deg2rad(atan(sind((1-N)*t)/((R/(E*N))-cosd((1-N)*t))))))+(E*sind(N*t));
  plot(X, Y, 'g.');
  hold on;
end
```

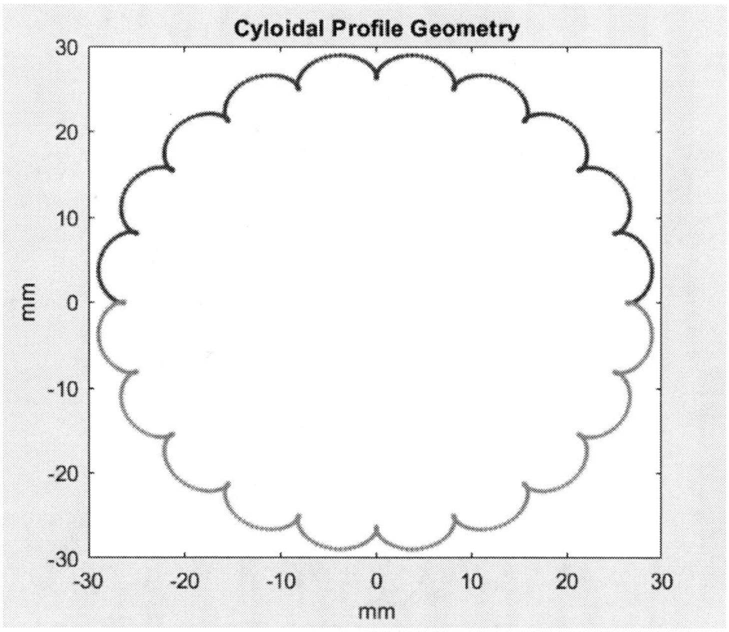

Figure 4. Cycloidal profile generated using code

2.3 Structural Analysis of Cycloidal Gears

Strength analysis of cycloidal rotor (disc) and casing is carried out. Working loads are applied on each of the parts to obtain maximum stress and deformations. Conditions: Fixed – backside of casing (where arm will be seated), force applied - 260 N, Mesh – tetrahedral
Material chosen - Polylactic acid (PLA)
Material properties of PLA

Density	:	1.25 g/cm^3
Young's modulus	:	3027 MPa
Passions ratio	:	0.32

From the analysis as seen in Fig.5, the maximum deformation is of 0.379 mm at the teeth of the output where the cycloidal gear and output completely meshes. The minimum deformation is zero where there is no contact between the cycloidal gear and the output. From the contour plot in Fig.6 , it can be seen that the maximum stress at completely meshed teeth because the force acting at that point is maximum and minimum stress where the gear is not in contact with the output. The minimum stress value is 298Pa and maximum value is 61.82 Mpa. Same is the case for the deformation of the cycloidal gear as it is for the out as observed in the Fig.7. The deformation is maximum at the teeth which is completely meshed with the output and the deformation goes on decreasing equally on both sides of the meshed teeth. The maximum deformation in this case is 0.355mm and minimum of 0mm about the centre of the gear. From the contour plot in Fig. 8 it can be observed that the maximum stress occurs at the root of one teeth that gets completely meshed which is 45.9Mpa. Also the stress occurs between the eccentric shaft and the cycloidal gear which is around 15 MPa.

IOP Conf. Series: Materials Science and Engineering 1272 (2022) 012017 doi:10.1088/1757-899X/1272/1/012017

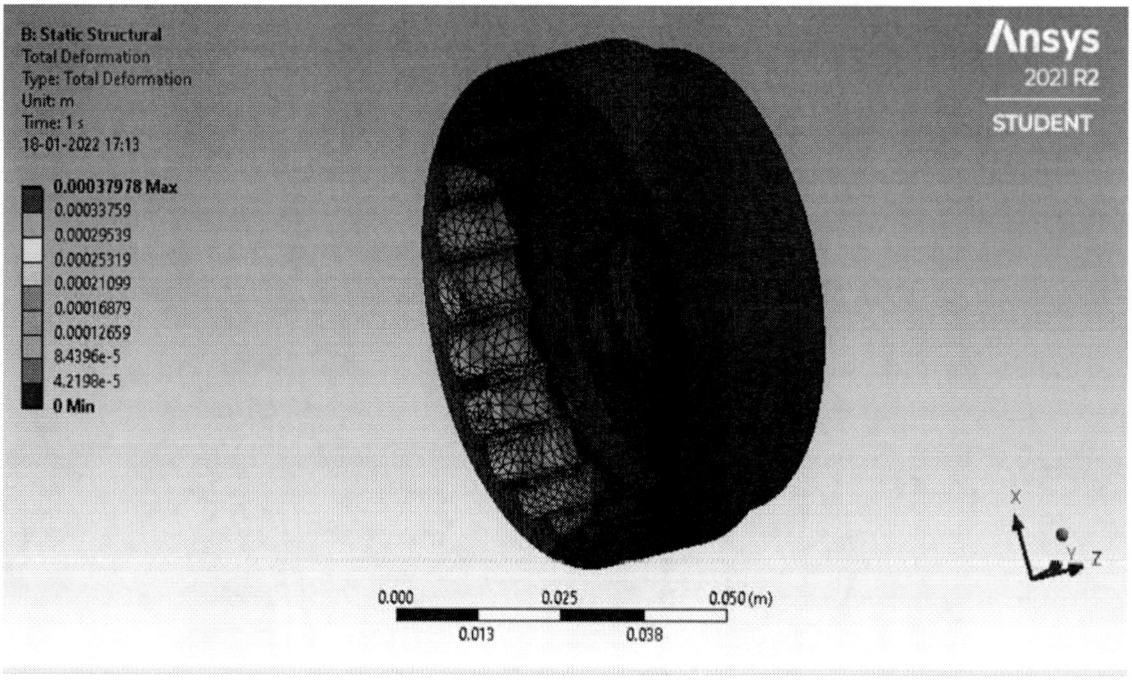

Figure 5. Deformation of annular gear of cycloidal gear box

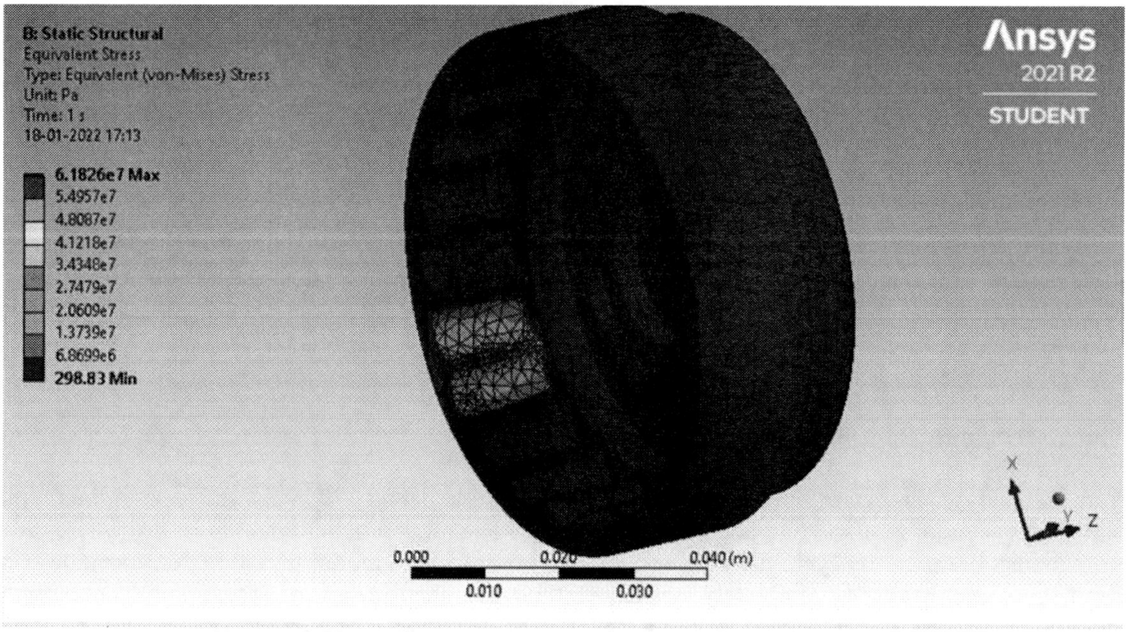

Figure 6. Von-Mises stress on annular gear of cycloidal gear box

Analysis was carried out in Ansys Student Version with following boundary conditions Fixed - Inner bearing face, Force applied - 260 N, Mesh – tetrahedral

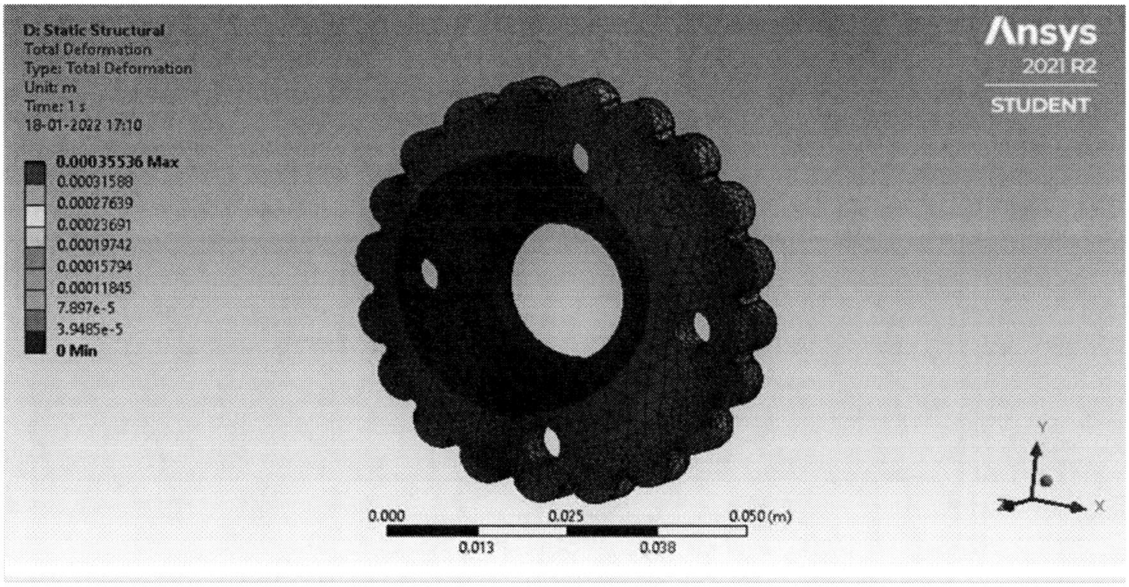

Figure 7. Deformation of cycloidal gear of cycloidal gear box

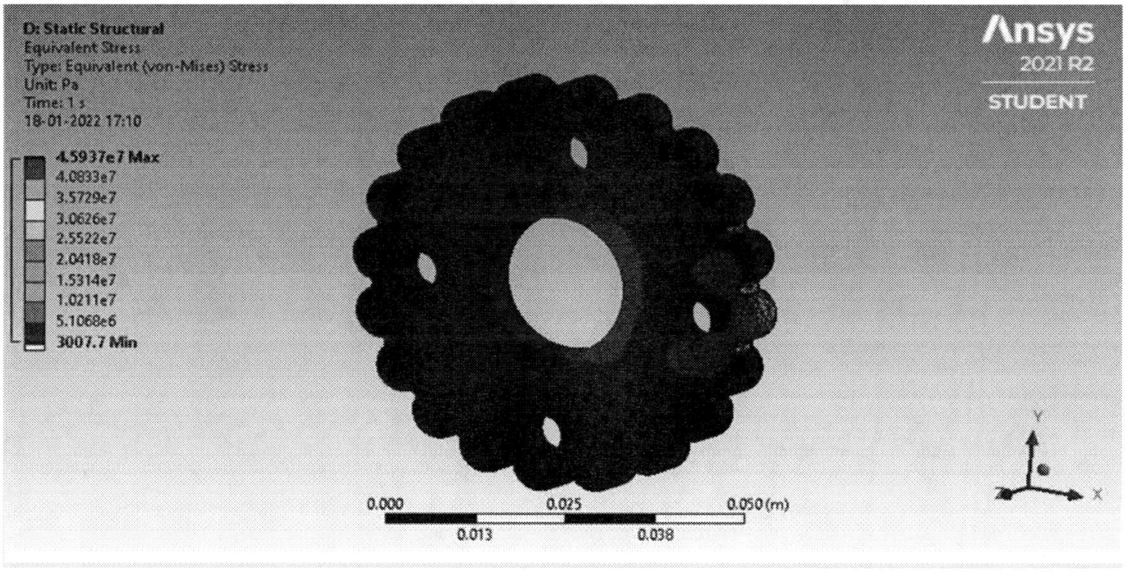

Figure 8. Von-Mises stress on cycloidal gear of cycloidal gear box

The stress levels and deformations found to be within the allowable limits as two out of phase cycloidal gears helps to reduce load distribution on individual gears. Also, contact stress and deformation of the proposed cycloid drive is much hence, proposed cycloid drive exhibits high load capability which is more suitable for robotic arm applications. Thus the structural strength was verified before the developments of prototype.

3. Motion Analysis

Motion study of CG was performed in computational tools. Fixed support boundary condition was given to the bolts and rotary input motion to the eccentric camshaft. Solid body contact assumed between the bearing and the discs, the eccentric shaft and the bearing, the cycloidal gear and the annular gear casing. Gravity condition also applied and a weight attached at the output side. The angular velocity of the output gear found to be one by twenty times of the input angular velocity. Similarly, the output torque which was derived by using the angular acceleration was also close to the desired values. The humanoid for which this gearbox is designed is going to an interactive entertainment bot with capability of holding 4.5 kg load at 0.2m distance which is equivalent to 7.8Nm. Fig. 9 shows the torque drawn by the motor over time. The values are fluctuating as gravity effect is considered and a single mass (for load) is given at output. Fig. 10 shows the angular velocity obtained at the output shaft. Input speed given to motor was 60 rpm, and the output speed computed was 3.13 rpm (0.3314 rad/s or 19 deg./s).The angular acceleration graph was also plotted shown in Fig. 11. The angular acceleration values obtained for each time step from motion analysis was exported to torque analysis. Here, these values were multiplied the inertia of the output part to calculate the torque that is required by the output (and the load) to rotate at each time step. The graph of torque required at each time step is shown below in Fig. 12.

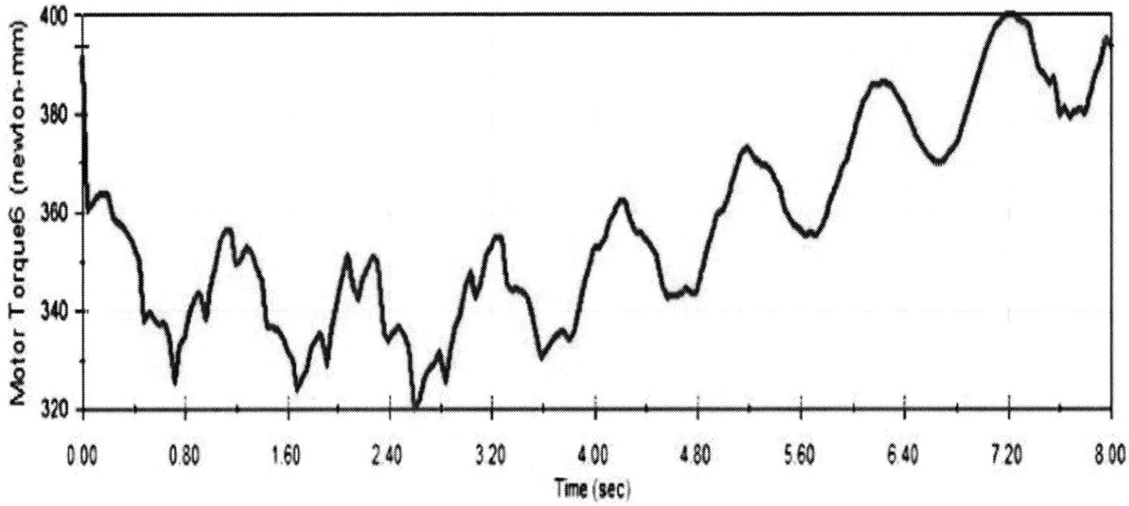

Figure 9. Motor torque vs time (sec.)

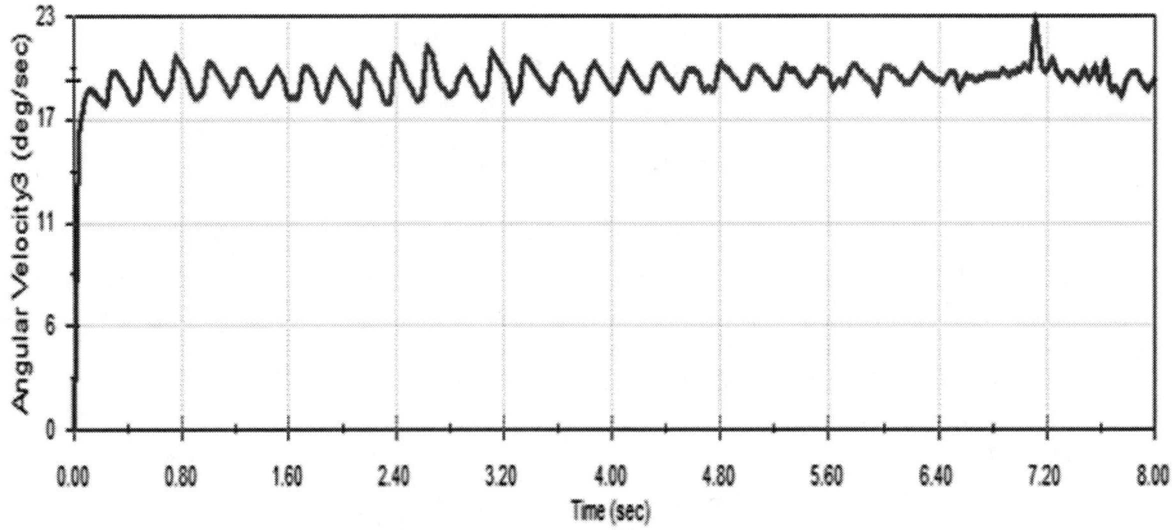

Figure 10. Angular velocity vs time (sec.)

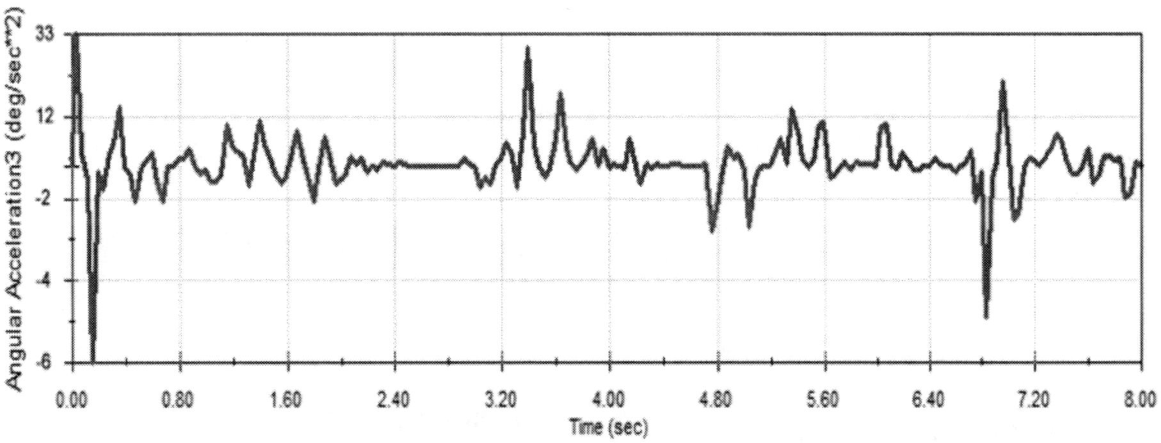

Figure 11. Angular acceleration vs time (sec.)

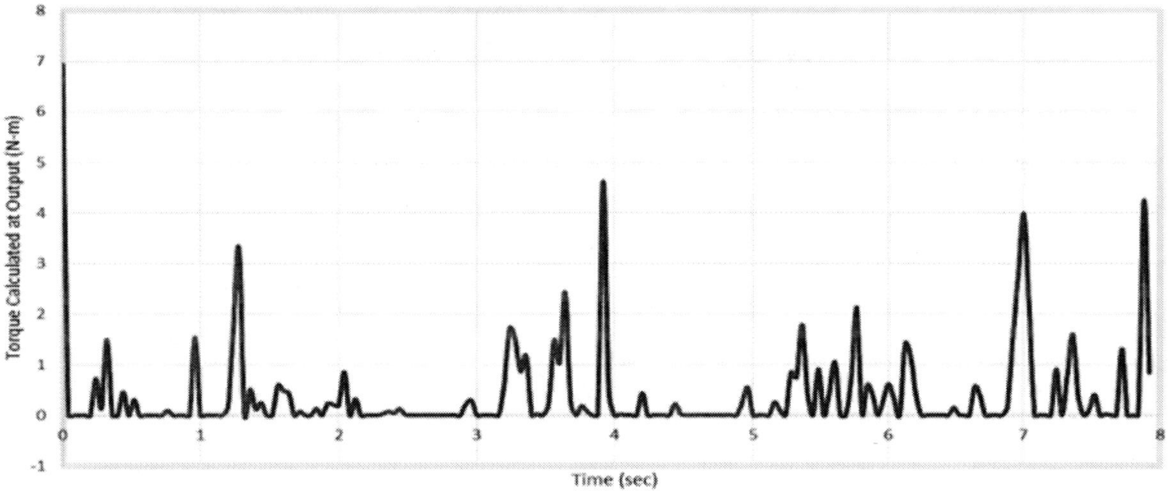

Figure 12. Torque output vs time (Sec)

4. Prototype and Experiments

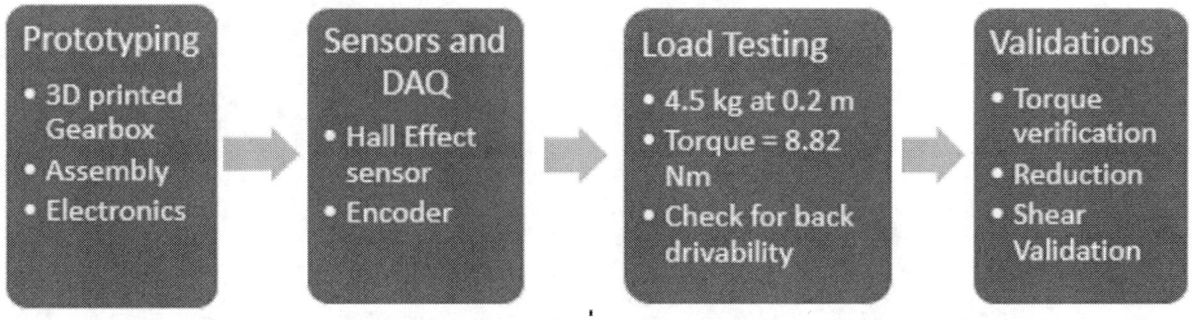

Figure 13. Design of experiment

4.1 Prototyping

Fig. 13 shows design of experiment plan to build and test the CG. As per design specifications a prototype of CG was 3D printed. The parts included the eccentric camshaft, cycloidal gears, annular gears, and casing. 3D printed components shown in Fig. 14. Special care to be taken to make accurate assembly of all components. Bearings are press fitted in the 3D printed cycloidal discs. Bolts were used as the connectors between the discs and the retainer which will allow the disc to wobble around it. Finally, the cycloidal discs are fitted into the cycloidal casing. Both the discs were mounted on the eccentric shaft in the phase difference of 180°. Input and output covers were also fitted and finally bolted to prepare the final gearbox assembly. Overall space dimensions of a cycloidal gear box was 60 x 60 x 20mm and mass assembly is 360g.

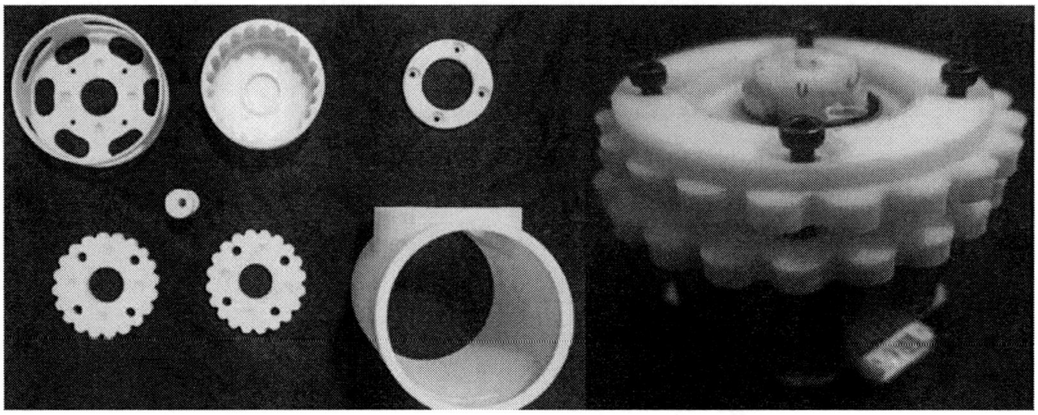

Figure 14. 3D printed components of cycloidal gear box

4.2 Testing setup

Input is given to gearbox using a 12 Volt, 0.45 N-m, NEMA 17 stepper motor. These motors come with an outer casing, which are easy to mount and ensure stable operation. This motor required a special motor driver due to its high ampere rating. A4988 motor controller was used along with the RUDRA board to drive this motor. This motor also allows to vary the speed. This could be done by changing the delay between the consecutive frequency pulses in the code. RUDRA board (which is very similar to the Arduino UNO Micro-controller). External power supply was given to the micro-controller using a 12 V, 2A adaptor.

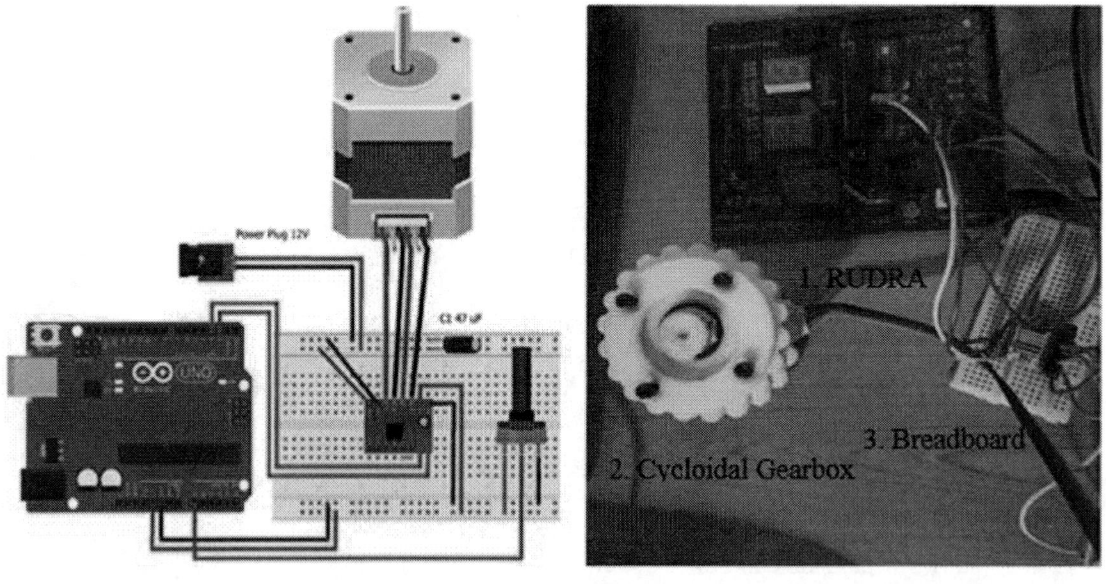

Figure 15. Test setup of cycloidal gear box

Fig.15 shows circuit to control NEMA 17 stepper motor with Arduino. Stepper motor is powered using a 12V power source, and the A4988 module is powered via Arduino.

4.3 Results

A 3D printed prototype is developed to validate motion analysis. It was actuated by NEMA 17 step motor (0.45 N-m) which is controlled by A4988 motor driver module on the Arduino platform. Tests are conducted on two prototypes as printing of hypocycloidal and epicycloid tooth profile is critical. In first prototype it was observed slipping during rotation and skipping some gear teeth in between and not rotating continuously. Also, the cycloidal disc was rotating out of plane. Gearbox was functioning properly when the motor and cycloidal disc were held at a certain position. One of the main reasons for the above-mentioned problems was the error in assembly that is backlash in meshing cycloidal teeth, this caused problems while rotation. Second prototype components precisely 3D printed and assembled accurately. Problems occurred in first prototype due to backlash are totally rectified. Many test trails are conducted on prototype of cycloidal gear box. Input and output speed were measured through rpm sensor. Displacement trials shows annular gear rotates by 18^0 during a complete rotation of the input shaft of 360^0. The output speed found to be 3 complete rotations in 60 seconds when the input speed of motor was 60 rpm. Motor torque 0.45Nm and output torque found to be 8.82Nm. Kinematic simulation results are very close to the experimental values of speed ratio and torque. Speed, torque and back drivability validated by running gear box at different speed and more than 50 trials. Overall, after 20 rotations of the input shaft, the output shaft completes one rotation. The transmission ratio of the gearbox is thus 20:1 in a very small space of 60 x 60 x 20mm.

5. Conclusion

Design of compact cycloidal gear box is proposed for the application such as humanoid shoulder joints. Research was focused on development of concepts and validation through design, modelling, motion analysis, 3D printing, prototyping and testing. Two out of phase cycloidal gears in assembly helps to reduce vibration and load distribution on individual gears. The output speed found to be 3 complete rotations in 60 seconds when the input speed of motor was 60 rpm. Motor torque 0.45Nm and output torque found to be 8.82 Nm. Also, motion analysis results are very close to the experimental values of speed ratio and torque. Speed, torque and back drivability validated by running gear box at different speed and more than 50 trials. Overall, after 20 rotations of the input shaft, the output shaft completes one rotation. The transmission ratio of the gearbox is thus 20:1 in a very small space of 60 x 60 x 20mm. Proposed cycloid drive exhibits high load capability which is more appropriate for robotic arm applications. Cycloidal gearbox satisfies requirement of humanoid shoulder joints such as high torque, light weight and compact assembly. Additional features noticed are back drivable, vibration and noise free transmission which are highly desirable for humanoid shoulder joints.

References

[1] Pablo López García, Stein Crispel, Elias Saerens, Tom Verstraten and Dirk Lefeber, "Compact gearboxes for modern robotics: A Review", Frontiers in Robot and AI, vol.7, Article 103, August 2020

[2] Jonathon W. Sensinger and James H. Lipsey, "Cycloid vs. harmonic drives for use in high ratio, single stage robotic transmissions," IEEE International Conference on Robotics and Automation, June 2012

[3] Biser Borislavov, Ivaylo Borisov, Vilislav Panchev, Design of a planetary-cyclo drive speed reducer cycloid stage; geometry; element analyses, Project work report, Linnaeus University, Sweden, May 2012

[4] Mirko Blagojevic, Nenad Marjanovic, zorica Djordjevic, Blaza Stojanovic and Aleksandar Disic, "A New Design of a Two-Stage Cycloidal Speed Reducer," Journal of mechanical design, ASME, vol. 133, no. 8, pp. 085001-12011, August 2011

[5] Logan C. Farrell, James Holley, William Bluethmann, Marcia K. O'Malley, "Cycloidal gear train in-use efficiency study," ASME, International Design Engineering Technical Conferences and Computers and Information in Engineering Conference, Quebec City, Canada, August 2018.

[6] Ta-Shi Lai, "Design and machining of the epicycloid planet gear of cycloid drives," The International Journal of Advanced Manufacturing Technology, vol. 28, no.7, pp.665-670, April 2006

[7] Wei Bo1, Wang Jiaxu1, Zhou Guangwu1, Yang Rongsong, Zhou Hongjun and He Tao, "Mixed lubrication analysis of modified cycloidal gear used in the RV reducer," Proc IMechE Part J: Journal of Engineering Tribology, Vol.230 no.2, pp. 121-134, February 2016

[8] Li Yawei and Wu Yuanzhe, "Design of a cycloid reducer Planetary stage design, shaft design, bearing selection design, and design of shaft related parts," Engineering, Corpus ID 203697900, 2012

[9] Abhi J. Rawal, Cycloidal-Drive Joint Design for Wearable Exoskeletons, Master Thesis, New Jersey Institute of Technology, December 2020.

[10] Hiroshi Matsuki, Kenta Nagano and Yasutaka Fujimoto, "Bilateral Drive Gear—A Highly Backdrivable Reduction Gearbox for Robotic Actuators," IEEE/ASME Transactions on Mechatronics, vol. 24, no. 6, pp. 2661-2673, December 2019.

[11] Benedetto Allotta, Lorenzo Fiorineschi, Susanna Papini, Luca Pugi, Federico Rotini and Andrea Rindi, "Redesigning the cycloidal drive for innovative applications in machines for smart construction yards," World Journal of Engineering vol.18 no.2 pp.302–315, March 2021

[12] Li, X., Chen, B., Wang, Y., Sun, G. et al., "Geometry Design of a Non-Pin Cycloid Drive for In-Wheel Motor," SAE Technical Paper 2015-01-2172, June 2015

[13] Wan-Sung Lin, Yi-Pei Shih and Jyh-Jone Lee, "Design of a two-stage cycloidal gear reducer with tooth modifications," Mechanism and Machine Theory, vol.79, 184–197, September 2014

[14] Nenad Petrovic, Mirko Blagojevic, Nenad Marjanovic, Milos Matejic, "Parametric drawing of a Cyclo drive shortened equidistant Epitrochoid gear," 7th International Quality Conference, Center for Quality, Faculty of Engineering, University of Kragujevac, May 2013.

[15] Mirko Blagojevic, Nenad Marjanovic, Zorica Djordjevic, Blaza Stojanovic, Vesna Marjanovic, Rodoljub Vujanac, Aleksandar Disic, "Numerical and Experimental analysis of the cycloid disc stress state, UDC/UDK 621.833.058:[519.6:004.421], May 2014.

[16] Ivan Pantić, Mirko Blagojević, "Kinematic analysis of single-stage Cycloidal speed reducer," Machine

[17] S.V. Thube and T.R. Bobak, "Dynamic Analysis of a Cycloidal gearbox using Finite Element Method," American Gear Manufacturers Association, ISBN: 978-1-61481-049-0, October, 2012.

[18] Wear assessment of 3–D printed parts of PLA (Polylactic acid) using Taguchi design and Artificial Neural Network (ANN) technique, Materials Research Express, 25 November 2020

Edge Blending to Enhance the Flow Over Double delta wing configuration during re-entry

Krishna Jaiswal, Pratik Ganorkar, Sunil Shinde

krishna.jaiswal20@vit.edu, pratik.ganorkar20@vit.edu , sunil.shinde@vit.edu

Abstract. During re-entry of space vehicle, the vehicle undergoes flow variation from hypersonic flow to subsonic flow, which causes huge heat transfer over the wing configuration. The double delta wing of configuration 76 degree - 40 degree is considered and effect of blending the wing at edges showed variation and enhancement of flow over the double delta wing configuration. Heat flux calculations were carried out at regions of shock- shock interactions and the effect of blend radius at edges were analyzed using simulations done in ANSYS Fluent. It was observed that aerodynamic efficiency increases with the edge blending which ultimately enhanced the flow field over the double delta wing.
Keywords. Double Delta Wings, Blended Wing, Re-entry Condition, Flow Enhancement

1. Introduction

During an aircraft design, selection of wing is one of the most important parameter that is to be taken into consideration. Wings are designed in taking into parameters that directly impact on the whole aircraft. These parameters include the lift force generation, fuel storage considerations, angle of attack, lift angle and maneuverability stall angle. A delta wing has greater advantages than other types of wings and are used in fighter planes and spaceship applications where the flow is either supersonic or hypersonic flows [1,2]. A delta wing sweeps sharply from the fuselage with an angle between the leading edge of the wing often high as 60 degrees and the angle between the fuselage and the trailing edge of the wing mostly around 90 degrees [3]. A double delta wing is essentially a wing with a kink in its leading edges that forms the support wherein the leading edge of the strake and main wing intersect.

Table 1. Nomenclature

Symbols	**Description**
α, φ	Angle of attack [deg]
δe	Elevon deflection angle [deg]
θ	Bent angle [deg]
$(C_D)_v$	Vortex Drag coefficient [n.d]
$(C_L)_v$	Vortex Lift coefficient [n.d] L
C_D	Drag coefficient [n.d]
C_L	Lift coefficient [n.d]
C_{Mx}	Rolling moment coefficient [n.d]
C_{My}	Pitching moment coefficient [n.d]
C_{Mz}	Yawing moment coefficient [n.d]

C_M	Pitching moment coefficient [n.d]
$C_{M\alpha}$	Pitching moment derivative [n.d]
C_N	Normal force coefficient [n.d]
C_Y	Side force coefficient [n.d]
h	Total height of vehicle cabin [m]
h_{ws}	Half-width of vehicle cabin [m]
l_t	Total length of vehicle [m]
l_{ref}	Wing mean aerodynamic chord [m]
M_∞ , M_o	Mach number [n.d]
m_{tot}	Total mass of the vehicle (TPS included) [kg]
r_n	Nose radius [m]
S_{ref}	Aeroshape planform area [m^2]
V_∞	Free-stream velocity [m/s]
V_{max}	Maximum landing speed [m/s]

2. Double delta wings

At the front end of the main wing, double delta wing has leading edge extensions, also called as strakes. These strakes have a higher swept angle in order to achieve vortex stability. Most importantly, the configurations that needed to be taken into consideration on double delta wing so as to achieve sufficient supersonic cruise performance is the sweep angle combination [3]. A monolithic wing body configuration with double delta wing having quite high sweep angle becomes more advantageous as it lowers the heat fluxes over the leading edge [4].

Investigations are carried out during the re-entry condition of vehicle. During re-entry of vehicle the vehicle experiences the flow change from supersonic to transonic then to subsonic flow. Previous study prevails that there must be improvisation of performance at low speed in order to achieve a safe horizontal landing. Moreover, safety is one of the important parameter to be considered [4].

Numerous study have been carried out on double delta wing and re-entry of vehicle. Antonio Viviani, Andrea Aprovitola, Luigi Iuspa and Giuseppe Pezzella [4] evaluated the low speed longitudinal aerodynamic performance of spacecraft which has a blended wing body aeroshape with unconventionality. The authors showed using body flap enhances longitudinal static stability, provides side slip control and also gives lateral directional manoeuvrability. To obtain high speed as well as low speed design requirements, multi-disciplinary design optimization (MDO) has been applied. Analytical results were compared with simulations keeping the spacecraft at low Mach number, it concluded that behaviour at landing angle of attack (AoA) exhibits the initializing and improvement of two vortex system which helps in vortex lift and thereby increases the lift coefficient. It drawbacks at ability for landing horizontally and also reduction of aerodynamic efficiency of 25%.

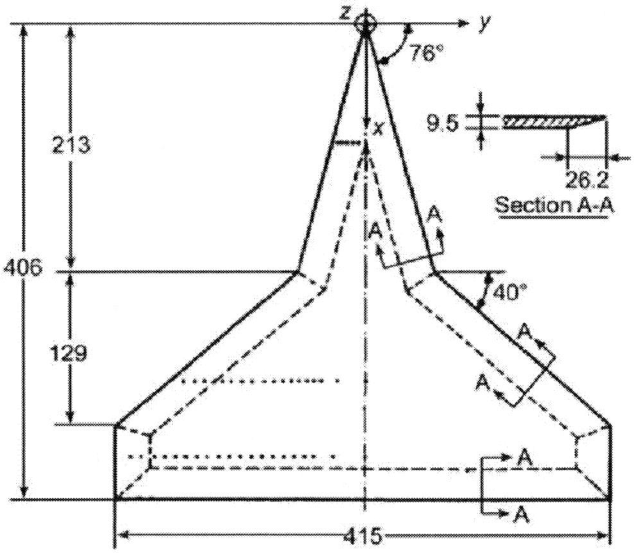

Figure 1. $76^0 - 40^0$ Double Delta Wing [10]

Gaoxiang, Chun, Honghui and Zonglin describes the Shock- Shock Interactions between the body and wings at supersonic flows with a wing considered as sharp wing and body as flat wedge [5]. The thermal protection of aircraft and design specifications are also detailed in their research. Wedge angle greatly affects the wave configuration as compared to swept angle. The flow structures were solved by flow dynamics, finding the geometrical relationships between the 3-dimensional steady and 2-dimensional unsteady problem with decomposed Mach numbers properly described in sections, the steady problem is considered as interactions of two incident waves, succeeding over the characteristic plane; for unsteady problem shock polar analysis was done [5].

P.R Vishwanath has investigated that skin friction drag reduction can be done by the use of riblets [6]. Riblets are micro-grooves that are applied over the surface in accordance with freestream direction. The author reviewed the performance of riblet when applied over the airfoil wings and wing- body configuration at varying speed. To determine the viscous drag reduction from riblets, three different methods were applied; the first approach is skin friction balance to directly measure the wall shear stress; the second approach is 2D boundary layer momentum integral [MI] and third approach is internal strain gauge balance [6]. Mohammad Ghalandari, Ibhrahim Mahariq, Farad Ghadak and Fahad Jarad described briefly the aileron mass parameter has an adverse effect on velocity and frequency of flutter issues [7]. Fluttering is the self-vibrations occurring at varying speeds, is considered to be most common problem, which can ultimately lead to structure failure if proper damping system is absent. The authors have applied the multi-disciplinary design optimization (MDO) technique [4] in order to achieve flutter wing analysis and optimization. The position and centre of mass are effective and no need to unify the mass moment of inertia. The MDO helped to investigate the best position for aileron of composite wing. The present work describes the flow over the double delta wing at re-entry condition, the variation of flow from supersonic to subsonic and the aerodynamic performance study over the wing is considered. Also the effect of edge blending to enhance the flow is done and simulations were done with help of ANSYS Fluent. The values of drag coefficient, lift coefficient and moment coefficient obtained from simulations are plotted against angle of attack. The values are compared with values obtained from Slender Wing Theory.

3. Numerical Methods

Considering the slender wing theory to obtain the relation between the drag & pressure coefficient along with lift coefficient at high angle of attack the following relations given by the theory:

$$(C_L)_p = \frac{\pi AR}{2} sin\alpha\, cos\alpha \qquad (1)$$

$$(C_L)_p = (C_L)_P \sin\alpha \qquad (2)$$

Then, the following vortex (v) lift terms:

$$(C_L)_V = (C_D)_{pf} \sin^2\alpha \cos\alpha \qquad (3)$$

$$(C_L)_V = (C_D)_{pf} \sin^3\alpha \qquad (4)$$

As a result, lift and drag coefficients become:

$$(C_L) = (C_L)_p + (C_L)_V = \frac{\pi AR}{2} \sin\alpha \cos\alpha + (C_D)_{pf} \sin^2\alpha \cos\alpha$$

$$(C_D)_p = (C_D)_p + (C_D)_V = \frac{\pi AR}{2} \sin^2\alpha \cos\alpha + (C_D)_{pf} \sin^3\alpha$$

Four main parameters that affect the wing configuration are the pressure or thrust, lift force, drag force and weight and shape of wing and body. Slender wing theory explains briefly the actual conditions that will be worked out at the peripheral of delta and double delta wings which gives the findings of aerodynamic parameters like relation of drag and lift coefficient along with angle of attack.

Now as to calculate the heat flux over the regions, basic numerical study and aerodynamic relations are to be considered as given [8]:-

$$To/T = 1 + ((\gamma-1)/2) * M^2 \qquad (5)$$

This will give the static temperature. Now further to find speed of object and speed of sound we use the below empirical relationships

$$Velocity\ of\ Sound = a = (YRT)^{1/2} \qquad (6)$$

and

$$Mach\ number = M = C/a = Object\ Speed / Speed\ of\ Sound \qquad (7)$$

Shock-shock interactions are considered for investigating the aerodynamic heating of wings, initial methods were to use the basic relations of pressure and heat in order to obtain aerodynamic heating but not applicable to complex geometry [5]. To obtain the flow deflections, the below relation is used from the oblique- shock interactions that is given by L. Laguarda [6] in its research as :-

$$\begin{aligned} v^i_{km} &= f(M_0, \varphi, \gamma) \\ &= \tan^{-1}[((2*(M^i_0 * \sin\varphi^i_{km})^2 - 1)/(Mi_0^2(\gamma + \cos 2\varphi^i_{km}) + 2)*\tan\varphi^i_{km})] \end{aligned} \qquad (8)$$

Where specific heat ratio is taken as 1.4.

Aerodynamic investigation was carried out with initial design of double delta wing [1] without giving blend radius at the edges and angle of attack given with enclosure created along with the tetrahedral meshing with mesh refinement over the wing, importing and solving the three dimensional RANS equation, considering the density based solver and enabling the energy equation; using the k-omega SST model; as flow is compressible and is computed using the ANSYS Fluent. Air is considered as perfect gas with density kept as ideal- gas and viscosity to be given as accordance to Sutherland's Law, thermal conductivity is kept at *0.0242W/mK* and the specific heat (at constant pressure) to be *1.006 kJ/kgK.* Boundary conditions were kept as inlet to be as pressure far-field giving initial Mach number to be 0.6 as considering the flow to be subsonic flow. Respective gauge pressure and temperature is given at inlet; the wall is kept at adiabatic condition and outlet to be pressure outlet with standard boundary conditions. The convergence condition was kept at decimal of 0.001. Convective fluxes were discretized using second order- Roe Flux Difference Splitting (FDS); while diffusive fluxes are discretized by considering a cell-centred scheme. The Courant Number (CN) is automatically numbered by the Solution Steering (SS) algorithm in Fluent wherein the flow can be kept as per our required conditions; moreover, the hybrid initialization is done because it uses the boundary conditions then solves the Euler problem and uses the solution to Euler problem as initial guess. Further keeping the same input condition, the wing was edge- blended and the simulations were carried out with all boundary conditions keeping same.

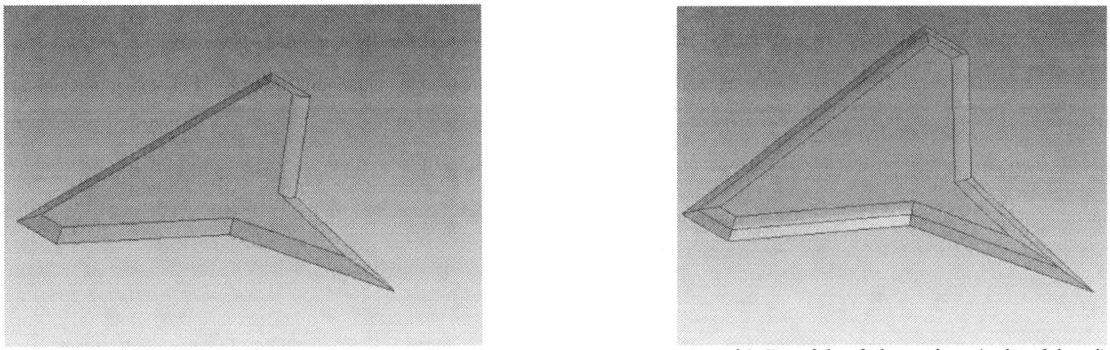

a) Double delta wing b) Double delta wing (edge blend)

Figure 2. Double Delta Wing (with edge blending)

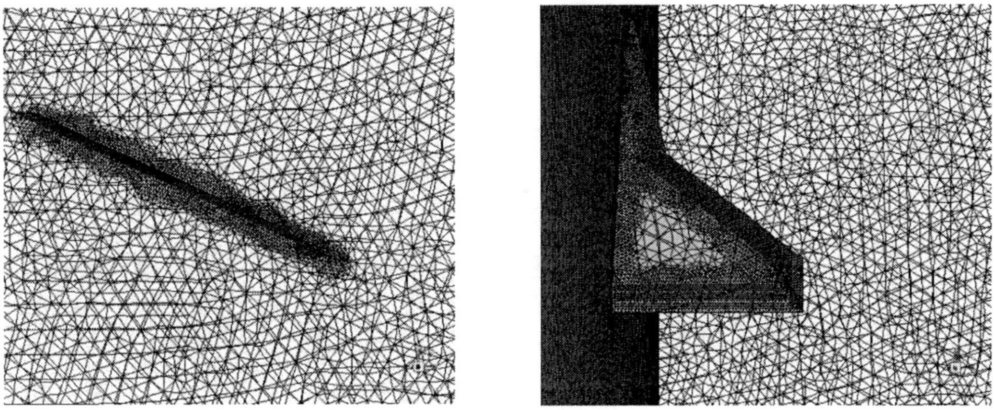

Figure 3. Meshing

4. Results

The results showed that by blending it by radius of 0.03m and increasing it to certain level, it was observed that the flow over the wing has enhanced that can be seen in pressure contour diagram.

The figures showed the simulations over the double delta wing at different angle of attack, the left shows the wing without blend and the right shows the pressure contour over the wing with edge blending. It can be observed that the blended wing helps in reducing the creation of spots over which the pressure gets debated. The values of drag, lift and moment coefficient were obtained, which was compared with Slender Wing theory.

The graphs were plotted with previous research data and were compared with the obtained results of edge blending. It was observed that the lift coefficient increases and drag coefficient decreases when edge blending is done, this exhibits that the flow has improvised over the wing since drag reduction causes friction to be emerged at lower value thereby increasing the smooth flow over the wing. Moreover, the lift to drag ratio increases which leads to increase in efficiency of aircraft. It was also observed that as angle of attack increases there is an increase in drag and lift to drag ratio tends to decrease. The maximum lift to drag occurs at initial 5^0 angle of attack during re-entry of spacecraft into the Earth's Orbit.

IOP Conf. Series: Materials Science and Engineering 1272 (2022) 012018 doi:10.1088/1757-899X/1272/1/012018

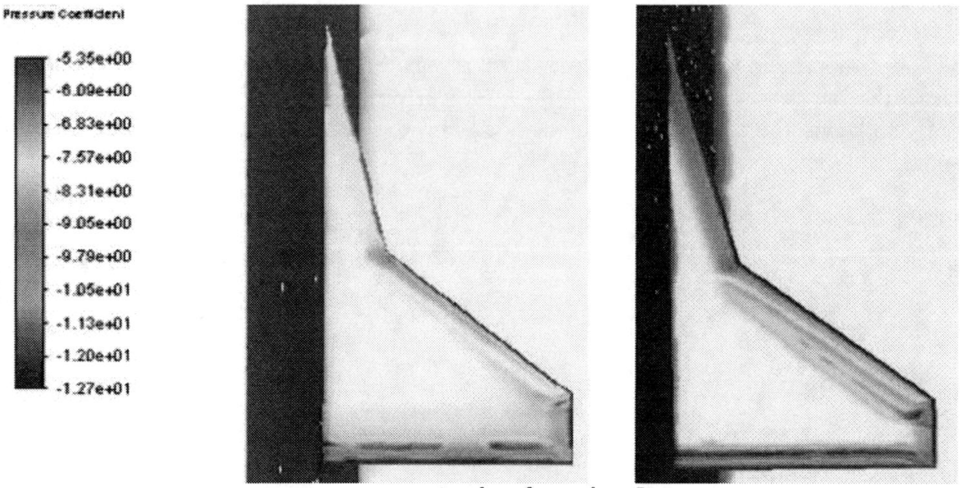

Angle of attack = 5°

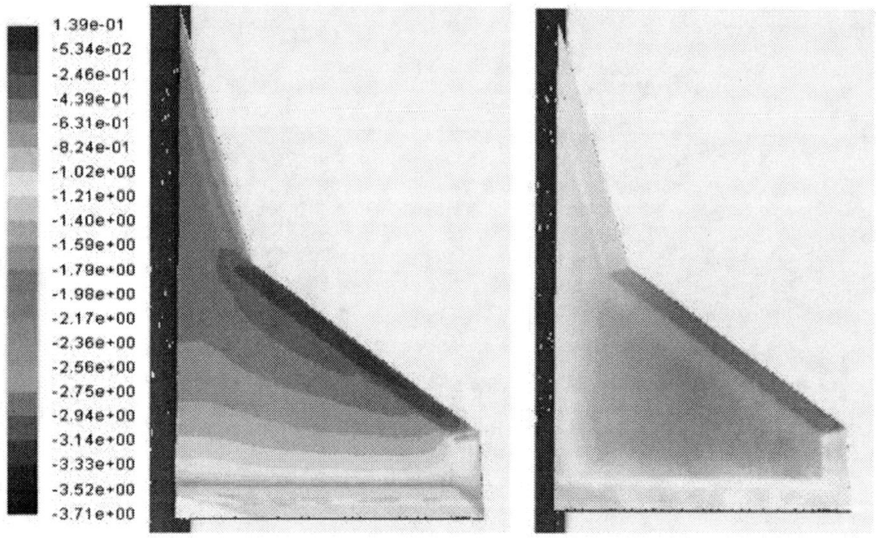

Angle of attack = 25°

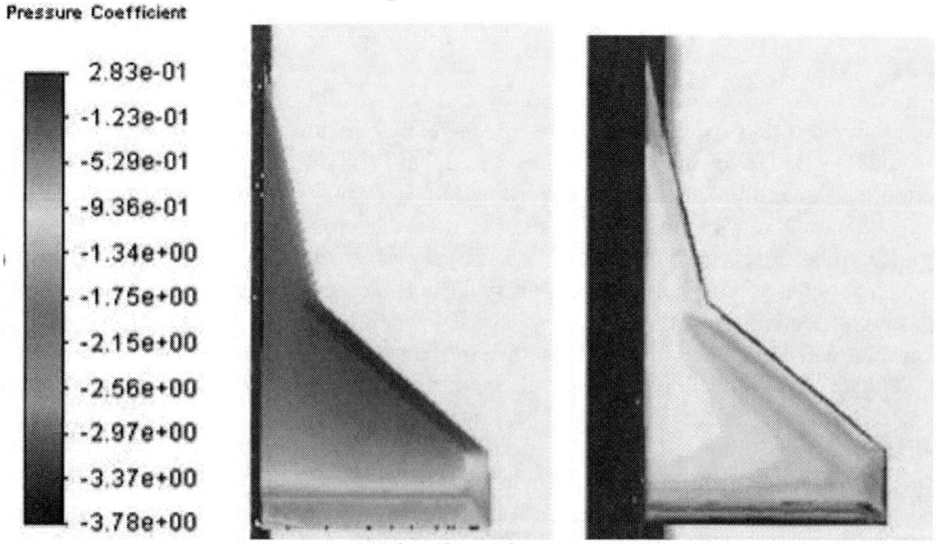

Angle of attack = 35°

Figure 4. Pressure Contours at different angle of attack (left: without blend; right: with edge blending)

The figure 5 describes the variation of drag coefficient with angle of attack, it was observed that at initial 5^0 angle of attack drag coefficient gets reduced as edge of wing is blended, thereby reducing the vortex generation at the end tips. But as angle of attack increases there is no much huge effect of blend as can be seen at angle of attack 45^0. The figure 6 describes the lift coefficient plotted against the angle of attack given, it was observed that lift coefficient increases at lower angle of attack for the blended wing, but at angle of attack at 45^0 the values seems to be almost same for normal and blended wing.

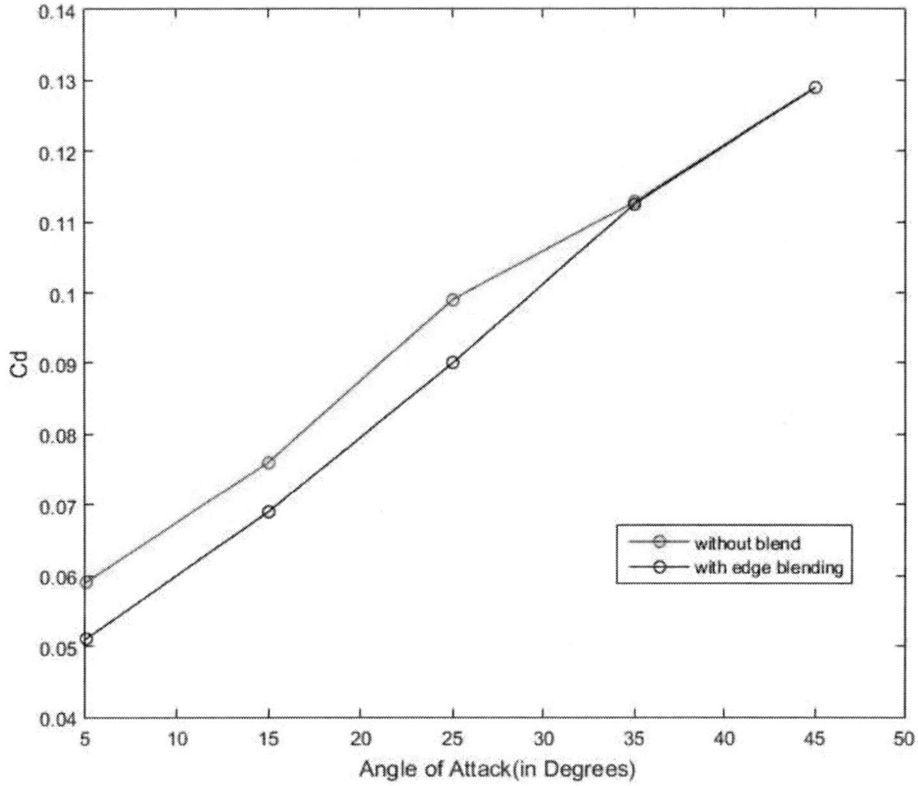

Figure 5. C_D v/s Angle of Attack

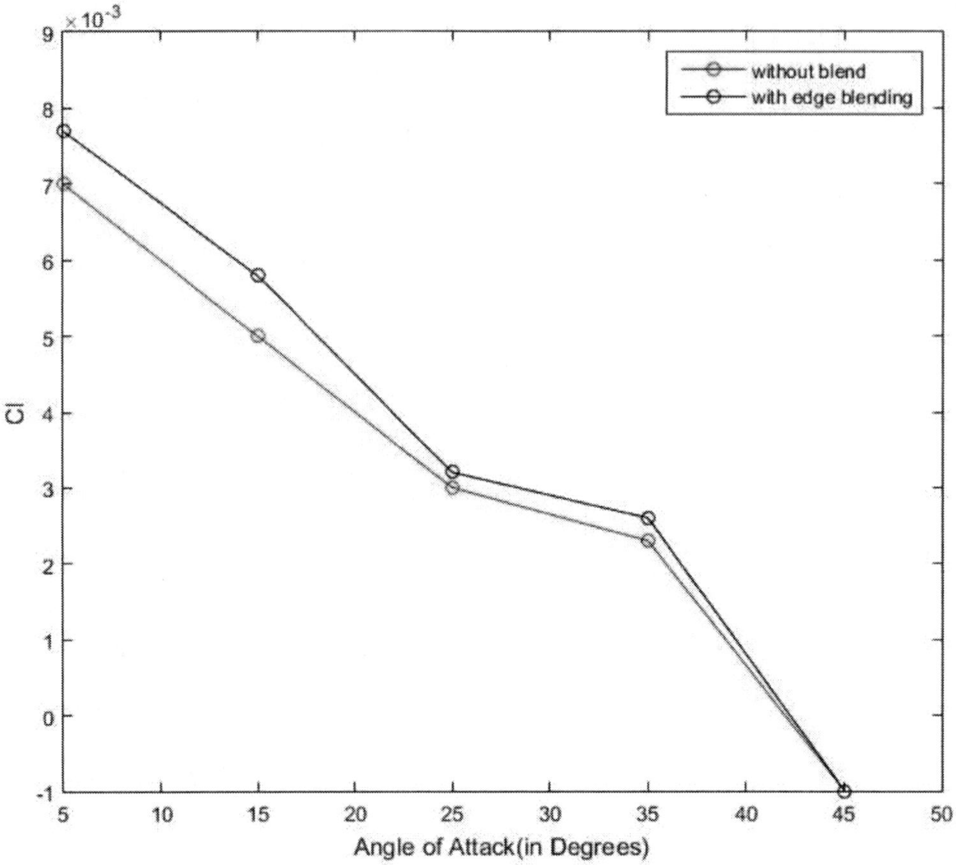

Figure 6. C_L v/s Angle of Attack

The figure 7 explains the lift to drag ratio versus the angle of attack and it can be thereby observed that the lift to drag ratio is much higher in case of blended wing as compared to normal double delta wing, the double delta wing with blend satisfies the subsonic re-entry conditions, with giving greater ability to accelerate and stabilize the wing since the flow gets enhanced as ratio increases. Moreover, the lift coefficient curve shows the non-linearity as it was expected.

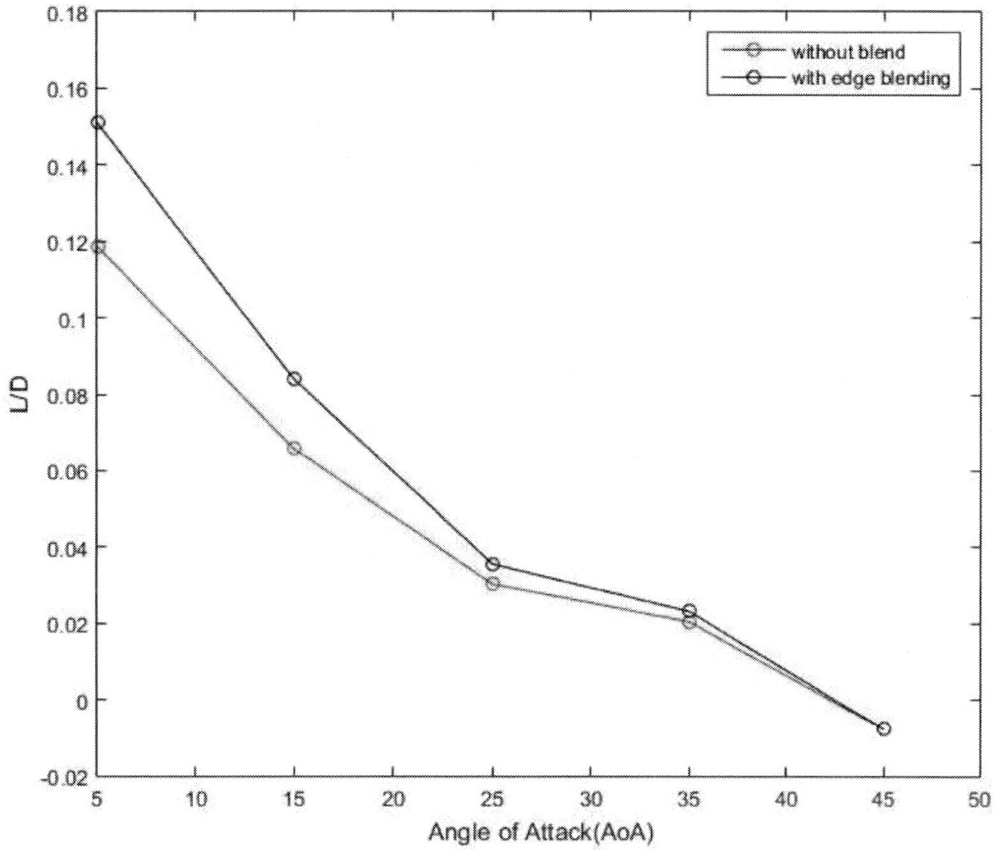

Figure 7. L/D v/s angle of attack

5. Conclusion

The edge blending exhibits greater aerodynamic efficiency as compared to normal $76^0 - 40^0$ double delta wing but limits at higher angle of attack from 45^0. The blended wing increases the stability and since the lower the drag and higher the lift more is the stability of wing. The pressure contours moreover show that the wing experiences lower pressure thereby enhancing the flow since it also cools the wing at the edges. The simulations were carried out considering the proper wind tunnel conditions and as per standard conditions the blended wing and the normal wing were simulated and the results were compared with previous research work, the blend concept showed greater advantages as compared to previous work.

Declaration of competing interest. The authors declare that they have no competing financial interest or personal relationship that could have influenced the paper.

CRediT authorship contribution statement.

Krishna Jaiswal: Conceptualization, Methodology, Analytic Calculations, Problem analysis, Investigation, Simulation, Writing- Original Draft.

Pratik Ganorkar: Investigation, Problem Defining, Software validation, Simulation, Writing- Original Draft.

Sunil Shinde: Project administration, Resources, Writing – Review and editing

References

[1] Li Qin; Sun Dong; Zhang HanXin : Detached-eddy simulations and analyses on new vortical flows over a 76/40^0 double delta wing; June 2013 Vol.56 No.6: 1062–1073, doi: 10.1007/s11433-013-5105-6

[2] Dr. P. Jeyajothiraj, Gopalsamy Murugesan;Theoretical and Experimental of Evaluation of Flow Field Over a Double Delta Wing at Supersonic Speeds

[3] LI Qin, SUN Dong& ZHANG HanXin :Detached-eddy simulations and analyses on new vortical flows over a 76/40° double delta wing

[4] Antonio Viviani, Andrea Aprovitola, Luigi Iuspa, Giuseppe Pezzella: Low speed longitudinal aerodynamics of a blended wing-body re-entry vehicle; Aerospace Science and Technology 107 (2020) 106303

[5] Gaoxiang XIANG, Chun WANG, Honghui TENG, Zonglin JIANG: Shock-Shock interactions between bodies and wings (2017) Chinese Journal of Aeronautics, (2018), 31(2): 255–261

[6] P.R Vishwanath: Aircraft viscous drag reduction using riblets, Progress in Aerospace Sciences 38 (2002) 571–600

[7] Mohammad Ghalandari, Ibrahim Mahariq, Farhad Ghadak, Oussama Accouche and Fahd Jarad: Aeroelastic optimization of High Aspect Ratio Wing with Aileron, Computers , Materials and Continua (2021)

[8] Aircraft Structures for engineering students Fourth Edition T. H. G. Megson

[9] M.R. Visball, R.E. Gordnier, Higher-order compact difference scheme applied to the simulation of a low sweep delta wing flow, in: 41st Aerospace Sciences Meeting and Exhibition, 6-9 January 2003, Reno, Nevada (AIAA Paper 2002- 620), 2003.

[10] N.G. Verhaagen, Effects of leading-edge radius on aerodynamic characteristics of 50 deg delta wings, in: Proc. of 48th AIAA Aerospace Sciences Meeting Including the New Horizons Forum and Aerospace Exposition, 4 - 7 January 2010, Orlando, Florida, AIAA 2013-0029, 2013

[11] A. Viviani, A. Aprovitola, L. Iuspa, G. Pezzella, Aeroshape design of reusable re-entry vehicles by multidisciplinary optimization and computational fluid dynamics, Aerospace Science and Technology 105 (2020) 106029

[12] Ansys fluent user's guide, 2019R1.

[13] Human Space Flight. NASA, Simulating dream chaser spacecraft aerodynamics: subsonic through hypersonic, https://www.nas.nasa.gov/SC19/demos/ demo1.htmlprettyPhoto, 2020

[14] Huang, R. et al., 2015. Design of active flutter suppression and wind-tunnel tests of a wing model involving a control delay. Journal of Fluids and Structures, 55, pp.409–427.

[15] F.X. Caradonna, C. Tung, Experimental and analytical studies of a model helicopter rotor in hover, , NASA TM-81232, 1981.

[16] Ahmed Z. Al-Garni and Farooq Saeed, "Experimental and Numerical Investigation of 65-deg Delta and 65/40-deg Double-Delta Wings in Sideslip" -28th AIAA Applied Aerodynamics Conference 2010.

Influence of Different Parameters on the Specific Wear Rate of PTFE Composites in the Steam Environment

A B Bhosale[1], M V Walame[1], Lathesh M[2]

[1]Vishwakarma Institute of Technology, Pune 411037, India

[2]Forbes Marshall Private Limited, Pune 411034, India

Abstract. Carbon filled PTFE material has been used in various engineering applications due to its properties of self-lubrication and minimal porosity. In Paper Industry carbon filled PTFE composite is used as a mechanical face seal to prevent the leakage of the steam. During the sealing application, the material is subjected to the wear hence, it is necessary to determine its wear rate to compute the life of the material when subjected to different loading conditions. Researchers have studied the tribological characteristics of carbon filled PTFE composites in dry sliding conditions as well as in a humid environment. However, wear behaviour of PTFE composite material in a steam environment is yet to be explored. In present study, the research is focused on the variation of specific wear rate of two grades of Carbon filled PTFE composites in a steam environment with the change in contact pressure and sliding velocity. In order to conduct the experiment in a steam environment, a pin on disc tribometer with a steam enclosure is designed and fabricated. The two grades used in this study are 25% Carbon filled PTFE and 35% Carbon filled PTFE. It was observed that the loss due to wear for both the materials increases with the rise in contact pressure. For 35% carbon filled PTFE, the specific wear rate decreases with the increase in sliding velocity whereas it increases with increase in sliding velocity for 25% carbon filled PTFE.

1. Introduction

In Paper Industry, the steam is passed through the rotating dryer cylinders over which the paper pulp rolls. The steam helps in maintaining the high temperature of the dryers. The high temperature dryers help in removing the moisture content from the produced paper. During the process, relative motion occurs between the rotating dryers and the stationary component. In order to restrict the steam leakage at this interface, mechanical face seals are used. The sealing material is expected to have a longer life cycle and at the same time should be able to restrict the leakage of steam. Hence, there is a need to study the tribological behaviour of the sealing material in a steam environment. Polytetrafluoroethylene (PTFE) is used in various applications due to its self-lubricating properties. PTFE is widely used for its property of low coefficient of friction however its resistance to the wear is low [1]. Studies show that the mass loss of PTFE and nylon 6 due to wear rises when the testing environment is changed from dry sliding environment to aqueous medium [2,3]. Hence polytetrafluoroethylene in its pure form cannot be used in the sealing application where the operating condition is dry saturated steam. The resistance to wear of the PTFE increases when the filler materials are added to it. Hence, the PTFE composites are used in the self-lubricating bearings and sealing applications [4,5]. K. Tanaka et al. [6] found that the carbon filled PTFE composites can be used in the applications with aqueous environment due to low water absorptivity and low coefficient of friction.

Researchers have studied the wear behaviour of materials in different environments such as hydrogen gas, inert gases, water vapour, dry sliding in air and steam. A friction and wear test rig was described by R. R. Paxton et al. [7] to study the lubrication of seal carbons. Factors affecting the wear rate such as face pressure, rubbing speed were studied and it was found that the rate of wear is significantly dependent on contact pressure as compared to that of increase in rubbing speed. PTFE-serpentine nanocomposite was slid against steel to study its self-lubricating properties by Zhining Jia et al. [8]. Parameters varied were applied load and sliding velocity and two mechanisms of wear were observed viz. adhesive and abrasive. Jaydeep Khedkar et al. [9] focused on the effect of addition of filler material (composites) of PTFE on its wear behaviour. Results showed that filler materials increase the hardness of PTFE and thereby improve its wear resistance. The transfer layer formation of PTFE composites due to different parameters like counter surface, sliding velocity was studied by HuLin Li et al. [10]. The material used in the study was PTFE polymer with copper as filler and as the contact pressure increased, the thickness of the transfer layer decreased. M. Conte et al. [11] emphasise on the wear behaviour of PTFE composites. Seven different composites of PTFE were studied and concluded that the use of composites improves the self-lubrication of the PTFE material. Anamol Sonawane et al. [12] worked upon the wear characteristics of Carbon filled PTFE. For dry sliding, an increased rate of wear was observed with the temperature rise. On the other hand, rate of wear decreased with the drop in contact pressure in case of 35% carbon filled PTFE. With the rise in the contact pressure, the rate of wear decreased for 25% carbon filled PTFE. The sliding wear characteristics of composites of PTFE was studied by H. Unal et. al [13] under dry sliding conditions. Four different filler materials were used for study of was 35% Carbon filled PTFE. It was observed that PTFE composites enhance the load carrying capacity of the pure PTFE which lowers the wear rate. Humid conditions in the presence of nitrogen gas environment was considered to study the influence of humidity on the wear of PTFE composites by Pontus Johansson et al. [14]. Wear loss of material was higher in dry conditions as compared to that of dry nitrogen. The comparison of wear loss of PTFE and its composites with change in temperature was studied by Ye Sujuan et al. [15] and concluded that wear rate increases with the rise in temperature. Yanzhen Wang et al. [16] utilised the glass fibre and carbon fibre as filler material to study the wear behaviour of the UHMWPE composite. The friction coefficient was less in water lubricated condition as compared to that of dry sliding condition. Water absorption phenomenon was also seen in the composites.

Many researchers have worked upon the tribological study of PTFE polymer with carbon as filler material; however, it is observed that less study has been done regarding the wear behaviour of carbon filled PTFE in the steam environment. This work mainly focuses on the determination of specific wear rate of two grades of Carbon filled PTFE in a steam environment with change in the contact pressure and sliding velocity. The sliding counter face used is lapped SS 431 discs.

2. Materials and Test Procedure

2.1 Materials

In this study, two grades of Carbon Filled PTFE are used viz. 35% CF-PTFE and 25% CF-PTFE. Table 1 shows the pin materials properties. The material properties were tested in a NABL approved laboratory. The tensile strength of 25% carbon filled PTFE used in this test ranges from 9.59 MPa to 10.36 MPa. The tensile strength for 35% carbon filled PTFE varies from 9.806 MPa to 17.61 MPa. The density of the materials is found to be 1.85 g/cc and 2.00 g/cc for 35% carbon filled PTFE and 25% carbon filled PTFE with shore D hardness of 65.

Table 1: Pin material properties

Material	Density (g/cc)	Tensile Strength (N/mm^2)
35% CF-PTFE (Material A)	2.0-2.14	9.806-17.61
25% CF-PTFE (Material B)	1.84-1.78	9.59-10.36

In the first stage, the carbon filled PTFE composites are made in the form of rods using the mould. These rods are then machined on CNC machines to form the pins. The pin samples are cylindrical with 10 mm diameter and 30 mm length. The counter face against which the pin slides is a stainless steel 431 disc with a lapped surface. The surface roughness of the disc is 0.268-0.495 micrometres.

2.2 Test Procedure

Figure 1: Pin on disc tribometer with steam enclosure

The testing environment for conducting the experiment is dry saturated steam conditions. Hence, it is difficult to conduct the experiments on a conventional pin on disc tribometer as it cannot withstand high steam pressure. In order to overcome this difficulty, a pin on disc tribometer with a steam enclosure is designed and fabricated for conducting the experiments as shown in figure 1. A torque sensor is used to monitor the torque readings due to the load acting upon the pin. Surface mounted RTD is used to measure the variation of temperature of the pin throughout the testing. Before the testing, the pin samples are washed and cleaned with acetone to ensure no impurities remain on the pin surface. The initial weight of the sample is measured on the Mettler Toledo weighing machine with least count of 0.1 micrograms. The pin is clamped with the pin holder and the surface mounted RTD which is inserted from top of the pin holder to measure the pin sample temperature. The disc is mounted over the disc holder using bolts and the pin is in contact with the disc at a diameter distance of 110 mm. The dead weights are put on the pin holder to get the desired contact pressure. The entire system is driven by a shaft and pulley assembly with the disc holder rotating with the driven shaft. Timing pulley is used to ensure homogeneous rotation. A Variable Frequency Drive (VFD) motor is used to regulate the rpm of the shaft. To store the temperature and the torque readings, a Data Acquisition system is installed. After the assembly of the samples on to the rig, the steam is allowed to flow into the chamber from the boiler. The required pressure is set using a Pressure Regulating Valve (PRV). The proximity sensor is installed to check the number of rotations of the disc. After the testing, the pin is removed from the pin holder and placed in the furnace at the temperature of 120°C for 2 hours to remove the adsorbed moisture from the pin. The sample is allowed to cool at ambient temperature and then the surface of the pin is cleaned with acetone. These samples after the test are weighed again. Hence the difference between the initial and final weight of the pin helps in determining the mass loss due to the wear. The test parameters for the testing in the steam environment are mentioned in table 2. During the tests of varying contact pressure, the steam

IOP Conf. Series: Materials Science and Engineering 1272 (2022) 012019 doi:10.1088/1757-899X/1272/1/012019

temperature was kept constant of 151°C (corresponding to 4 bar pressure), sliding distance of 7775 m and sliding velocity of 0.960 m/s. While varying the sliding velocity, the contact pressure was set at 0.427 N/mm^2.

Table 2: Parameters for test

Parameter	Stage 1	Stage 2	Stage 3	Stage 4	Stage 5	Stage 6
Contact Pressure (N/mm^2)	0.173	0.427	0.5523	0.753	1.372	2.748
Sliding Velocity (m/s)	0.738	0.960	1.477			

Calculation for specific wear rate is given by equation 1,

Specific Wear Rate =Loss of Volume/(Normal Load *Sliding Distance) (1)

And Volume Loss is given by equation 2,

Loss of Volume = Loss of Mass/Density (2)

Wear Loss per unit area per unit sliding distance=Mass Loss/(Contact Area*Sliding Distance) (3)

3. Results and Discussion

3.1 Results for contact pressure

Fig 2 shows the trendline for the change in specific wear rate by varying the contact pressure and maintaining constant steam temperature of 151.39 °C for material A and material B. The graph is obtained from the experimental results tabulated in table 3 and table 4. In case of material A, specific wear rate decreases with increase in the contact pressure at constant steam temperature. Non uniform transfer layer is observed on the wear track at low contact pressure indicating more abrasive wear. At higher contact pressure, a thick and uniform transfer layer is seen which acts as lubricant and hence the specific wear rate drops. The transfer layer of material A is shown in figure 4 (a) and (c). The empirical relation for the specific wear rate in terms of the contact pressure for material A is obtained as

$$W = 3.42E - 5 * P^{-0.202}$$

R^2=0.681

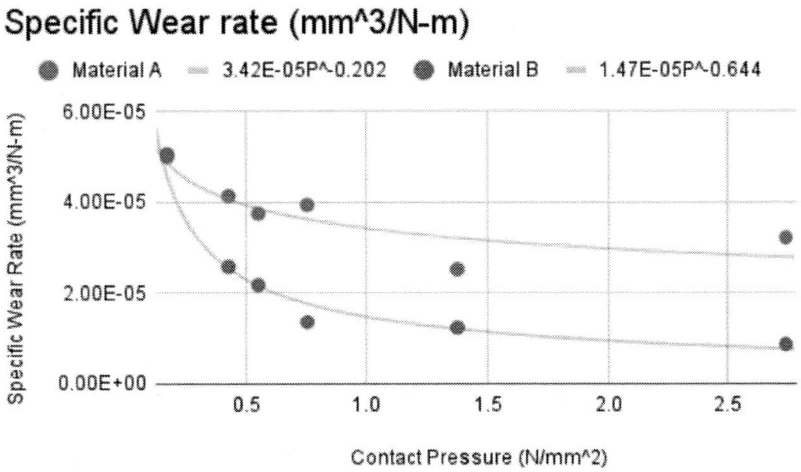

Figure 2: Variation of specific wear rate with contact pressure

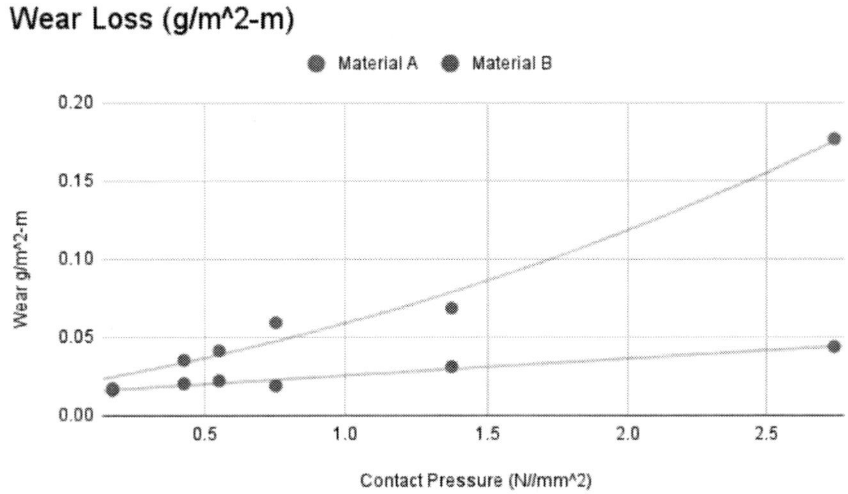

Figure 3: Variation of loss due to wear per unit area per unit sliding distance with contact pressure

In case of material B, specific wear rate decreases as the contact pressure increases from 0.173 N/mm^2 to 2.748 N/mm^2 at constant saturated steam temperature corresponding to 4 bar pressure. Transition in the formation of the transfer layer is seen as the contact pressure goes on increasing. At low contact pressure, the stick and slip type of transfer layer is seen on the wear track whereas at high contact pressure the uniform thick black film is seen. The thick transfer layer avoids the direct contact of the pin and the disc. Hence the rate of wear is decreased. Figure 4 (b) and (d) shows the transfer layers of material B. The empirical relation for the specific wear rate in terms of the contact pressure for material B is obtained as

$$W = 1.47E - 5 * P^{-0.644}$$

R^2=0.953

IOP Conf. Series: Materials Science and Engineering 1272 (2022) 012019 doi:10.1088/1757-899X/1272/1/012019

Table 3: Results for specific wear rate of Material A

Contact Pressure (N/mm^2)	Steam Pressure (bar)	Saturated steam temperature (°C)	Sliding Speed (m/s)	Sliding Distance (m)	Mass Loss (gm)	Specific Wear Rate (mm^3/N-m)	Wear g/m^2-m
0.173	4	151.39	0.960	7776	0.01059	5.00E-05	0.01733
0.427	4	151.39	0.960	7776	0.02157	4.13E-05	0.03531
0.5523	4	151.39	0.960	7776	0.02527	3.75E-05	0.04137
0.7538	4	151.39	0.960	7775	0.03624	3.94E-05	0.05934
1.3727	4	151.39	0.960	7776	0.04190	2.52E-05	0.06860
2.748	4	151.39	0.960	7775	0.10793	3.22E-05	0.17673

Table 4: Results for specific wear rate of Material B

Contact Pressure (N/mm^2)	Steam Pressure (bar)	Saturated steam temperature (°C)	Sliding Speed (m/s)	Sliding Distance (m)	Mass Loss (gm)	Specific Wear Rate (mm^3/N-m)	Wear g/m^2-m
0.173	4	151.39	0.960	7776	0.00992	5.06E-05	0.01624
0.427	4	151.39	0.960	7776	0.01246	2.58E-05	0.02040
0.5523	4	151.39	0.960	7776	0.01359	2.18E-05	0.02225
0.7538	4	151.39	0.960	7775	0.01176	1.36E-05	0.01900
1.3727	4	151.39	0.960	7776	0.01895	1.23E-05	0.03103
2.748	4	151.39	0.960	7775	0.02674	8.61E-06	0.04378

With increase in contact pressure, wear loss per unit area per unit sliding distance increased steadily up to contact pressure of 1.3727 N/mm^2 in case of material A after which a steep rise was observed in the graph. In the case of material B, a very slight increase was observed with increase in contact pressure from 0.173 N/mm^2 to 2.748 N/mm^2.

IOP Conf. Series: Materials Science and Engineering 1272 (2022) 012019 doi:10.1088/1757-899X/1272/1/012019

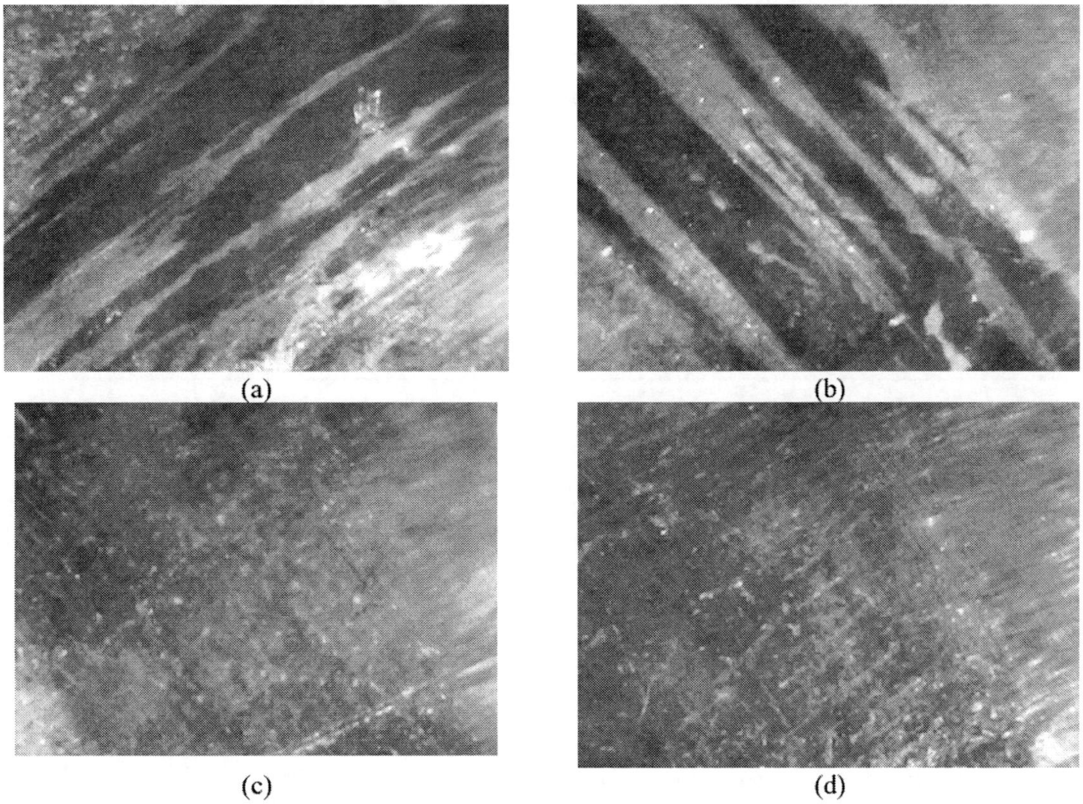

Figure 4: (a) Non uniform transfer layer: material A (b) Non uniform transfer layer: material B (c) Uniform transfer layer: material A (d) Uniform transfer layer: material B

3.2 Results for sliding velocity

Table 5 and 6 shows the variation of specific wear rate with change in sliding velocity for material A and material B respectively. The graph is plotted in figure 5 based on the data obtained in table 5 and 6. Both the materials exhibit opposite behaviour with the change in sliding velocity. For 35% carbon filled PTFE, the specific wear rate decreases with the increase in sliding velocity. On the other hand, specific wear rate increases with increase in sliding velocity for 25% carbon filled PTFE. An empirical relationship is obtained with specific wear rate as the response variable and the sliding velocity as the independent variable.

For 35% carbon filled PTFE, the equation is,

$$W = 4.52E - 5 * V^{-0.725}$$

$R^2 = 0.857$

For 25% carbon filled PTFE, the equation obtained is,

$$W = 2.73E - 5 * V^{1.13}$$

$R^2 = 0.999$

Table 5: Results for specific wear rate of material A (sliding velocity)

Contact Pressure (N/mm^2)	Steam Pressure (bar)	Saturated steam temperature (°C)	Sliding Speed (m/s)	Sliding Distance (m)	Mass Loss (gm)	Specific Wear Rate (mm^3/N-m)
0.427	4	151.39	0.738	7776	0.03165	6.06E-05
0.427	4	151.39	0.960	7776	0.02157	4.13E-05
0.427	4	151.39	1.477	7776	0.01859	3.56E-05

Table 6: Results for specific wear rate of material B (sliding velocity)

Contact Pressure (N/mm^2)	Steam Pressure (bar)	Saturated steam temperature (°C)	Sliding Speed (m/s)	Sliding Distance (m)	Mass Loss (gm)	Specific Wear Rate (mm^3/N-m)
0.427	4	151.39	0.738	7776	0.00949	1.95E-05
0.427	4	151.39	0.960	7776	0.01246	2.58E-05
0.427	4	151.39	1.477	7776	0.02062	4.27E-05

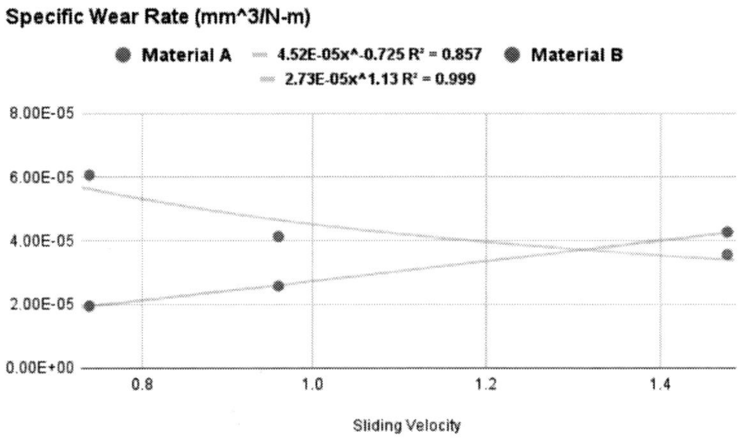

Figure 5: Variation of specific wear rate with sliding velocity

4. Conclusion

The wear characteristics of two different grades of Carbon filled PTFE composites sliding against SS 431 lapped disc was studied in the steam environment. Influence of contact pressure and the sliding velocity on the specific wear rate of PTFE composites was studied. The experimental results conclude that:

1. In the steam environment, for 35% carbon filled PTFE, decrease in the specific wear rate is observed with the rise in contact pressure to the power of -0.202. At the corresponding value of contact pressure, the loss mass due to wear for material A is more when compared to the material B.

2. For 25% carbon filled PTFE, the specific wear rate goes on decreasing to the power of -0.644. Minimum specific wear rate was observed at 2.748 N/mm^2.
3. An empirical relation for 35% carbon filled PTFE was obtained as

$$W = 3.42E - 5 * P^{-0.202}$$
$$W = 4.52E - 5 * V^{-0.725}$$

4. For 25% carbon filled PTFE the empirical relation obtained is

$$W = 1.47E - 5 * P^{-0.644}$$
$$W = 2.73E - 5 * V^{1.13}$$

5. In the steam environment, with the increase in sliding velocity, specific wear rate decreased to the power of 0.725 in the case of 35% carbon filled PTFE whereas for 25% carbon filled PTFE, rise in the specific wear rate was observed.
6. At lower contact pressure in the steam environment, the stick and slip type of transfer layer is observed for 25% as well as 35% carbon filled PTFE which further increases the specific wear rate.
7. In the steam environment, an increase in the carbon percentage from 25% to 35% as a filler material, reduces the resistance to wear of the PTFE composites.

References

[1] S. K. Biswas, Kalyani Vijayan, “Friction and Wear of PTFE- a review”, Wear, Vol-158, Issues 1-2, Elsevier, October (1992). https://doi.org/10.1016/0043-1648(92)90039-B

[2] Steinbuch, R. T., "Nylon 6 as a Bearing Materials," Wear, Vol. 5, No. 6, Nov./Dec. 1962, pp. 458-466.

[3] Pratt, G. C, "Plastic-Based Bearing," in Braithwait, E. R. ed., Lubrication and Lubricant, Elsevier, Amsterdam, 1967, pp. 377-426.

[4] A. Harnoy, Bearing design in machinery: engineering tribology and lubrication, Marcel Dekker, Basel, New York (2002). https://doi.org/10.1201/9780203909072

[5] M.W.J. Lewis, “Friction and Wear of PTFE-Based Reciprocating Seals, Lubrication Engineering”, Volume 42, Issue 3, Pages 152-158, March 1986

[6] K. Tanaka, “Friction and Wear of semicrystalline polymers sliding against steel under water lubrication”, Journal of Lubrication Technology, 102, 526 (1980). https://doi.org/10.1115/1.3251590

[7] R. R. Paxton, W. R. Shobert, Testing High Speed Seal Carbons, Asle Transactions, Vol-5, Issue-2, https://doi.org/10.1080/05698196208972474

[8] Zhining Jia, Yulin Yang, Self-lubricating properties of PTFE/serpentine nanocomposite against steel at different loads and sliding velocities, Composites: Part B, 43 (2012) 2072-2078, https://doi.org/10.1016/j.compositesb.2012.01.014

[9] Jaydeep Khedkar, Ioan Negulescu, Efstathios, I. Meletis, Sliding wear behaviour of PTFE composites, Wear, 252 (2002) 361-369. https://doi.org/10.1016/S0043-1648(01)00859-6

[10] HuLin Li, ZhongWei Yin, Dan Jiang, Liyong Jin, YuQing Cui, A study of the tribological behaviour of transfer films of PTFE composites formed under different loads, speeds and morphologies of the counterface, Wear, 328-329 (2015) 17-27, https://doi.org/10.1016/j.wear.2015.01.028

IOP Conf. Series: Materials Science and Engineering 1272 (2022) 012019 doi:10.1088/1757-899X/1272/1/012019

[11] M. Conte, A. Igartua, "Study of PTFE composites tribological behavior", Wear, 296 (2012) 568-574, https://doi.org/10.1016/j.wear.2012.08.015

[12] Anamol Sonawane, Abhijeet Deshpande, Satish Chinchanikar, Yashwant Munde, "Dry sliding wear characteristics of carbon filled polytetrafluoroethylene (PTFE) composite against Aluminum 6061 Alloy", Materials Today: Proceedings. 44 (2021) 3888-3893, https://doi.org/10.1016/j.matpr.2020.12.929

[13] H. Unal, A. Mimaroglu, U. Kadioglu, H. Ekiz, "Sliding friction and wear behavior of polytetrafluoroethylene and its composites under dry sliding conditions", Materials & Design, Vol-25, Issue-3, May 2004, 239-245, https://doi.org/10.1016/j.matdes.2003.10.009

[14] Pontus Johansson, Par Marklund, Marcus Borling, Yijun Shi, "Effect of humidity and counterface material on the friction and wear of carbon fiber reinforced PTFE composites", Tribology International, https://doi.org/10.1016/j.triboint.2021.106869

[15] Ye Sujuan, Zeng Xingrong, "Tribological Properties of PTFE and PTFE Composites at Different Temperatures", Tribology Transactions, 57:3, 382-386, (2014). http://dx.doi.org/10.1080/10402004.2013.812759

[16] Yanzhen Wang, Zhongwei Yin, Hulin Li, Gengyuan Gao, Xiuli Zhang, "Friction and wear characteristics of ultrahigh molecular weight polyethylene (UHMWPE) composites containing glass fibers and carbon fibers under dry and water-lubricated conditions", Wear, 380-381 (2017) 42-51, https://doi.org/10.1016/j.wear.2017.03.006

[17] Alan O. Lebeck, "Principles and Design of Mechanical Face Seals".

[18] Harish Hirani, "Fundamentals of Engineering Tribology with Applications".

IOP Conf. Series: Materials Science and Engineering 1272 (2022) 012020 doi:10.1088/1757-899X/1272/1/012020

Design and analysis of energy efficient wind turbine blades

Umesh Chavan[1], Dhiraj Ghode[2], Ranveer Ghorpade[2], Hritika Aacharya[2], Vaibhav Ubale[2], Kedar Urunkar[2] Nitin Satpute[3]

[1, 2] Department of Mechanical Engineering, Vishwakarma Institute of Technology Pune, Maharashtra, India
[3] Department of Mechanical Engineering, Vishwakarma University, Pune, Maharashtra, India

Abstract. Inspired by humpback whale flippers, this study presents the effect of leading-edge tubercles on the wind turbine blade. Force models for lift and drag are developed on the blade which have an important role in the wind turbine performance. The study compares the CFD simulation of fluid flow over blade with leading edge tubercles and the conventional blade. Present work also covers strength and random vibration analysis for both the blades. Furthermore, different tubercle blades have been compared for lift and drag coefficients by varying tubercle amplitude. Structural analysis shows tubercled blade is suitable from structural and strength point view. In tubercled blade overall lift to drag-coefficient ratio increased up to 8.5, whereas in conventional this ratio is about 1.7 which is found to be very low. This ratio clearly indicates that in the presence of tubercles on the blade, performance improving even in lower wind speed areas. Overall tubercled blade found to be more suitable for modern wind turbines.

Keywords: Wind energy, HAWT, humpback whale, tubercles, CFD, lift to drag ratio

1. Introduction

Turbine is a device that converts the wind energy into electrical energy. Blade is a very essential part of HAWT (horizontal axis wind turbine). It is well known that the tubercles on the whale flippers enhance its manoeuvrability by increasing the lift and reducing the drag. Betz's law states maximum power that a wind turbine can generate 59% wind kinetic energy [1]. Laurens E. Howle and M. Murray [2] reported blades with smooth leading edges, blades with leading-edge tubercles demonstrated enhanced performance. Watts and Fish [3] examined the performance of a low aspect ratio wing with leading-edge protuberances. A two-dimensional aerofoil with a humpback whale flipper results in a significant loss of lift and increase in drag [4, 5]. J.H. Chen, Li, V.T. Nguyen [6] through their experimental tests and CFD analysis concluded that addition of protuberances does not have a significant effect of the lift but the drag decreased drastically. Goruney and Rockwell [7] studied the effect of TLE on a swept delta wing. H. Johri, C. Henoch, D. Custodio, A. Levshin [8] studied the effect of leading edge protuberances considering variation of amplitude and wavelength. Aftab SMA, Ahmad KA [9] compared spherical and sinusoidal tubercle leading edge. They reported that that smaller amplitude tubercles performed better for both sinusoidal and spherical tubercle leading edge ,there is significant reduction in maximum lift/drag, but the reduction in maximum lift and drag is less in case of spherical tubercle leading edge than sinusoidal tubercle leading Edge. Halefom Kidane, Gebremeskel Tekle [10] studied design and analysis of 5 kW wind turbine blade for rural and remote areas. Norén-Cosgriff, Karin and Kaynia, Amir [11] presented estimation of natural frequencies and damping using dynamic field data from an offshore wind turbine.The main objective of this work is to analyse the conventional wind turbine blade as well as the turbine blade with tubercles inspired by the humpback whale. Force models for lift and drag are developed on blade which have an important role in the wind turbine performance. The study compares the CFD simulation of fluid flow over blade with leading edge tubercles and the conventional blade.

2. Design of blade

2.1 Design Parameters

The blade is designed for a horizontal axis wind turbine having a power capacity of 4 KW. The rotor radius has been calculated using the parameters such as power coefficient (Cp = 0.4), drive train coefficient (nd = 0.9), Velocity of the wind in flow (V = 7.89m/s), generator efficiency (ng = 0.9). The number of blades are assumed to be 3 for the wind turbine. The efficiency of a 3-blade turbine is much more than the efficiency of a 2-blade turbine. But for turbines with blades more than three, the efficiency does not increase drastically whereas the overall weight increases. Therefore, an optimum number of 3 blades is taken for the design. The wind velocity is calculated to be 7.89 m/s. the average velocity proposed at a height of 10m is 4.65 m/s, turbine height 24m. V1= wind speed measured at the reference height H1 (known wind velocity), V2= wind speed estimated at height H2=24m (unknown velocity). α=Ground surface friction coefficient = 0.23

$$V(24) = V(10) * \left[\frac{H2}{H1}\right]^{0.23} \quad \ldots\ldots(1)$$

Tip speed ratio (TSR), for a water-pumping windmill, for which greater torque is needed, TSR lies in the range $1<\lambda<3$. For electrical power generation, the TSR lies in the range $4<\lambda<10$. The higher speed machines use less material in the blades and have smaller gearboxes, but require more sophisticated aerofoils [10].The aerofoil NACA 63-215 was analysed in the software Q blade. Following graph shown in fig. 1 obtained and input parameters are mentioned in table 1.From a graph of Cl/Cd vs. α, the highest ratio of Cl/Cd was obtained at angle of attack equal to 6.5°. The ratio Cl/Cd being highest is a desirable condition. Similarly, the power coefficient was found out from the results of Q blade which is 0.4.

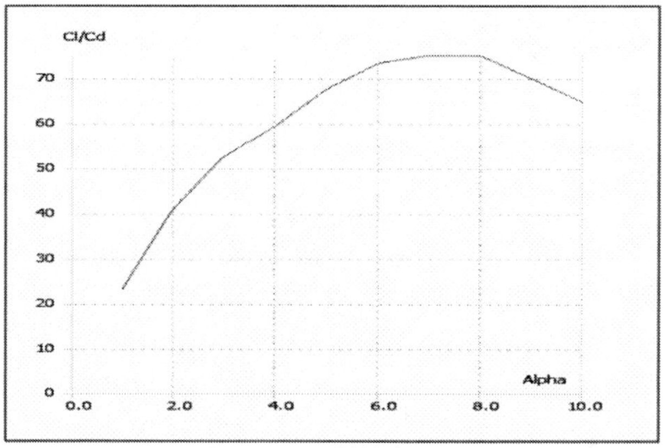

Figure 1: Cl/Cd vs α

Table no. 1: Design input parameters

Parameter	Value	Parameter	Value
Power	4 kW	Tower height	24 m
Rotor Radius	3.6 m	Generator efficiency	0.9
Number of blades	3	Drive train efficiency	0.9
Wind Velocity	7.89 m/s	Coefficient of lift design	0.9
Tip Speed Ratio (TSR)	7	Angle of attack	6.5
Aerofoil type	NACA 63-215	Power coefficient	0.4

The turbine blade was divided into 10 elements. For each element, the chord length and twist angle were calculated. Using these values, the turbine blade was modelled in in Solid works
Rotor diameter:

$$R = \left(\frac{2P}{Cp*nd*ng*V^3*\rho}\right)\text{^}2 \text{...................(2)}$$

Cp = Power coefficient or capacity factor, ŋd = Drive train coefficient
V = velocity (The velocity of the wind in flow), ŋg = Generator efficiency
Substituting in above formula

$R = \left(\frac{2*4000}{0.4*0.9*0.9*7.89^3*1.225}\right)\text{^}2$, R= 3.6m

Blade is divided in 10 elements
Cord length:

$$\lambda r = \lambda\left(\frac{ri}{R}\right) \text{ ....(3)}$$

Where, λ= local tip speed ratio, R= rotor radius, ri = midpoint radius
Ψi = the angle of relative wind of the aero foil

$$\Psi i = (2/3)*tan^{-1}(1/\lambda r) \text{ ....(4)}$$

$$\Psi i = \alpha + \theta \text{ .... (5)}$$

Where, α = angle of attack, ɵ = twist angle
Cord length

$$C = \left(\frac{8*\pi ri}{\beta*Cid}\right)(1 - \cos \Psi i) \text{ ....(6)}$$

C_{id} = Coefficient of lift design, β = Number of blades
Sample Calculation:
For element 1:

$$ri = R/(2*10) = 0.18m$$
$$\lambda r = 7*(0.18/3.6) = 0.35$$
$$\Psi i = (2/3)*tan^{-1}(1/0.35) = 47.13^0$$
$$C = \left(\frac{8*3.14*0.18}{3*1.3}\right)(1 - \cos 47.13) = 0.37m$$
$$\theta = \Psi i - \alpha = 47.13 - 6 = 41.13^0$$

Similarly, the calculations are done for remaining elements and the data is given in table 2

Table no 2: Blade element calculations

Element	ri (m)	λr	Ψi (⁰)	C (m)	θ (⁰)
2	0.54	1.05	29.06	0.43	22.5
3	0.9	1.75	19.83	0.34	13.3
4	1.26	2.45	14.80	0.26	8.30
5	1.62	3.15	11.74	0.21	5.24
6	1.98	3.85	9.70	0.18	3.20
7	2.34	4.55	8.26	0.15	1.76
8	2.7	5.25	7.189	0.13	0.60
9	3.06	5.95	6.36	0.12	-0.13
10	3.42	6.56	5.70	0.10	-0.79

2.2 Bending Moment

The lift force that propels the turbine round is distributed along the blade roughly proportionally to the local radius, i.e., there is more force near the root than near the tip. The blade bends as a result of the lift force [3].
Flap wise bending moment:

IOP Conf. Series: Materials Science and Engineering 1272 (2022) 012020 doi:10.1088/1757-899X/1272/1/012020

The flap-wise bending moment on single blade, M_B for a turbine with B blades is:

$$M_B = T * 2/B * 3*R \text{ ....(7)}$$

Where R = radius, T = torque, B =number of blades
The mean torque (T) is the power divided by the rotational speed.

$$T\ (Nm) = Power\ /Angular\ velocity \text{ ....,,,......(8)}$$

Angular velocity (ω) = Tip Speed Ratio (TSR * Wind speed/rotor radius)
Edgewise Moment
The bending moment in the edgewise direction of single blade (M_Z), is torque divided by the number of blades.

$$M_Z = torque\ /Number\ of\ blades \text{ ....(9)}$$

The flap wise bending moments of various blade elements: The aerodynamic loads cause the flap wise bending moment, which can be calculated using BEM theory [10].

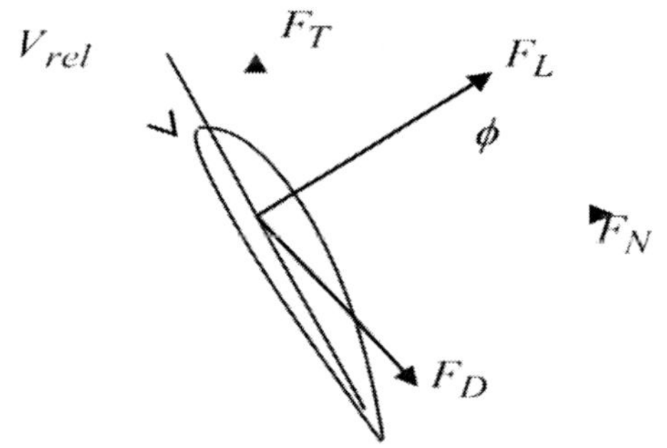

Figure 2: Forces acting on wind turbine blade [1]

EWBM = Edge wise bending moment,
FWBM = flap wise bending moment.
From the fig.2 free body diagram, the forces can be calculated by the following formulas

$$F_N = F_L \cos\phi + F_D \sin\phi \text{ ....(10)}$$

$$F_T = F_L \sin\phi - F_D \cos\phi \text{ ....(11)}$$

$$FL = ½ [Cl\rho R^2V^2cl] \text{ .... (12)}$$

c = Chord length and l = Length of the blade element

$$FT = ½ [C_T\rho\pi\ R^2V^2] \text{ ....(13)}$$

2.3 Model of blade

The turbine blade is modelled in Solid works software fig.3 and fig.4. The aerofoil shape NACA 63-215 was imported in the software. For element 1 the aerofoil shape was scaled to obtain a chord length of 0.37m. The twist angle for the element is 41.13°. The aerofoil was rotated by the respective twist angle. Similarly, the aerofoil for all the other elements was imported for their respective chord length and twist angle. The loft feature in Solid works was used to connect all the aerofoil and create a 3D structure.

Figure 3: Arrangement of aerofoils

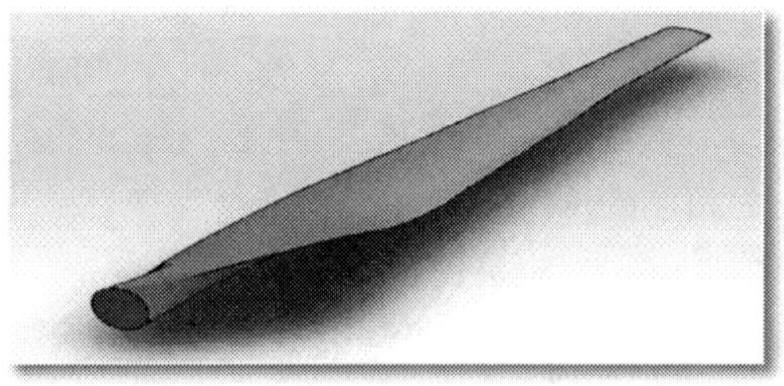

Figure 4: Wind turbine blade with straight edge

2.4 Tubercle blade modelling

The CAD models shown in fig.5 are modelled in Solid works software. The tubercles are designed on previously modelled conventional turbine blade. The amplitude of tubercles is 14% of the chord length. Ideally inter tubercular distance on whale fins is 8-10% of the span length of the wing (ref). The tubercles designed in proposed model has inter tubercular distance as 10% of the span length of the blade. A tubercle is added between each element except for last one.

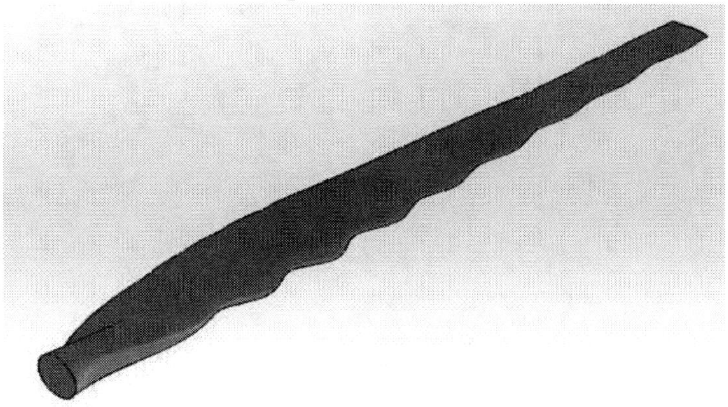

Figure 5: Wind turbine with leading edge tubercles

3. Analysis

Fluid flow analysis has been conducted over both the blades in Ansys software as shown in fig.6. The k-ξ SST model of velocity which represents the equations of turbulent flow has been used for CFD analysis. It is assumed that the blade is enclosed in a cubical enclosure, which acts as a wind tunnel. Boundary conditions are that the inlet velocity is 7.89 m/s and the outlet pressure is the atmospheric pressure.

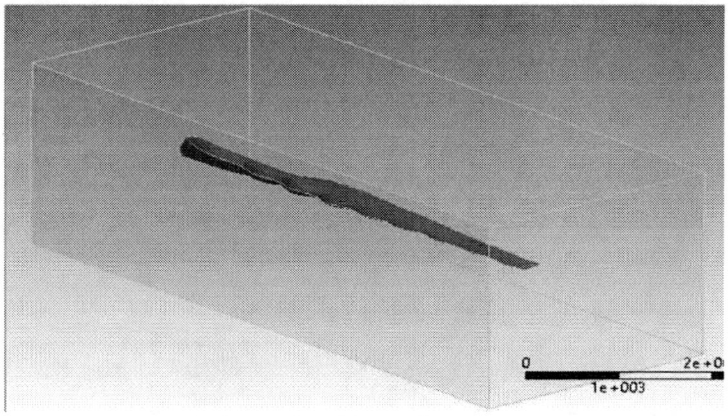

Figure 6: CFD Analysis setup

Both the blades have been analysed for equivalent stress and deformation in Ansys software. The material chosen for blades is E glass fibre. The blade is loaded with edgewise bending moment, flap-wise bending moment and pressure acting which has been found out through CFD analysis. Dynamic characteristics of wind turbine blades are largely affected by environmental loads from wave and wind action. The nature of this wind excitation is random. The fatigue load due to this random vibration is one of the important factors to consider. Hence, random vibration analysis has been carried out on both the blades. The loading data that is PSD acceleration, has been taken from Noren-Cosgriff's [11] on estimation of natural frequencies of offshore wind turbine.

4. Results and discussions

The drag coefficient showed a tremendous dip on the tubercled blade as shown in fig.7. The drag coefficient was about 1.2 for the conventional blade, whereas in the tubercled blade shown in fig. 8 stopped down to 0.2.The overall lift-coefficient to drag-coefficient ratio increased from 1.667 for conventional blade to 8.5 for the tubercled blade, nearly 5x increase shown in fig. 9 and fig.10.

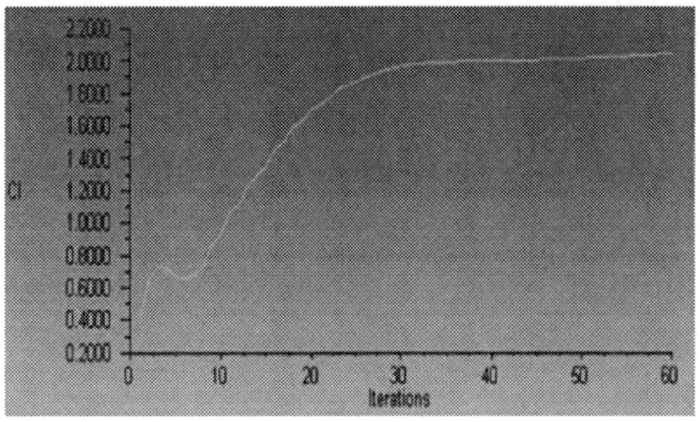

Figure 7: Lift Coefficient for conventional blade

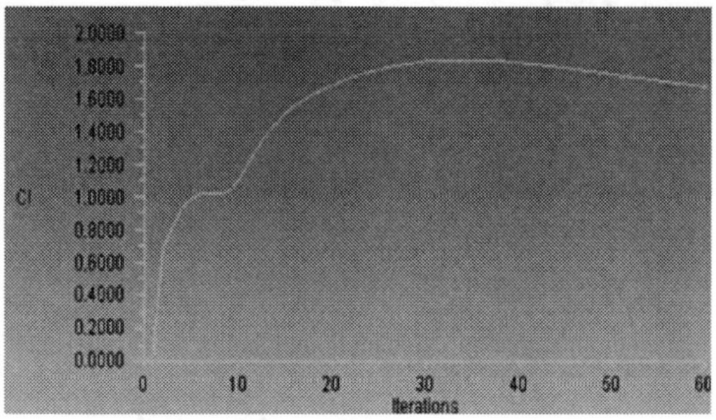

Figure 8: Lift Coefficient for tubercled blade

There are several factors which influence the drag and lift forces, but the lift coefficient-to-drag coefficient ratio clearly indicates that the overall lift on the blade increases and the drag decreases in the presence of tubercles on the blade, thereby improving its performance even in lower wind speed areas. When the air flow is divided, the air flowing on the top of the aerofoil has to cover more distance than the air flowing on the bottom side in the same amount of time. This implies that the air flowing above the aerofoil has a greater velocity and that is verified through the results fig.11 and fig. 12. The lift coefficients of the two types of blades were found out using CFD analysis in Ansys, for the given angle of attack. The lift coefficient for the conventional blade was about 2, obtained over 60 iterations, whereas for the tubercled blade, it was found to be about 1.7.

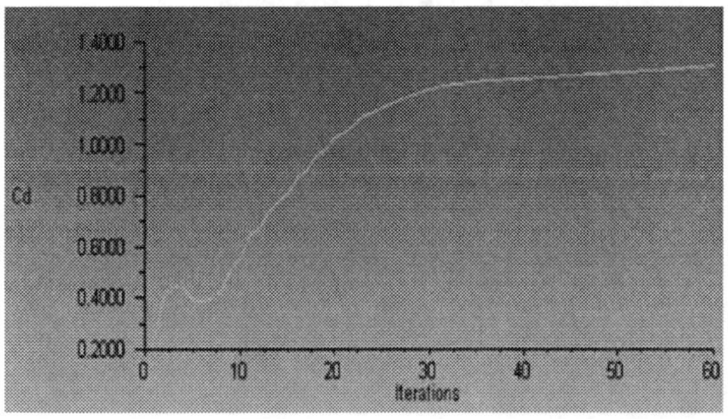

Figure 9: Drag coefficient for conventional blade

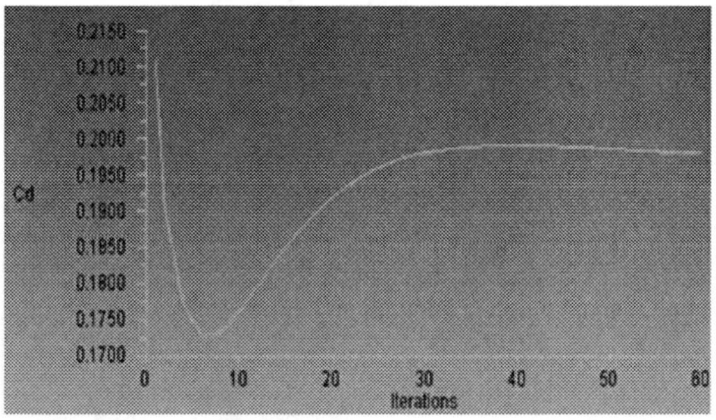

Figure 10: Drag Coefficient for tubercled blade

Figure 11: Velocity streamline conventional blade

Figure 12: Velocity streamline tubercled blade

Bernoulli's principle states that fast moving air has a lower pressure as compared to slow moving air. This implies that the pressure on the bottom surface should be higher as compared to the lower surface. This distribution of pressure is the cause for the lift and drag forces. Pressure contours shown in fig.13 and fig. 14 implies that there is a significant pressure difference between the pressure on upper and lower sides of the wind blade and pressure on the lower side is significantly high. The average overall pressure acting on the upper profile of the conventional blade is about 0.786 Pa, as visible from the pressure contours. Similarly, the overall pressure on the tubercled blade was about 12.4 Pa. This pressure is relative to the working pressure, which in this case, is atmospheric pressure. These values indicate that upper profile of the tubercled blade has an even lower pressure than conventional blade, which assists in increasing the Lift force due to a larger pressure difference between the two profiles.

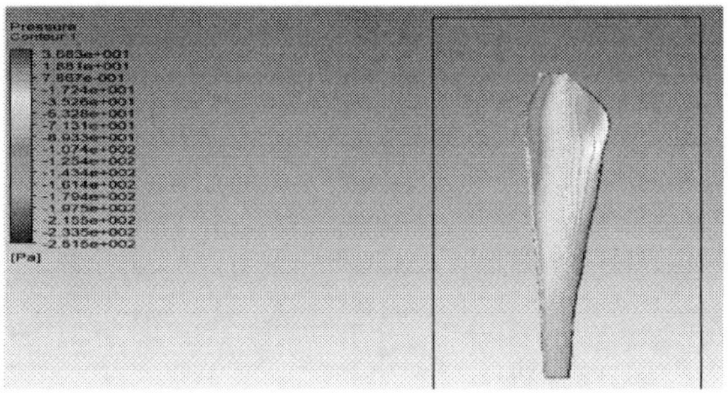
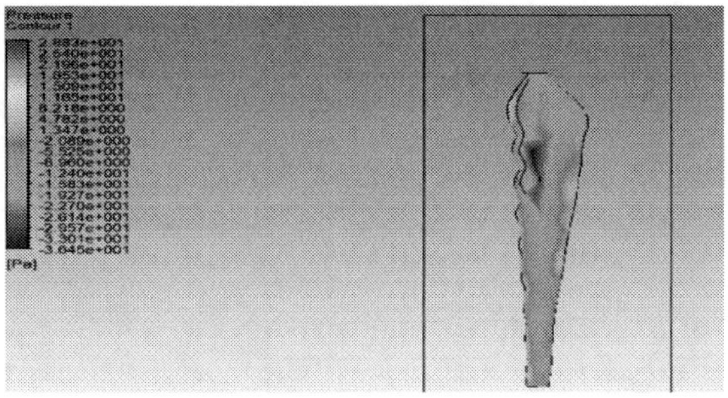

Figure 13: Pressure contours on upper profile of conventional blade **Figure 14:** Pressure contours on upper profile of tubercled blade

The average overall pressure acting on the upper profile of the conventional blade is about 0.786 Pa, as visible from the pressure contours. Similarly, the overall pressure on the tubercled blade was about -12.4 Pa. This pressure is relative to the working pressure, which in this case, is atmospheric pressure. These values indicate that upper profile of the tubercled blade has an even lower pressure than conventional blade, which assists in increasing the lift force due to a larger pressure difference between the two profiles.

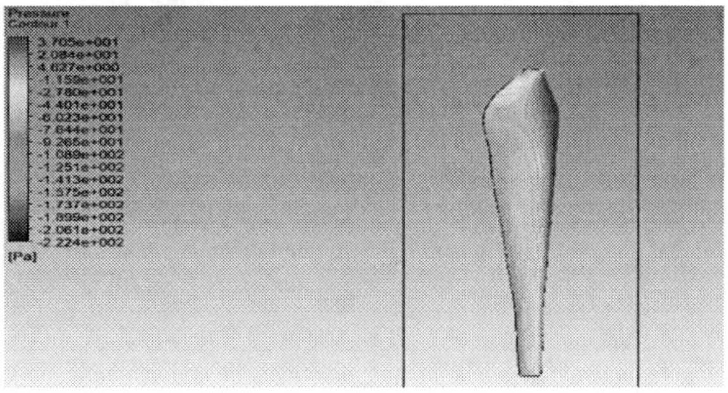
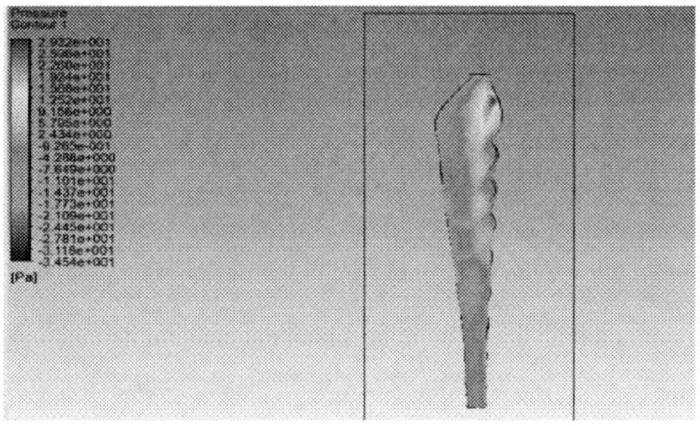

Figure 15: Pressure contours on lower profile of conventional blade **Figure 16:** Pressure contours on lower profile of tubercled blade

Fig. 15 and fig. 16 shows average overall pressure acting on the lower profile of the conventional blade is about 4.627 Pa, as visible from the pressure contours. The maximum pressure being 37.05 Pa, on the leading edge. Similarly, the overall pressure on the tubercled blade was about 5.79 Pa, maximum pressure being 29.32 Pa over a small patch near the tubercle. Even though the maximum pressure on the tubercled blade is lower than conventional blade, the overall pressure acting on the lower profile is higher. This in turn, increases the overall pressure gradient between the two profiles, giving an overall increased lift in the case of tubercled blade. Furthermore, five different tubercled blades were designed with amplitude modification, with scale factors 0.9, 0.95, 1, 1.02, 1.05 respectively. The original blade with scale factor 1 had an amplitude which is 14% of the cord length, from scale factor 0.9 to 1.05 the amplitude % goes on decreasing shown in fig.17 and fig.18.

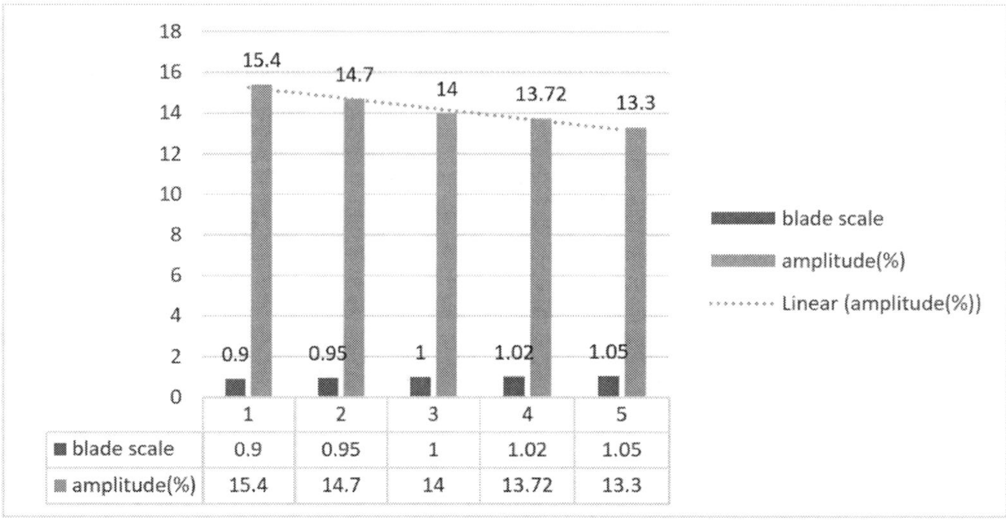

Figure 17: Corresponding amplitude and scale factor of tubercles.

Analysis was done on all five blades; it's drag and lift coefficient was found out. The lift coefficient decreases as the amplitude of the tubercles increase whereas the drag coefficient increases with increase in tubercle amplitude.

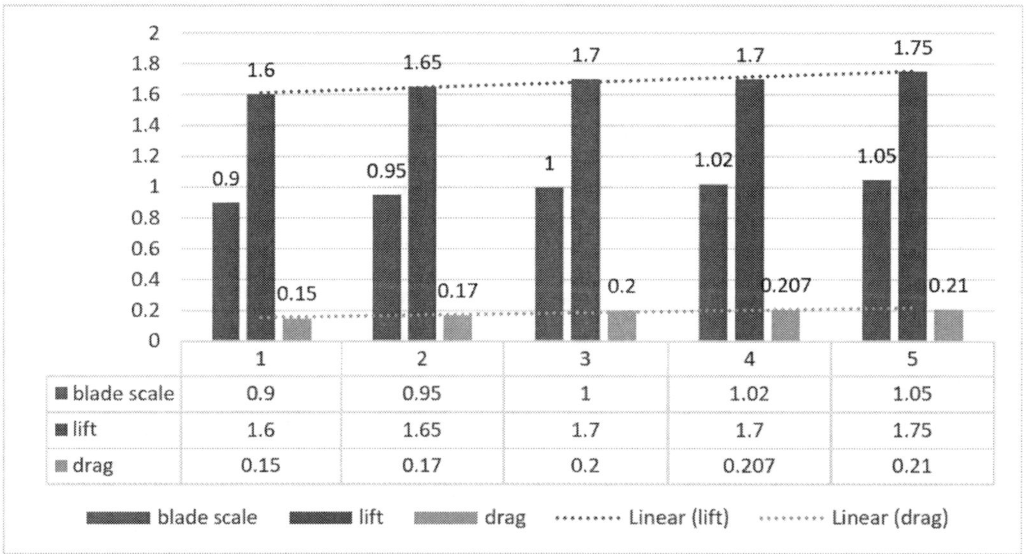

Figure 18: Lift and drag coefficients for tubercle blade with increasing amplitude

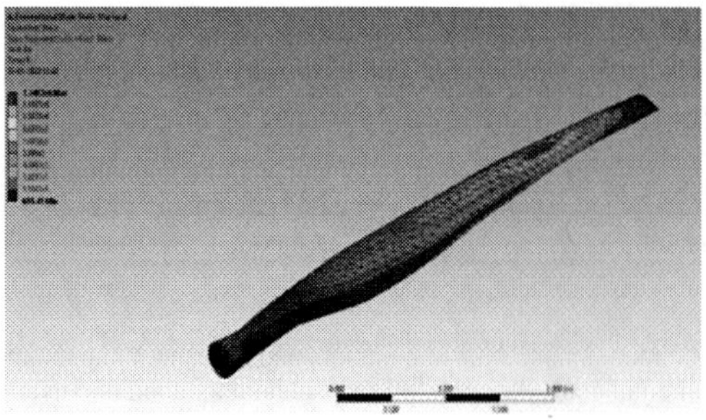

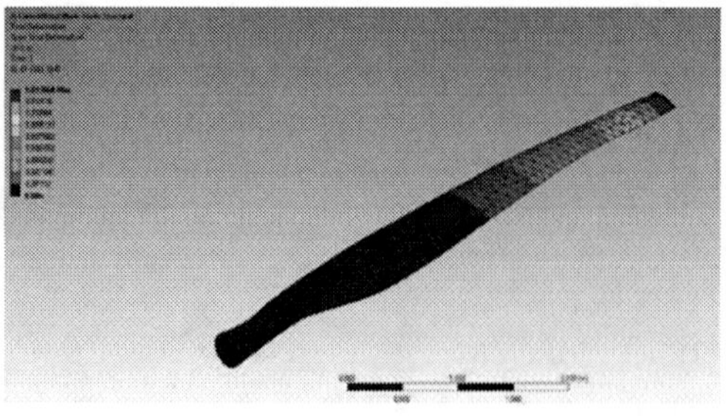

Figure 19: Equivalent stress conventional blade **Figure 20:** Total deformation conventional blade

IOP Conf. Series: Materials Science and Engineering 1272 (2022) 012020 doi:10.1088/1757-899X/1272/1/012020

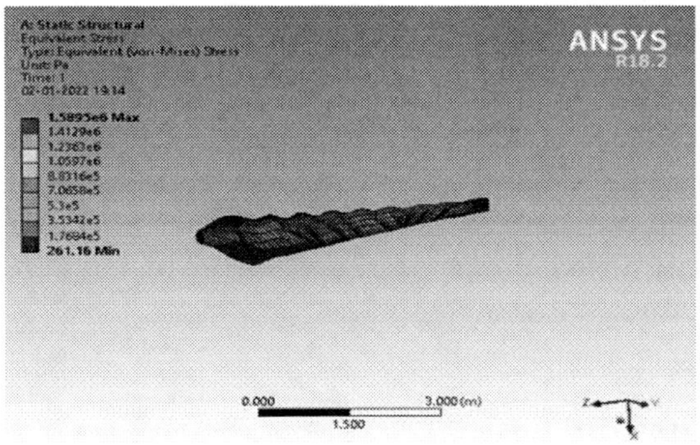

Figure 21: Equivalent stress on tubercled blade

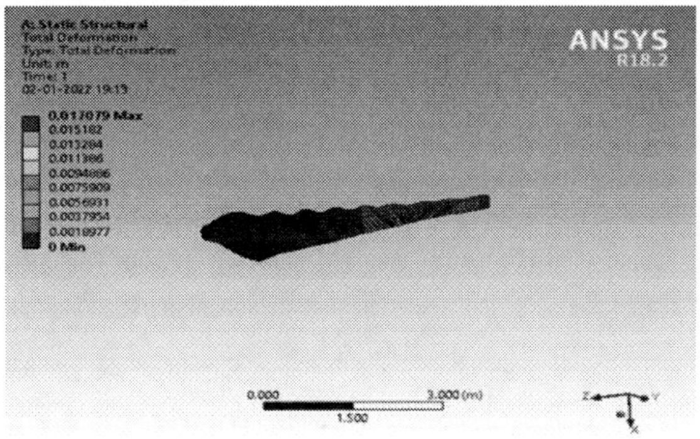

Figure 22: Total deformation of tubercled blade

Both the blades are analysed for static loads shown in fig. 19 to fig. 26. The stress developed in tubercled blade in comparison with conventional blade is on slightly higher level but well within the yield stress value of the material. The Maximum stress developed in conventional blade due to EDBM, FWBM and acting pressure is 1.36 Mpa and that in tubercled blade is 1.58 Mpa. The overall stress in conventional turbine blade is 0.44 Mpa and in tubercle blade is 0.53Mpa. The deformation in both the blades is almost same. The maximum stress in tubercled blade is found near tubercle trough and can be accounted due to decrease in blade span. The stress values in both the blades are well under yield limit with a factor of safety of 3. Below are the results showing the dynamic stress and deformation on conventional and tubercled blade.

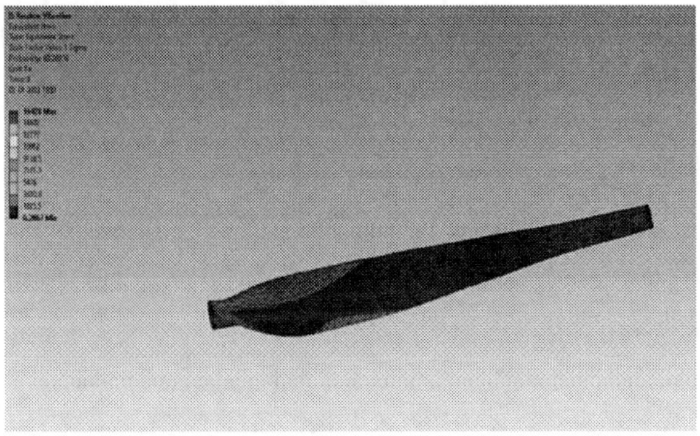

Figure 23: Equivalent stress conventional blade

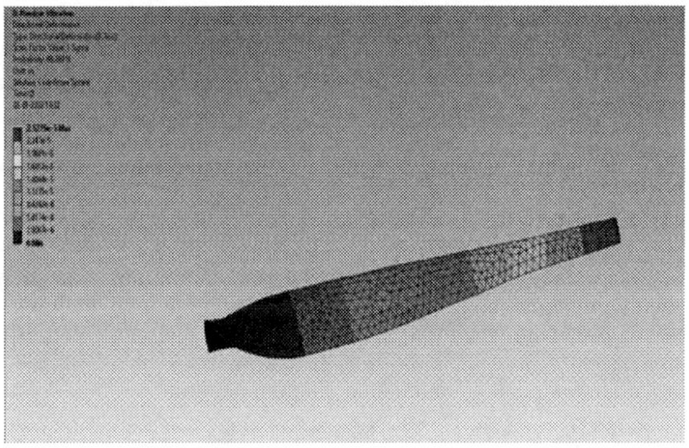

Figure 24: Total deformation conventional blade

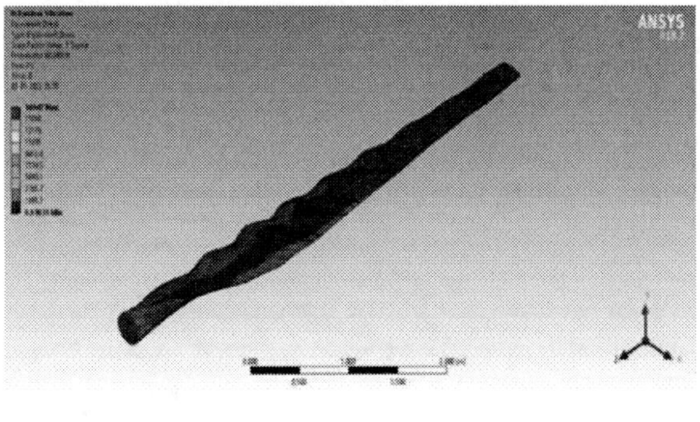

Figure 25: Equivalent stress tubercled blade

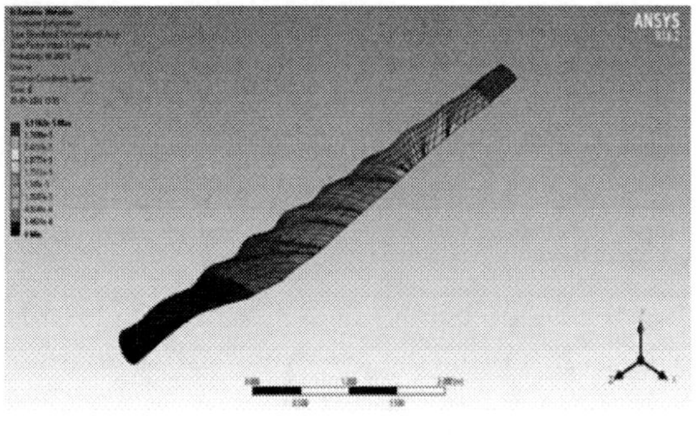

Figure 26: Total deformation tubercled blade

The stress due to random vibration in conventional blade is on lower side with overall stress throughout the span of blade being around 3.6 Kpa and with maximum stress of 16.42 Kpa and that in tubercle blade is 3.76 Kpa and maximum stress reaching value of 16.94 Kpa

5. Summary of results and discussion:

Initially, blade with straight leading edge had been designed and later tubercles of aspect ratio 1.5 have been added to the blade. Fluid flow simulation has been conducted over both the blades. The difference in results has been observed because the fluid flow becomes turbulent and vortices are formed as the flow passes over the tubercles. As flow becomes turbulent much earlier, the boundary layer thickness decreases and it stays attached to the blade for a longer period of time. Hence, the wake region formed is much lesser due to which there is significant decrease in the drag coefficient.

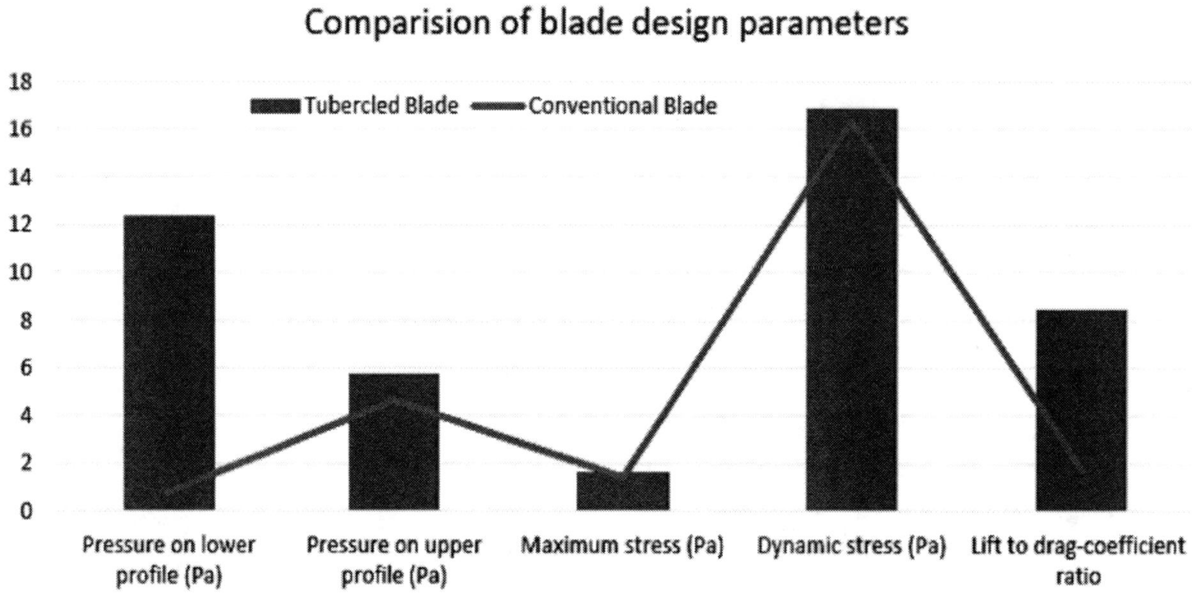

Figure 27: Comparison of blade design parameters

Fig. 27 shows the comparison of conventional and tubercled blade design parameters. The average pressure in conventional blade upper and lower profile found to be 0.8 Pa and 4.6 Pa respectively. Similarly, average pressure in tubercled blade upper and lower profile are about 12.4 Pa and 5.79 Pa respectively. The maximum stress developed in conventional blade is 1.36 Mpa, whereas in tubercled blade it is 1.58 Mpa which is below the allowable limit. Stresses under dynamic loads and random vibrations in conventional and tubercled blade found to be almost same 16.42 Kpa and 16.94 Kpa respectively. In tubercled blade overall lift to drag-coefficient ratio increased to 8.5, whereas in conventional blade lift to drag-coefficient ratio is about 1.7 which found to be very low. This analysis shows tubercled blade is more suitable.

6. Conclusion

The effects of tubercles on the leading edge of NACA 63-215 aerofoil have been studied through CFD analysis. In tubercled blade overall lift to drag-coefficient ratio increased to 8.5, whereas in conventional blade lift to drag-coefficient ratio is about 1.7 which found to be very low. The results of simulation showed less difference in lift coefficient but the drag coefficient reduces considerably due to the tubercles. The lift to drag ratio has increased by about 4 times after the addition of tubercles. This ratio clearly indicates that the overall lift on the blade increases and the drag decreases in the presence of tubercles on the blade, thereby improving its performance even in lower wind speed areas. The results also indicate that the amplitude of the tubercles has a significant role as compared to the wavelength in reducing the drag. The wavelength plays a major role in flow separation. As amplitude of the tubercles increases the drag coefficient decreases but this also results in decrease in blade width which results in decrease in strength of the blade. This analysis shows tubercled blade is suitable from structural and strength point view. Overall tubercled blade is more suitable for modern wind turbines.

References

[1] Martin O. L. Hansen, Aerodynamics of Wind Turbines, second ed., Earthscan, 2008.
[2] Murray, M. M., Miklosovic, D. S., Fish, F. E., and Howle, L., "Effects of Leading Edge Tubercles on a Representative Whale Flipper Model at Various Sweep Angles," Proceedings of Unmanned Untethered Submersible Technology (UUST), UUST05, Autonomous Undersea Systems Inst., Lee, NH, Aug. 2005.
[3] Watts, P., and Fish, F. E., "The Influence of Passive, Leading Edge Tubercles on Wing Performance," Proceedings of the Twelfth International Symposium on Unmanned Untethered Submersible Technology (UUST), UUST01, Autonomous Undersea Systems Inst., Lee, NH, Aug. 2001.
[4] Miklosovic, D. S., Murray, M. M., Howle, L. E., and Fish, F. E., "Leading-Edge Tubercles Delay Stall on Humpback Whale (Megaptera novaeangliae) Flippers," Physics of Fluids, Vol. 16, No. 5, pp. L39–42. doi:10.1063/1.1688341, May 2004
[5] Stein, B., and Murray, M. M., "Stall Mechanism Analysis of Humpback Whale Flipper Models," Proceedings of Unmanned Untethered Submersible Technology (UUST), UUST05, Autonomous Undersea Systems Inst., Lee, NH, Aug. 2005.
[6] J.H. Chen, S.-S. Li, V.T. Nguyen, The effect of leading edge protuberances on the performance of small aspect ratio foils, 15th International Symposium on flow visualization, ISFV15-040-S21, June 25-28, 2012, Minsk, Belarus
[7] Goruney T. and Rockwell D., "Flow past a delta wing with a sinusoidal leading edge: near-surface topology and flow structure", Experiments in fluids, vol. 47, no. 2, pp. 321–331, 2009.
[8] H. Johari, C. Henoch, D. Custodio, A. Levshin, Effects of leading-edge protuberances on airfoil performance, AIAA Journal, 45 (2007), 2634–2642, 2007
[9] Aftab SMA, Ahmad KA CFD study on NACA 4415 airfoil implementing spherical and sinusoidal Tubercle Leading Edge. PLoS ONE 12(8): e0183456. https://doi.org/10.1371/journal.pone.0183456 PMID: 28850622, August 2017
[10] Halefom Kidane, Gebremeskel Tekle, Design and Analysis of 5 kW wind Turbine Blade for Rural and Remote Areas Institutions in Ethiopia: Case of Degua Warren Kebele, Journal of Applied Mechanical Engineering, Vol. 9 Iss. 3 No: 334 DOI: 10.35248/2168-9873.20.9.334, October 2020
[11] Norén-Cosgriff, Karin & Kaynia, Amir. (2021). Estimation of natural frequencies and damping using dynamic field data from an offshore wind turbine, Marine Structures, Volume 76, 102915, https://doi.org/10.1016/j.marstruc.2020.102915, March 2021

IOP Conf. Series: Materials Science and Engineering 1272 (2022) 012021 doi:10.1088/1757-899X/1272/1/012021

Study of active magnetic bearings (AMB) on non–synchronous rotors

Nabam Teyi*, Sandeep Singh

Department of Mechanical Engineering, North Eastern Regional Institute of Science & Technology (NERIST), Nirjuli–791110, Arunachal Pradesh, India.

**nbt@nerist.ac.in*

Abstract. A rotor is never free from vibration owing to constant centrifugal force acting on it while rotating. A practical rotor has inherent mass unevenness flaw, or bearing misalignment flaw, or self-weight bow flaw, or presence of discontinuity or cracks flaw. If the vibrations are not controlled within limits, the rotor's purpose of existence is lost. These vibrations can be synchronous if the frequency of vibration is equal to, or multiple of, the speed of rotation of the rotor. Or may be non-synchronous if the frequency of vibration of the rotor is not a multiple of its rotational frequency. Active magnetic bearings (AMBs) have been in use for vibration control, vibration suppression, rotor stability and condition monitoring of rotordynamic systems, since many years. A lot of authors have performed various numerical and experimental works in an AMB–rotor configuration and reported their findings. The goal of this paper is to report and present the effect of AMB solely on non–synchronous rotors for the benefit of the scientific community. Firstly, this paper briefly introduces the constructional features on AMB–rotor system. Then, the gist of every paper is written, as well as specification of the work provided in tabular form. Finally, some comments on limitations and future applications are made. The paper is separately sectioned into numerical investigations, and experimental investigations.

Keywords: Active Magnetic Bearing (AMB), Non–Synchronous Rotors, Rotor Flaws, Active Vibration Control, Condition Monitoring.

1. Introduction

Any recognizable machinery is essentially a collection of a few links that move relative to one another and are joined together by certain kinematic pairings. The system's mechanism is established by the formation of connections via which specified and constrained motion transfer occurs, and the exchange of energy at the input and output of such systems transforms them into machines. Every big mechanical machine we are familiar with is, like automobiles, turbines, and ships are nothing more than a collection of motion transmission systems, with spinning shafts playing a significant part. Smaller appliances, such as fans, washing machines, and mixer grinders, are simple applications of a spinning shaft with the requisite blades on its surface. A rotor by definition is a revolving shaft with any linked unit on its surface. That means, a rotor unit is a shaft that has an extension, attached to its axis. And since all machineries include motion transfer from rotary to linear or vice versa, a rotating shaft or a rotor is fundamental to every machine system. Despite the fact that machines are carefully engineered and are robust, shafts might acquire defects without warning. Teyi & Singh (2022) reported that rotor shafts that are defective have mismatched bearings, shaft bow, mass imbalance, or cracks. Engineers are always concerned with a rotor's operational integrity and conformity to the central axis, both at rest and in motion. Ideally, the disc or shaft should not deviate off its rotational axis under all conditions, which does not occur in actuality. As a result, bearing support for a spinning shaft is an essential design issue for any engineer when building a rotor system. Because of the extremely fast rotation of rotors in most applications, contacts forming conventional bearings have given way to non-contact active magnetic bearings. Active magnetic bearing is a form of magnetic bearing that uses closed–loop control to provide non–contact support for a shaft. These bearings employ magnetic forces that are at least partially created by actively controlled electromagnets. Section 2 goes into further insight about AMBs. A typical rotor is seen in figure 1.

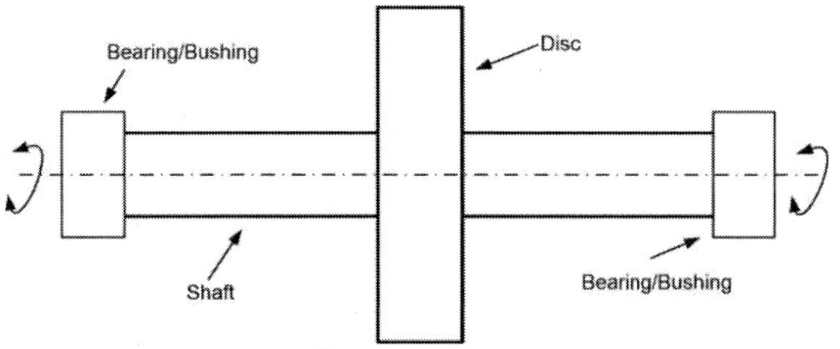

Figure 1. Rotor.

There are two types of rotor vibrations: synchronous and non–synchronous vibrations. The frequency of synchronous vibrations corresponds to the rotating frequency of the rotor. A shaft spinning, let's say at 3600 rpm (60 Hz) will have a synchronous vibration of 60 Hz. Synchronous vibration is mostly generated by imbalance. The force produced by an imbalance grows in proportion to the square of the shaft rotational speed. $F = mr\omega^2$, where F is the imbalance force, m is the imbalance mass, r is the radius of the imbalance mass, and ω is the rotational speed of the shaft. The orbit of a synchronous vibration depicts the rotor rising to a peak after exactly one spin. Non–synchronous orbits do not have the same frequency as the rotating frequency. Sub–synchronous vibrations occur at frequencies lower than the rotational frequency. The speed of a spinning shaft may be denoted as the 1x, or "one time". This is the first frequency on a vibration spectrum graph. Mechanical faults apart from imbalance are also of 1× harmonics (one time), such as bow, misalignment, looseness, and so on. Synchronous peaks are harmonics of the driving speed in whole numbers (2x, 3x, 12x, etc). In contrast non–synchronous peaks show as decimal values rather than whole numbers since they are not connected to the shaft (7.3x, 22.7x, 51.8x, etc).

Every substance has a natural frequency that is constant. If the shaft speed approaches its natural frequency, resonance occurs, and the amplitude of vibration becomes very enormous, causing the rotordynamic system to lose stability. This is referred to as the system's critical speed. The rotordynamic system returns to its normal state just beyond this speed (i.e. when the shaft speed increases further). After a while at a greater speed, it achieves its second critical speed, and so on. A rigid rotor is one in which the rotor is considered to be stiff, and the rotor characteristics are derived from the bearing's behaviours; it often works at substantially less than its initial bending critical speed. These rotors can be balanced in any two arbitrary directions at low speeds. A flexible rotor, on the other hand, has a rotor shaft that has both stiffness and flexibility and can bend at or above its critical speed. A rotor with a flexible shaft experiences dynamic changes in the distribution of imbalance as it spins. This necessitates balancing at high speeds on a variety of balance planes at different critical speeds.
With all of the aspects described above, it is apparent that researchers have been exploring novel approaches, tools, algorithms, analyses, and solutions to ensure the shaft's good health, theoretically and experimentally. And, for the benefit of the scientific community, a regular review of such

collection of work must be reported. For example, Teyi and Singh (2022) discussed the use of data science methods for analysis rotors with cracks.

Numerous studies have looked at the effects of integration of standard bearings with active magnetic bearings with a rotor system that may vibrate synchronously or non–synchronously. This paper seeks to provide a specific assessment of the literature encompassing numerical and experimental studies of the effect of active magnetic bearings on non–synchronous rotors.

2. Active magnetic bearing construction

AMBs have really no touch between the stator, which is a stationary housing, and the rotor, which is a rotating target. The rotor is raised into the air by AMBs, acting as a pure bearing, and then rotated up to the necessary operating speed by the drive. AMBs operate as a controller during run–up, limiting undesirable high levels of vibration, particularly during disturbances and excitations. Using a closed–loop controller with additional hardware components is important since AMBs can only generate attractive forces. AMBs have several benefits, including a high achievable rotational speed, configurable bearing rotordynamic characteristics, no surface rubbing and no lubrication, minimal losses and a broad working temperature range, decreased power use, and electronic control as reported by Haberman & Liard (1980) and Maslen et al. (1996).

There are four basic components in a magnetic bearing system: the actuator, the controller, the power amplifier, and the proximity sensor. The controller receives a signal from the non–contact proximity sensor when the rotor position changes. A correction signal is sent from the controller to the power amplifier, which then adjusts the magnetic coils' current accordingly. Electromagnets are used to create magnetic flux and restore equilibrium to a rotating machine. Levitation of the rotating disc is accomplished in real time. AMBs generate attractive forces on the rotor using electromagnets. Figure 2 depicts a schematic depiction of the components of AMBs.

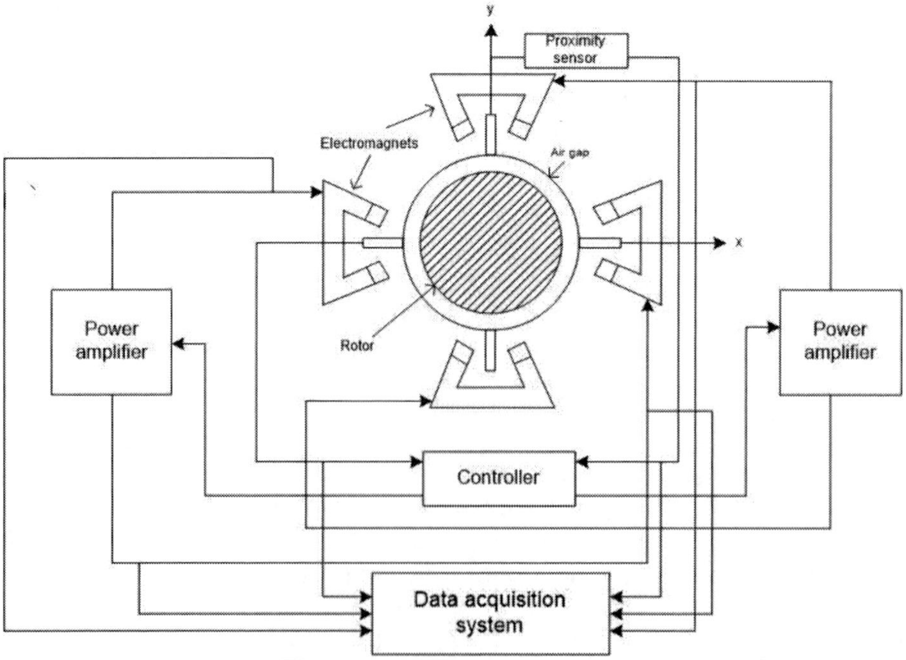

Figure 2. AMB–rotor system.

IOP Conf. Series: Materials Science and Engineering 1272 (2022) 012021 doi:10.1088/1757-899X/1272/1/012021

3. Abbreviations and their full forms

The review begins in the following section, which is followed by a Table that summarizes the specifications of the articles. Here, it is necessary to specify the complete forms of the abbreviations that will be used in the Tables.

Nomenclature

AAEMS	Actively Actuated Eccentric Mass System
ABD	Auto–Balancer Device
AC	Axial Compressor
ACC	Accelerometer
AFSFD	Active Floating Ring Squeeze Film Damper
AMB	Active Magnetic Bearing
AAMB	Axial Active Magnetic Bearing
RAMB	Radial Active Magnetic Bearing
TAMB	Thrust Active Magnetic Bearing
AMD	Active Magnetic Damper
AVC	Active Vibration Control
BB	Ball Bearing
BTS	Blade Tip Sensor
CAD	Computer Aided Drafting/Design
CMG	Control Moment Gyro
DP	Displacement
EMA	Electromagnetic Actuator
FBLMS	Fast Block Least Mean Square Algorithm
FDOF	Five Degree of Freedom
FEA	Finite Element Analysis
FF	Feed Forward
FIR	Finite Duration Impulse Response
GPM	Gain Phase Modifier
HB	Hybrid Bearing
HERMB	Heteropolar Radial Magnetic Bearing
HFMB	Hybrid Foil Magnetic Bearing
HS	Hall Sensor
JB	Journal Bearing

IOP Conf. Series: Materials Science and Engineering 1272 (2022) 012021 doi:10.1088/1757-899X/1272/1/012021

LQR	Linear Quadratic Regulator
MIMO	Multi Input Multi Output
MOROLAC	Multi Objective Recursive Open Loop Adaptive Control
NAF	Non Adaptive Feedback
ODE	Ordinary Differential Equation
PD	Proportional Derivative
PDF	Proportional Derivative Feedback
PID	Proportional Integral Derivative
PP	Proximity Probe
REB	Rolling Element Bearing
RMS	Root Mean Square
ROLAC	Recursive Open Loop Adaptive Control
SDOF	Six Degree of Freedom
SG	Strain Gauge
SISO	Single Input Single Output
SPHS	Schroeder Phased Harmonic Sequences
SR	Slip Ring
THO	Transient Harmonic Oscillator
TPB	Tilting Pad Bearing
TS	Temperature Sensor
UMP	Unbalance Magnetic Pull

4. Review of numerical investigation

This section examines significant research articles released in the last 15 years on numerical studies of the influence of active magnetic bearings on suppression and associated condition monitoring for non–synchronous vibration rotor systems.

Amer et al. (2006) explored the dynamics of a basic rigid disk–rotor hung by AMBs without gyroscopic effects. Multiple scales were used to analyze system stability and resonance instances. Using frequency response equations, system stability was explored. Numerical results focused on both parameter effects and primary parametric resonance. The results helped minimize vibration in rotor–AMB systems. The major parametric resonance example had the worst system behavior. The system featured multi–valued solutions, jump, and initial–condition sensitivity. Steady–state amplitudes were monotonically decreasing in frequency and damping functions. Marx & Nataraj (2007) discussed magnetic bearing–supported rotor systems with harmonic base excitation. Conventional magnetic bearings with nonlinear properties functioned poorly in such conditions. The suggested compensation approach combined conventional, proportional, and differential feedback control. Nonlinear expressions were utilized to represent the rotor system's bearings. Each disturbance frequency produced a multiharmonic system response, a nonlinear system characteristic. They used optimum control to minimize a performance index, which optimized the corrective current function's trigonometric coefficients. The control approach suppressed rotor vibration to amplitudes much less than disturbance amplitudes for all disturbance frequencies.

When constructing magnetic suspension systems, Lei and Palazzollo (2008) presented a way for employing large flexible rotordynamics models to analyze and build magnetic suspension systems. A flexible shaft rotordynamics model for a flywheel with magnetic bearings was developed by them. The magnetic suspension compensator was built using a finite element model of the rotor system. Closed–loop rotordynamic properties were optimized using eigen value analysis, Campbell plots, waterfall charts, and mode shapes. Large rotor order and high spin speed complicated magnetic suspension control. Their work was an example of a physical controller in a flywheel system that provided steady rotor suspension and good disturbance rejection in all operational modes. Eissa et al. (2008) studied the nonlinear response of a basic rigid disk–rotor supported by AMB without gyroscopic effects. Multi– excitation primary resonance and Coupled second–order nonlinear ODEs described the spinning shaft. Multiple scales provided approximations. Using the frequency response function approach, numerical simulations highlighted the steady–state response and stability of various solutions. System characteristics affected the rotor's nonlinear response differently. Multiple–valued solutions and jumps arose for steady–state response. In their work, Das et al. (2008) discovered that electromagnetic exciters provided non–contact control force of actuation, which reduced rotor vibration owing to imbalance and enhanced rotor–shaft stability limit speed, resulting in safer operations at higher speeds. The control action required no maintenance, wear, or friction. Increased exciter coil turns and pole–face area decreased responsiveness and current regulation. There were appropriate exciter positions on the stator along the shaft for lowering imbalance response amplitude, increasing stability limit speed, and decreasing control current. The authors identified best locations for exciters for the current situation, while reducing support forces for longer bearing life.

Hussain's (2009) research of a flexible rotor in magnetic bearings with a significant geometric coupling parameter revealed nonlinear dynamical phenomena. To observe how the spindle's response to speed parameter x in the 0.05–5 Hz range affected vibration patterns with periods of 2, 3, 6, 9, and 17, they looked for both periodic and quasi–periodic behavior. The rotor's operation speed influenced the number of attractors present at the same x value. When the rotor–bearing system was subjected to preloads, fluid forces, or other external excitation factors that moved the rotor's beginning conditions from one basin of attraction to another, synchronous rotor vibration became non–synchronous or chaotic. According to Das et al. (2010), parametric excitation from base stiffness and rotor–shaft internal damping may occur in rotor–bearing systems subjected to base motion. A PD feedback control on the actuator's rotor shaft displacement could eliminate both types of instability. Parametric instability zones were moved away from excitation zones by manipulating the actuator's control force on the rotor's stiffness. The initial critical speed and the stability limit speed were both raised by proportional feedback, which improved rotor performance. External stationary dampening provided by the actuator improved the rotor–bearing system's stability limit speed and reduced response amplitude. The ability to express linear force expressions was made possible owing to careful selection of actuator and control parameters. Hussain (2010) affirmed that statically misaligned flexible rotor supported by active magnetic bearings exhibited nonlinear behavior. Responses of periods 2, 3, 4, 6, 8, 12, and 16 were highly influenced by the geometric coupling parameter, as were quasi–periodicity and chaos. Magnetically supported spinning machinery became non–synchronous or chaotic if external influences caused the rotor's location to change basins of attraction. Non–synchronous and chaotic vibrations should be avoided in rotating machinery because they produced alternate strains on its key components, which could lead to fatigue failure.

Sahinkaya et al. (2011) discussed rotor vibration control and base force management in rotating machinery. Using magnetic bearings, an adaptive multi–objective technique was devised. Weighting techniques were devised using two stages, combining the system's receptance matrices' basic weightings and a single weighting parameter that may be adjusted to balance the two target functions. Open–loop adaptive technique avoided any additional measurements for this new real–time controller.

Before conducting an experiment, the suggested multi–objective adaptive controller was tested using simulations of an existing experimental apparatus. Bouaziz et al. (2011) investigated the dynamic behavior of a misaligned rotor installed in two identical AMBs. The rotor had x, y, and x, z degrees of freedom. A bias current was thought to provide an electromagnetic bearing field. Eight linearized spring and damping factors modelled AMBs' dynamic support characteristics. As the stator–shaft gap rose, they decreased. A theoretical model quantified shaft misalignment's influence. The results of this study showed that angular misalignment had a two and four fold effect on the dynamic behavior of a misaligned rotor, respectively, on the influence of angular misalignment on dynamic behavior. As the number of magnets increased, so did their magnitudes, and the second harmonic reduced as the air gap widened. Four, six, and eight–magnet AMBs were compared. This study found that adding magnets helped reduce misalignment–caused vibration.

Bouaziz et al. (2013) performed numerical analysis to observe the effect of angular misalignment on rotor– AMBs transient dynamics. A spatial model of a misaligned rotor installed in two identical AMBs accounted for angular misalignment using a flexible connection. Transient electric motor drove the driving rotor. Four, six, and eight–magnet AMBs were utilized. Bias and control current powered them. Modeling stiffness and damping matrices supported AMB's dynamics. These were calculated using electromagnetic theory and magnetic forces. Using Newmark time integration, the driven rotor's general equations of motion were obtained and solved. Simulation findings determined frequency and transient misalignment responses. Misaligned rotor dynamic responses demonstrated the transient regime's effect. Simulations showed that angular misalignment–induced vibrations were more significant in four–magnet AMBs. Spectral components of the system response were studied to diagnose the malfunction in transient mode. Kumar et al. (2016) created a 12–pole AMB and excitation method to regulate rotor vibration to generate circular force. AMB employing AC stimulation, was proposed to counterbalance. Switching AMB teeth excitation currents generating a spinning force synchronized with rotor unbalance but in the opposite direction. The suggested technique required real–time estimation of the unbalanced force's direction. Proper excitation switching produced a rotating force with the same frequency as the excitation.

One of the newer hybrid imbalance vibration control strategies proposed by Jung and DeSmidt (2018) was for an ABD–rotor system that included a passive bearing and an AMB. When operating at supercritical speeds, the rotor's steady–state imbalance was compensated by an ABD with eccentric balancing weights. Due to the ABD's nonlinearity, non–synchronous limit–cycle behavior caused rotor vibration. The suggested hybrid control approach guaranteed worldwide asymptotic stability of synchronous balancing. A "Luenberger–like" observer was used to constantly estimate the status of an ABD balancer ball. Before this technique, controllers had to modify rotor speed to get the desired performance after achieving the goal speed. As part of his research, Hussain (2020) studied how flexible a rotor's response to magnetic and auxiliary bearing load sharing was affected by its shaft's flexibility. When it came to damping off rotor response vibrations, numerical simulations showed that auxiliary bearing contact stiffness (expressed by the stiffness ratio) was more effective than the Coulomb sliding friction coefficient. For periods 2, 4, 8, and 16, the rotor responded with sub–synchronous vibrations. When the stiffness parameter was raised from 0 to 5.0, a period–doubling cascade occurred in the rotor response, resulting in chaos. Jung (2020) introduced a new AAEMS control approach to cancel rotor imbalance vibration which could provide virtually zero–imbalance rotor vibration even with presumed artificial disturbances and errors. The author proposed three AAEMS functions to attain great performance; first, two Kalman filters that estimated the unknown rotor unbalance and coulomb friction of AAEMS; second, effective angular orientation management of the AAEMS employing LQR–based optimum control that reduced rotor uneven vibration and actuator power utilization; and third, fine–tuning angular position control with AAEMS for additional imbalance suppression in the event of estimation mistakes and disturbances.

Zhang et al. (2021) analyzed the bifurcation properties of an externally excited AMB–rotor system at high speeds. An approximate analytical approach was given to evaluate bifurcations remote from harmonic resonances in forced vibration systems. Stability was examined using Floquet theory after it was applied to the AMB–rotor system and several static equilibrium and periodic solutions were found. Numerical integration and extended cell mapping digraph methods validated the approximate analytical technique. AMB–rotor system showed supercritical pitchfork bifurcation of static equilibrium. External excitation and controller settings affected dynamical properties. Controller settings were also changed based on study results to prevent nonlinear behaviours and increase system performance. Wang (2021) studied a vibration control active floating squeeze film damper concept. Active floating squeeze film damper design allowed fluid stiffness and damping changes. In this device, an AMB and a squeezing film damper were employed to obtain oil film forces and magnetic forces. The lumped mass model was solved using explicit Newmark– β, while the finite element rotor system employed explicit and implicit Newmark– β. According to the simulation results, frequency shifts were observed by adjusting the proportional gain, but the amplitude reductions by adjusting the derivative gain appeared to be inconsistent due to the relative shifting position of node and force acting points of active magnetic bearing for different modes, as well as the nonlinear strength of floating ring squeezed film damper. Liu et al. (2021) proposed a single–bearing HGMB. Gas bearing's bushing was the closed magnetic poles of the AMB, which shared the same air gap. System dynamics were adjusted with electromagnetic forces. Numerical findings showed that the suggested bearing had superior rotational precision, reduced sub–synchronous vibration, lesser power expenditure than rigid gas bearings and AMB, and fewer air friction losses than traditional HGMBs.

Table 1 sums up the bearing specifications, operating speed, control technique, and paper objective for each paper.

Table 1. Outline of numerical studies.

Author (Year)	Bearing	Speed (rpm)	Control/Sensors	Aim
Amer et al. (2006)	2 RAMB	–	–	Rotors with a parametric impulse
Marx & Nataraj (2007)	HERMB 8 poles	30000	PDFB Off	Suppression of base excitation
Lei & Palazzollo (2008)	–	20000	–	Design of large order, flexible rotors
Eissa et al. (2008)	2 RAMB	–	–	Non-linear response without gyroscopic effect
Das et al. (2008)	1 AMB, 2 REB, HERMB, 8 poles	900	PD	Vibration suppression
Hussain (2009)	2 RAMB, 8 poles	–	–	Misaligned rotor dynamics
Das et al. (2010)	1 RAMB, HERMB, 8 poles	900	PD	Coupled vibration control
Hussain (2010)	2 RAMB, 8 poles	–	–	Misaligned rotor

				dynamics
Sahinkaya et al. (2011)	2 RAMB	2400	PD, ROLAC, MOROLAC, PP	Vibration suppression
Bouaziz et al. (2011)	2 RAMB, HERMB, 8, 12 and 16 poles	8000	–	Vibration suppression due to misalignment
Bouaziz et al. (2013)	2 RAMB, HERMB, 8, 12 and 16 poles	8000	PID, PP	Cause of misalignment
Kumar et al. (2016)	1 AMB, 2 REB, 12 Poles	–	–	Unbalance counteraction
Jung & DeSmidt (2018)	AMB with ABD	–	–	New hybrid imbalance vibration control approach
Hussain (2020)	AMB with Auxilliary bearing	–	–	Magnetically flexible rotor bearing response
Jung (2020)	AAEMS	–	–	Cancelling the rotor imbalance vibration
Zhang et al. (2021)	2 AMB	27000	SDOF model with PD controller.	Analysis of supercritical pitchfork bifurcation
Wang (2021)	AMB with AFSFD		PD	Vibration reduction effectiveness by controlling the magnetic forces
Liu et al. (2021)	AMB, HFMB, HGMB	180000	PID, PD	Reducing the low–frequency vibrations and instability caused by the fluid,

5. Review of experimental investigation

This section summarizes the experimental studies conducted on AMBs and associated bearings using various control strategies to study their impact on non–synchronous rotordynamic systems. The pertinent papers are considered from the last 30 years.

Kasarda et al. (1991) showed that a between–bearing AMD reduced sub–synchronous rotor vibrations by adding damping and rigidity. They observed that in one case, an AMD installed at two– thirds rotor span, reduced sub–synchronous and uncompensated synchronous vibrations by 98% and 18%, respectively. And in another case, an AMD at two–thirds rotor span reduced sub–synchronous vibrations by 93% and synchronous vibrations by 17%. The first case had a larger decrease since AMD stiffness shifted the initial critical speed away from the disturbance frequency. Lang et al. (1995) used AMBs to manage nonlinear and non–synchronous vibrations of a stiff rotor powered by non– conservative cross–coupling methods. Unfamiliar cross–coupling characteristics of a rotor were approximated online using an AMB–adaptive controller's control technique and a time–dependent forgetting factor. The AMB controller's settings were adjusted to reduce cross–coupling that stabilized the rotor system. Simulation of a stiff rotor supported by two AMBs calculated transient cross– coupling response characteristics. This control technique guaranteed system stability for substantially larger cross–coupling coefficient values than non–adaptive feedback controllers constructed least square algorithms or pole placement. Johnson et al. (1998) developed a closed loop technique that reduced multi–frequency rotor vibration. The control algorithm needed identifying the partial receptance matrix for all relevant frequencies. Initial estimates of system dynamics determined each component's amplitude. Each component's phasing was optimized using SPHS, which assured a low peak value signal in the time domain and defined frequency domain amplitude characteristics. This test signal eliminated the need to apply a distinct signal for each frequency, speeding up identification. Using an SPHS–based test signal, multi–frequency identification of system dynamics was successful in experiments. Calculating the optimal closed loop controller gain reduced non–synchronous component vibration and maintains synchronous vibration control.

Matsushita et al. (2002) evaluated AMB–equipped flexible rotors for seismic response severity. The lowest natural frequency of stiff modes was 20–30 Hz to minimize resonance or rubbing induced by seismic agitation. Even if AMB rotors were constructed this way, reaction severity was still higher than oil–lubricated rotors. In addition to PD control, FF excitation was added to the AMB rotor to reduce seismic vibration response severity. Comparing the response vibrations of PD control and PD with FF control proved the control's efficiency. Experimentally, FF reduced vibration by 60%. Johnson et al. (2002) experimented with magnetic bearing actuators in a FF active control strategy to reduce gear noise and vibration. Gear meshing induced noise and vibration that aggravated strain mechanical components. The authors employed AMBs as actuators to impart control forces directly onto the spinning shaft. Proximity to the source allowed for great control with a limited number of actuators. Jung & DeSmidt (2018) AMB with ABD – – New hybrid imbalance vibration control approach Hussain (2020) AMB with Auxilliary bearing – – Magnetically flexible rotor bearing response Jung (2020) AAEMS – – Cancelling the rotor imbalance vibration Zhang et al. (2021) 2 AMB 27000 SDOF model with PD controller. Analysis of supercritical pitchfork bifurcation Wang (2021) AMB with AFSFD PD Vibration reduction effectiveness by controlling the magnetic forces Liu et al. (2021) AMB, HFMB, HGMB 180000 PID, PD Reducing the low–frequency vibrations and instability caused by the fluid, The control approach was tested using a four–square gear setup. In order to decrease vibration and pressure error sensor outputs, an X–LMS control system with a time domain filtered proximity sensor near the gear teeth was used as a reference sensor to drive two magnetic bearing actuators.. At the gear mesh frequency, one microphone reduced acoustic levels by 20 dB, while four microphones reduced levels by 6 dB. Keogh et al. (2002) implemented transient vibration attenuation in a MIMO flexible rotor/magnetic bearing system. The approach used measured rotor harmonic feedback. Unlike prior harmonic controllers, the new one added transient dynamics. The controller was developed using observed data and target transient vibration responses from disturbance step changes. The controller's stability bounds were tolerant to measurement inaccuracy. Experimental validation of the controller in a flexible rotor/magnetic bearing system showed quick vibration reduction and near removal of transient overshoot.

AMB and conventional support bearings were proposed by Kasarda et al. (2004) as an active control system. In place of standard support bearings, the AMD was used as an AMB. Using an AMD to control sub–synchronous rotor vibrations in a high–speed single–disk laboratory rotor was demonstrated as a proof of concept. AMD was found to diminish sub–synchronous vibrations by 93 % in a recent study. A closer comparison of the critical speed to the operating speed showed that AMD improved synchronous vibration response by increasing system rigidity. Keogh et al. (2004) investigated the dynamics and control of rotor vibration when auxiliary bearings or bushes were induced. There were significant phase shifts in the detected vibration components when an enormous unbalanced mass was applied to the rotor. Each sensor measurement had similar phase changes, up to 1800 at some speeds. For linear non–contacting controllers, applying control forces would be ineffectual and aggravate rotor dynamic state. A redesigned controller converted a contacting rotor from a non–contacting state. Keogh & Cole. (2006) studied rotor displacement signal vibrations in a magnetic bearing system with auxiliary bearings. An additional magnetic bearing was employed to connect the rotor dynamic reactions created by auxiliary bearings and bushes to the circular forces induced by the circular motion. In order to overcome the practical difficulties of manufacturing direct imbalance reactions at higher speeds, magnetic bearing responses were generated at an artificial low operating speed. Comparable results were found for an aligned and misaligned end bush by 50% radial clearance. Complex orbit features including asymmetric bounce and rub were produced without detailed nonlinear analysis and modelling. Matras et al. (2006) examined a magnetic–bearing–supported rotor system's adaptive disturbance rejection mechanism. The method was comparable to Sacks' and Bodson's method for rejecting persistent excitations in large space structures, but it was better for MIMO systems. The approach suppressed persistent excitations of known frequencies within a range. The authors discussed some implementation challenges, including determining acceptable weighting matrix values and needing current feedback.

Anegawa et al. (2008) explored the connection of shaft bending vibration with rotor blade in– plane vibration. Because of coupling and the coupled resonance of an eight–blade system, there is an unstable zone. They observed that at a certain rotating speed, the shaft's vibration was confined to tiny amplitude owing to anti–resonance points, but the blades' resonance was large. The blades' dynamic absorber effect caused a violent coupled resonance at a different speed. De Moraes & Nicoletti (2010) performed experiments with a hydrodynamic tilting–pad bearing with integrated electromagnetic actuators that demonstrated that the shaft's lateral vibration may be controlled via PD. The optimal condition reduced vibration RMS amplitude by 54%. Limitations on the actuating system's force frequency response reduced control efficiency at high shaft speeds. An examination of the pad shape and its connection with the shaft suggested design potential to optimize magnetic flux channeling and magnetic forces on the shaft. Jiang & Zhu (2011) proposed an adaptive FIR filter–based approach for reducing multi–frequency periodic vibration in AMB–rotor systems. The procedure was theoretically feasible. Adaptive filter used FBLMS algorithm to increase convergence and minimize computational complexity. The FBLMS algorithm's computing cost was unaffected by the amount of frequency components analyzed. Each frequency component's convergence rate was modified by setting its step size parameter. The reciprocating simulated disturbance test and the rotating harmonic vibration test revealed that the proposed adaptive filter using the FBLMS algorithm suppressed multi–frequency periodic vibration. Yang (2011) employed permanent magnet bearings for radial rotor levitation and electromagnetic actuators for axial stability in small magnetically levitated rotating devices. No controlled force was applied in these directions, causing rotor radial oscillations. Near the rotor's resonance frequency, oscillations became extreme. This article described an electromagnetic actuator's implementation and control. The actuator created axial and radial rotor forces. Axial force-controlled rotor axial direction, whereas radial force suppressed resonance–related oscillation.

Viveros & Nicoletti (2014) presented analytical and experimental investigation of rotor vibration attenuation employing a tilting–pad journal bearing with integrated electromagnetic actuators. Experimental results in the frequency domain demonstrated shaft–vibration amplitude reductions of 11% for 600 rpm and 18% for 1100 rpm, agreeing with the mathematical model. Comparison of the actuation capacity of the studied bearing to that of other active bearings showed that it was competitive in terms of particular control capacity. Fang et al. (2015) provided an AVC method for attaining maximum vibration force and torque. Notch filters, feed–forward controllers, and GPM made up the proposed AVC controller. Results showed that the GPM could perfectly adjust power amplifier flaws and changes and managed the rotation axis according to AVC requirements. The suggested technique was extremely appropriate for space applications of magnetically suspended moment exchange devices (CMGs and flywheels). Lyu et al. (2016) used an existing rotor–AMB test setup with exciter AMBs to mimic the workings of energy storage flywheels. The generator's negative stiffness and the flywheel disk's gyroscopic effects were replicated using the two AMBs at the shaft's mid and quarter spans. The generator's negative stiffness had a negligible effect on the rotor dynamics, according to simulation data. When using the exciter AMBs, the simulated gyroscopic forces were similar to those anticipated by a previous finite element study of the gyroscopic matrix. While the rotor orbits under various additional forces and replicated excitation forces displayed similar orbit tilting characteristics, this suggests that the emulation technique presented in this research was successful.

For the flexible rotor system with auxiliary bearings, Ebrahimi et al. (2018) studied the correlation between the theoretical model and the tests for the system, in order to provide an overview of the global system nonlinear dynamics. Various auxiliary bearing stiffness levels were tested. Gyroscopic moments, shaft flexibility, and a magnetic actuator geometric connection were all accounted for in the model's math. Bifurcation diagrams, system trajectories, power spectra, Poincaré maps, and the maximal Lyapunov exponent were used to study the dynamic behaviour of the system. The first chaotic area appeared at lower rotational speed as auxiliary bearing stiffness decreased. A high–speed two–pole induction generator using AMBs was used by Kim et al. (2020) to prove the existence of UMP effects. With the help of an analytical model and FEA data, the authors were able to predict the UMP's force. The simulation model coefficients were determined using a detailed CAD model and experimental modal analysis data. AMB linearized coefficients were determined by the stiff body whirling mode of the rotor. For further investigation, UMP effects were studied by performing an experimental rotordynamic simulation with mixed eccentricity and comparing the findings to vibration data during ramp–down operation of the machine. Both the rotor's intrinsic frequency and the additional vibration induced by supply frequency stimulation multiplied by two were found to be reduced in UMP experiments on induction machines. Kalista et al. (2021) proposed a solution for a genuine test rig fitted with AMBs and rotor vibration sensors to test rotor seals. The authors experimentally and statistically determined seal qualities need the appropriate rotor orbit shape. The generalized notch filter approach reduced harmonic vibrations. A modified generalized notch filter for harmonic vibration creation was also introduced. Injecting harmonic current into each AMB axis' control loop excited the rotor's harmonic vibration in an AMB system. Coupled rotor motion in the AMB axis made signal amplitude and phase adjustments laborious. The broad notch filter technique automatically provided rotor harmonic vibration. Measured results showed that AMBs may be employed to generate exact rotor orbit shapes.

Table 2 outlines each paper's bearing characteristics, operating speed, control mechanism, and paper objective.

Table 2. Outline of experimental studies.

Author (Year)	Bearing	Speed (rpm)	Control/Sensors	Aim
Kasarda et al. (1991)	1 RAMB, HERMB, 4 poles	4000	PP	1st mode vibration suppression
Lang et al. (1995)	2 RAMB	20000	NAF	Non synchronous vibration suppression
Johnson et al. (1998)	1 RAMB, 1 BB, 4 poles	10000	PD, SG, ACC, BTS	Testing the blades for vibration
Matsushita et al. (2002)	2 RAMB	6000	PD, ACC, PP	Seismic waves' effect on vibration suppression
Johnson et al. (2002)	2 RAMB, 2 BB, HERMB, 8 poles	51000	PID, PP, Microphone	Gear noise reduction
Keogh et al. (2002)	2 RAMB, 2JB, HERMB, 8 poles	9000	PID, PP	Vibration suppression in mass loss test
Kasarda et al. (2004)	1 RAMB, HERMB, 8 Poles	900	PD	Sub synchronous vibration suppression
Keogh et al. (2004)	2 RAMB, HERMB, 8 poles	1740	PID, PP, THO	Vibration suppression in mass loss test
Keogh & Cole. (2006)	2 RAMB, 2 bushes	1800	PID, PP	Bringing nonlinear effects
Matras et al. (2006)	2 RAMB	72000	PD, PP	Adaptive disturbance rejection
Anegawa et al. (2008)	1 RAMB, 1 BB	1800	DP, SG, SR	Resonance of couple blade shaft
De Moraes & Nicoletti (2010)	EMA in TPB, 1 BB, 4 poles	1200	PD, PP	Vibration suppression
Jiang & Zhu (2011)	2 RAMB, 1 PMA	18000	PP, FBLMS	Multiharmonic vibration suppression
Yang (2011)	2 PMB, 1 EMA, HERMB, 4 poles	1800	PID, Laser meter, HS	Vibration suppression
Viveros & Nicoletti (2014)	EMA in TPB, 1 BB, HB, 4 poles	1100	PD, PP, TS	Vibration suppression
Fang et al. (2015)	2 RAMB, 1 TAMB	12000	FF, PP	Vibration suppression
Lyu et al. (2016)	4 RAMB, HERMB, 16 poles	7600	PP, μ–synthesis	Imitate gyroscopic effect of flywheel

Ebrahimi et al. (2018)	2 AMB, 2 Auxiliary bearings	3000	STM32F103RE T6 microcontroller,	Design recommendations for AMB systems with auxiliary bearings
Kim et al. (2020)	2 RAMB, 1 RAMB, 2 poles	12500	Magnetizing and leakage inductance	UMP effects on the rotordynamics of induction machine
Kalista et al. (2021)	2 RAMB, 1 RAMB, 8 poles	3000	FDOF, SISO	Generating precise rotor orbit during rotordynamic experiments

6. Conclusion

The study on AMBs in non–synchronous rotordynamic systems is summarized in this review article. Among the key priorities were vibration reduction, stabilization, and condition monitoring.

In AMB-supported rotors, the control strategy to suppress vibration plays a bigger role than the disturbance amplitudes. Hence, control design for a rotor under fault is a major area of research. Research indicates that the optimization of the pole face area, coil turn numbers, and mass of the magnetic system with respect to the vibration suppression obtained is a grey area. It is a suitable area of investigation and research.

The number of active magnets used in the AMB set up also plays a role in the stability of the system. While 4 magnet AMB leads to greater vibrations in rotors as compared to 6 pole or 8 pole AMBs under angular misalignment, the influence of the number of poles of AMB magnets in the presence of other flaws like rub and rotor crack has not been reported. This field is open to investigation.

It is a fact well known now that the rotors are more stable in the spin speed range between the 1 st and 2nd critical speeds. But, to reach the range of a better stable rotor, the first critical speed has to be passed through. Most of the literature available deals with the rotor–AMB behaviour operating away from the critical speeds. More analytical and experimental work is required to gain the real benefit of the AMB in a rotor system by examining the system at critical speeds. The control strategy and design for AMB running rotors at critical speeds will be another investigation of great use in the industry.

Declaration of competing interest

There are no known financial or personal interests of the authors that could have affected this paper.

References

[1] Teyi, N. & Singh, S. (2022). A decadal review of various modelling and analysis of cracked rotors. *Procedia Struct. Integr, 39*, 333–346 https://doi.org/10.1016/j.prostr.2022.03.103.
[2] Teyi, N. & Singh, S. (2022). A review of application of data science tools in crack identification and localization. *Procedia Struct. Integr. 39,* 608–623. https://doi.org/10.1016/j.prostr.2022.03.134.
[3] Haberman, H. & Liard, G. (1980). An active magnetic bearing system, *Tribol. Int. 2 (3,)* 85–89.
[4] Maslen, E.H, Allaire, P.E, Noh, M.D. & Sortere, C.K. (1996). Magnetic bearing design for reduced power consumption, *J. Tribol.. 118*, 839–846. https://doi.org/10.1115/1.2831617.
[5] Amer, Y. A., Eissa, M. H., Hegazy, U. H., & Sabbah, A. S. (2006). Dynamic behavior of an

AMB/supported rotor subject to parametric excitation. *J. Vib. Acoust. Trans. ASME*, *128*(5), 646–652. https://doi.org/10.1115/1.

[6] Marx, S., & Nataraj, C. (2007). Suppression of base excitation of rotors on magnetic bearings. *Int. J. Rotating Mach*, https://doi.org/10.1155/2007/91276.

[7] Lei, S., & Palazzolo, A. (2008). Control of flexible rotor systems with active magnetic bearings. *J. Sound Vib.*, *314*(1–2), 19–38. https://doi.org/10.1016/j.jsv.2007.12.028.

[8] Eissa, M. H., Hegazy, U. H., & Amer, Y. A. (2008). Dynamic behavior of an AMB supported rotor subject to harmonic excitation. *Appl. Math. Model.*, *32*(7), 1370–1380. https://doi.org/10.1016/j.apm.2007.04.005.

[9] Das, A. S., Nighil, M. C., Dutt, J. K., & Irretier, H. (2008). Vibration control and stability analysis of rotor–shaft system with electromagnetic exciters. *Mech Mach Theory*, *43*(10), 1295–1316. https://doi.org/10.1016/j.mechmachtheory.2007.10.007.

[10] Inayat–Hussain, J. I. (2009). Geometric coupling effects on the bifurcations of a flexible rotor response in active magnetic bearings. *Chaos Soliton Fract.*, *41*(5), 2664–2671. https://doi.org/10.1016/j.chaos.2008.09.041.

[11] Das, A. S., Dutt, J. K., & Ray, K. (2010). Active vibration control of unbalanced flexible rotor–shaft systems parametrically excited due to base motion. *Appl. Math. Model.*, *34*(9), 2353–2369. https://doi.org/10.1016/j.apm.2009.11.002.

[12] Inayat–Hussain, J. I. (2010). Nonlinear dynamics of a statically misaligned flexible rotor in active magnetic bearings. *Commun Nonlinear Sci Numer Simul.*, *15*(3), 764–777. https://doi.org/10.1016/j.cnsns.2009.04.020.

[13] Sahinkaya, M. N., Abulrub, A. H. G., & Burrows, C. R. (2011). An adaptive multi–objective controller for flexible rotor and magnetic bearing systems. *J Dyn Syst Meas Control.*, *133*(3). https://doi.org/10.1115/1.4003421.

[14] Bouaziz, S., Belhadj Messaoud, N., Mataar, M., Fakhfakh, T., & Haddar, M. (2011). A theoretical model for analyzing the dynamic behavior of a misaligned rotor with active magnetic bearings. *Mechatronics*, *21*(6), 899–907. https://doi.org/10.1016/j.mechatronics.2011.05.001.

[15] Bouaziz, S., Messaoud, N. B., Choley, J. Y., Maatar, M., & Haddar, M. (2013). Transient response of a rotor–AMBs system connected by a flexible mechanical coupling. *Mechatronics*, *23*(6), 573–580. https://doi.org/10.1016/j.mechatronics.2013.05.002.

[16] Kumar, G., Choudhury, M. D., Natesan, S., & Kalita, K. (2016). Design and Analysis of a Radial Active Magnetic Bearing for Vibration Control. *Procedia Eng.*, *144*, 810–816. https://doi.org/10.1016/j.proeng.2016.05.090.

[17] Jung, D. Y., & DeSmidt, H. (2018). A new hybrid observer based rotor imbalance vibration control via passive autobalancer and active bearing actuation. *J. Sound Vib.*, *415*, 1–24. https://doi.org/10.1016/j.jsv.2017.11.024.

[18] Inayat–Hussain, J. I. (2020). Nonlinear Dynamics of a Magnetically Supported Flexible Rotor in Auxiliary Bearings. *J. Phys. Conf Ser.*, *1489*(1). https://doi.org/10.1088/1742–6596/1489/1/012015.

[19] Jung, D. Y. (2020). A control strategy of actively actuated eccentric mass system for imbalance rotor vibration. *Actuators*, *9*(3). https://doi.org/10.3390/ACT9030069.

[20] Zhang, X., Sun, Z., Zhao, L., Yan, X., Zhao, J., & Shi, Z. (2021). Analysis of supercritical pitchfork bifurcation in active magnetic bearing–rotor system with current saturation. *Nonlinear Dyn.*, *104*(1), 103–123. https://doi.org/10.1007/s11071–021–06220–w.

[21] Wang, H. (2021). A modeling method for a rotor system with an active floating ring squeeze film damper *Proc. Inst. Mech. Eng. C: J. Mech.*, *235*(4), 627–638. https://doi.org/10.1177/0954406220939112.

[22] Liu, Q., Wang, L., Li, Y. & Lei, G. (2021). Single–structured hybrid gas–magnetic bearing and its rotordynamic performance. *Nonlinear Dyn.*, *104*, 333–348 https://doi.org/10.1007/s11071–021–06295–5.

[23] Kasarda, M. E. F., Allaire, P. E., Humphris, R. R., & Gunter, E. J. (1991). The effect of an electromagnetic damper on vibrations in rotating machinery. *ASME Des. Eng. Tech. Conf, Part F168436–4*, 245–251. https://doi.org/10.1115/DETC1991–0341.
[24] Lang, O., Wassermann, J., & Springer, H. (1995). Adaptive vibration control of a rigid rotor supported by active magnetic bearings. *Proceedings of the ASME Turbo Expo*, *5*(October 1996). https://doi.org/10.1115/95–GT–181.
[25] Johnson, D., Brown, G., & Mehmed, O. (1998). A magnetic suspension and excitation system for spin vibration testing of turbomachinery blades. *American Institute of Aeronautics and Astronautics, AIAA*, https://doi.org/10.2514/6.1998–1851.
[26] Matsushita, O., Imashima, T., Hisanaga, Y., & Okubo, H. (2002). A seismic vibration control of flexible rotors using active magnetic bearing. *J. Vib. Acoust. Trans. ASME., 124*(1), 49–57. https://doi.org/10.1115/1.1423633.
[27] Johnson, M., Kasarda, M., & Bash, T. (2002). Active control of gear noise using magnetic bearings for actuation. *ASME International Mechanical Engineering Congress and Exposition, Proceedings*, 177–182. https://doi.org/10.1115/IMECE2002–33051
[28] Keogh, P. S., Cole, M. O. T., & Burrows, C. R. (2002). Multi–state transient rotor vibration control using sampled harmonics. *J. Vib. Acoust. Trans. ASME.*, *124*(2), 186–197. https://doi.org/10.1115/1.1448321.
[29] Kasarda, M. E. F., Mendoza, H., Kirk, R. G., & Wicks, A. (2004). Reduction of subsynchronous vibrations in a single–disk rotor using an active magnetic damper. *Mech. Res. Commun.*, *31*(6), 689–695. https://doi.org/10.1016/j.mechrescom.2004.04.004.
[30] Keogh, P. S., Cole, M. O. T., Sahinkaya, M. N., & Burrows, C. R. (2004). On the control of synchronous vibration in rotor/magnetic bearing systems involving auxiliary bearing contact. *J Eng Gas Turb Power*, *126*(2), 366–372. https://doi.org/10.1115/1.1689362.
[31] Keogh, P. S., & Cole, M. O. T. (2006). Contact dynamic response with misalignment in a flexible rotor/magnetic bearing system. *J Eng Gas Turb Power*, *128*(2), 362–369. https://doi.org/10.1115/1.2056530.
[32] Matras, A. L., Flowers, G. T., Fuentes, R. J., Balas, M., & Fausz, J. (2006). Suppression of persistent rotor vibrations using adaptive techniques. *J. Vib. Acoust. Trans. ASME*, *128*(6), 682–689. https://doi.org/10.1115/1.2345668.
[33] Anegawa, N., Fujiwara, H., & Matsushita, O. (2008). Blade–shaft coupled resonance vibration by using active magnetic bearing excitation. *Proceedings of the ASME Turbo Expo*, *5*(PART B), 1183–1190. https://doi.org/10.1115/GT2008–50904.
[34] De Moraes, D. C., & Nicoletti, R. (2010). Hydrodynamic bearing with electromagnetic actuators: Rotor vibration control and limitations. *Proceedings of ISMA 2010 – International Conference on Noise and Vibration Engineering, Including USD 2010*, 3715–3721.
[35] Jiang, K., & Zhu, C. (2011). Multi–frequency periodic vibration suppressing in active magnetic bearing–rotor systems via response matching in frequency domain. *Mech Syst Signal Process.*, *25*(4), 1417–1429. https://doi.org/10.1016/j.ymssp.2010.11.012.
[36] Yang, S. M. (2011). Electromagnetic actuator implementation and control for resonance vibration reduction in miniature magnetically levitated rotating machines. *IEEE T Ind Electron*, *58*(2), 611–617. https://doi.org/10.1109/TIE.2010.2046000.
[37] Viveros, H. P., & Nicoletti, R. (2014). Lateral vibration attenuation of shafts supported by tilting–pad journal bearing with embedded electromagnetic actuators. *J Eng Gas Turb Power*, *136*(4). https://doi.org/10.1115/1.4026038.
[38] Fang, J., Xu, X., & Xie, J. (2015). Active vibration control of rotor imbalance in active magnetic bearing systems. *JVC/J. Vib. Control*, *21*(4), 684–700. F.
[39] Lyu, X., Di, L., Yoon, S. Y., Lin, Z., & Hu, Y. (2016). A platform for analysis and control design: Emulation of energy storage flywheels on a rotor–AMB test rig. *Mechatronics*, *33*, 146–160. https://doi.org/10.1016/j.mechatronics.2015.12.007

[40] Ebrahimi, R., Ghayour, M., & Khanlo, H. M. (2018). Nonlinear dynamic analysis and experimental verification of a magnetically supported flexible rotor system with auxiliary bearings. *Mech Mach Theory*, *121*, 545–562. https://doi.org/10.1016/j.mechmachtheory.2017.11.018

[41] Kim, H., Sikanen, E., Nerg, J., Sillanpää T., Sopanen, J.T. (2020). Unbalanced Magnetic Pull Effects on Rotordynamics of a High–Speed Induction Generator Supported by Active Magnetic Bearings – Analysis and Experimental Verification. *IEEE Access*, *8*, 212361–212370, doi:10.1109/ACCESS.2020.3039629.

[42] Kalista, K.; Liska, J.; Jakl, J. (2021). A Vibration Sensor–Based Method for Generating the Precise Rotor Orbit Shape with General Notch Filter Method for New Rotor Seal Design Testing and Diagnostics. *Sensors, 21*, 5249. https://doi.org/10.3390/ s21155249.

Processing and tribological behaviour of carbon fiber reinforced polylactic acid composites

B. Suresha[1]*, Vikas Hanamasagar[1] and A. Anand[1]

[1]Department of Mechanical Engineering, The National Institute of Engineering, Mysuru-570 008, India

*Corresponding Author: B. Suresha, sureshab@nie.ac.in

ABSTRACT

Thermoplastic composites are acquiring more importance in various tribological applications due to their self-lubricity, low melting, and ease of fabrication. In this research article, processing via fused deposition modelling and dry sliding wear behaviour of neat and short carbon fiber reinforced polylactic acid composites have been investigated. A series of wear tests using a pin-on-disc wear apparatus was carried out by varying tribo-parameters namely applied load (10, 20, and 30 N), sliding velocity (0.5, 1.0, and 1.5 m/s), and sliding distance (1000, 2000, and 3000 m) following the response surface methodology. Analysis of variance was performed for optimizing the operating parameters with respect to minimum specific wear rate. The experimental outcomes showed decrease in specific wear rate of polylactic acid composites by 70 % in comparison to neat polylactic acid, due to reinforcement of short carbon fibers. Worn surface morphology of neat polylactic acid and carbon fiber reinforced polylactic acid show fine groves in the sliding direction, thinning of fibers and net work of microcracks and few pull-out of fibres from the matrix material.

Keywords: Short carbon fiber reinforced polylactic acid; Response surface methodology; Dry sliding wear; Analysis of variance; Worn surface morphology

1 INTRODUCTION

Thermoplastics belong to a group of polymers that are acquiring a lot of interest due to their ability to soften and melt by the application of heat and can be processed easily either in liquid or heat softened state making them reusable. Thermoplastics are being employed in many different applications due to its ability to improve the desired mechanical and thermal characteristics by suitable addition of reinforcements. The most used manufacturing processes for fabrication of thermoplastics are thermoforming, extrusion, blow moulding, injection moulding and many more [1]. Among the various thermoplastics, polylactic acid (PLA) based thermoplastics are being widely employed in automotive and aerospace applications such as, sliders, bearings, cams and gears because of their better mechanical and thermal properties [2]. Additive manufacturing (AM) has been recently developed as a rapid prototyping technique with its ability to develop complex designed structures with minimum material waste, low cost and higher accuracy as compared with other conventional methods [3].

Due to excessive demand from industries, a wide variety of 3D printing technologies have been developed such as selective laser sintering (SLS), stereo lithography, selective laser melting (SLM) and fused deposition modelling (FDM) [4]. Fused deposition modelling is an AM technique wherein raw material in filament form is passed through the rollers and the heating unit assists in melting of filament, that is layered on to the build platform as per required dimensions and this process continues until the desired structure is obtained [05]. Tymrak et al., [6] found that printing parameters have notable influence on the mechanical characteristics of FDM fabricated parts. The results show that the average tensile strength of acrylonitrile butadiene styrene and PLA specimens are 29 MPa and 57 MPa respectively. Addition of fibers/fillers is the most convenient way to enhance the mechanical, tribological and thermal characteristics of composites [7]. Lopes et al., Wang et al., and Sonsalla et al., [8-10] confirmed that reinforcing of ABS and PLA composites with carbon fibers resulted in improvement of both mechanical and thermal properties. Aqzna et al., [11] investigations showed an improvement in Young's modulus, impact strength, and Rockwell hardness by 21%, 18% and 93% respectively by the addition of zinc ferrite nanoparticles as compared to neat ABS polymer.

Friction, wear and fatigue are the most typically encountered problems in industries resulting in engineering part damage and replacements [12]. Recently, the attention of research is being focussed on study of tribological behaviour of thermoplastics fabricated by AM method. Abdualla et al., [13] studied the results due to different infill geometries on tribological behaviour of ABS specimens fabricated by FDM technique using Pin-on-disc apparatus. Results indicated that the coefficient of friction (CoF) value ranged from 0.27 to 0.51 and wear rate ranged from 0.27×10^{-6} to 0.24×10^{-6} mm^3/Nm. It was noted that both solid and triangular infill pattern showed lower CoF and specific wear rate in comparison to other infill geometries. Sahar et al., [14] looked into the impact of layer orientation and layer thickness on wear and friction properties of PLA samples. Analysis of variance findings proved the influence of layer thickness was insignificant in comparison to normal load and build orientation. The worn surface morphology studies revealed adhesive wear was more dominant as compared to fatigue wear. Kumar et al., [15] illustrated that specific wear rate of ABS specimens raised as the applied load and sliding velocity increased, whereas CoF reduces with higher sliding velocity and sliding distance. Pawlak et al., [16] compared the friction characteristics of PLA fabricated by injection moulding and FDM technique. The results revealed that CoF of specimen fabricated with FDM and injection moulding were similar in the range of 0.35 to 0.55. Pramendra et al., [17] showed that specific wear rate and CoF of PLA composites can be reduced significantly by using natural fibers as reinforcements. Response surface methodology (RSM) serves as a good statistical tool for analysis and optimization of responses efficiently with minimum number of trials [18].

From the literature review, addition of short carbon fibers into various thermoplastic polymers had shown considerable improvement in mechanical, thermal and tribological properties fabricated by conventional methods namely compression moulding, extrusion, injection moulding and so on. The present article focusses on the effect of short carbon fiber reinforcement on dry sliding wear behaviour of PLA composites fabricated by FDM technique.

2 MATERIALS AND METHODS

Materials

Filaments namely Polylactic acid (PLA) and 20 wt. % short carbon fiber reinforced polylactic acid (CF-PLA) composites having 1.75 mm fiber diameter were purchased from eSUN 3D, Karnataka, India. The fiber loading of 20 wt. % in PLA was limited due to optimum strength and stiffness of the final composites by 3D printing. Furthermore, the 20-wt. % carbon fiber reinforcement is found to be optimum for better interfacial bonding between the fiber and matrix, and addition of carbon fiber > 20 wt. % makes the filament brittle and not favourable to produce composites by 3D printing. Table 1 represents the physical and mechanical properties of PLA and CF-PLA composite.

Table 1. Physical and mechanical properties of PLA and CF-PLA composites

Parameters	PLA	CF-PLA
Melting point (°C)	195 – 215	215-225
Glass transition (°C)	60 – 65	60-70
Density (g/cm^3)	1.21-1.43	1.29 -1.5
Tensile strength (MPa)	48	59
Flexural strength (MPa)	89	106

Fabrication of PLA and CF-PLA composites

Figure 1 represents the process flow chart for preparation of PLA and CF-PLA composites using FDM technique. The first step is preparation of the CAD model of required samples using mechanical modelling software's, which is than sliced using an Ultimaker Cura slicing software for development of G-codes. The process parameters considered for printing of samples is as shown in table 2.2.1. The G codes developed are provided as an input for Creality Ender-3 Pro FDM printer. The material in the form of filament is supplied through the rollers to the heating unit, which is heated to a pre-determined temperature at the nozzle and then moved according to the G-codes, depositing the material on the build platform. This procedure is repeated until required samples are obtained and are removed after cooling of samples.

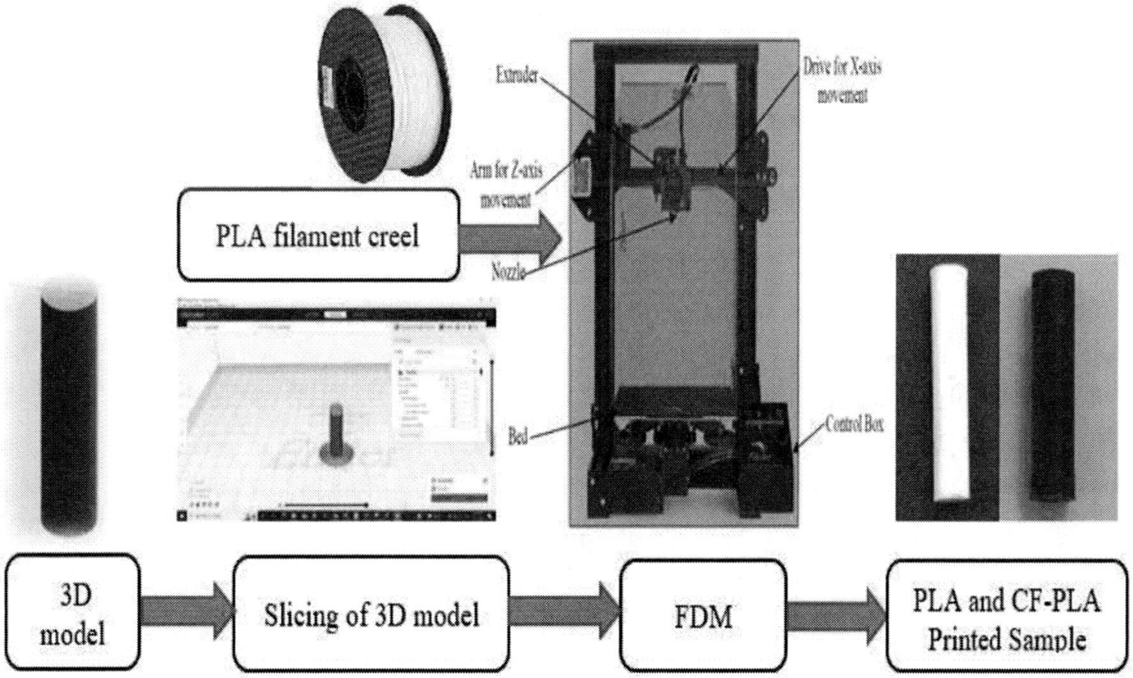

Figure 1. Process flow chart of PLA and CF-PLA composites using FDM

Table 2. Printing parameters for fabrication of PLA and CF-PLA composite

Material / Parameters	PLA	CF-PLA
Nozzle Temperature (^{0}C)	205	210
Bed Temperature (^{0}C)	60	60
Layer thickness(mm)	0.12	0.12
Infill Density (%)	100	100
Printing Speed (mm/sec)	70	70
Build Orientation	Vertical	Vertical
Infill Pattern	Lines	Lines

Density measurement

Density of neat PLA and CF-PLA composite samples was measured using high precision electronic weighing machine following Archimedes' principle as per ASTM D792 standards [19]. The mass of sample in air was recorded and mass of sample submerged in distilled water at room temperature was noted. Based on equation (1), the sample density was determined experimentally. The theoretical density of samples was determined by rule of mixtures using equation (2).

$$\rho_{ex} = \frac{M_a}{M_a - M_w} S \quad (1)$$

where, ρ_{ex} is measured density of sample (g/cm^3), M_a is the mass of sample in air (g), M_w is the mass of sample in distilled water (g), and S= specific gravity of water (g/cm^3).

$$\frac{1}{\rho_{th}} = \frac{M_m}{\rho_m} + \frac{M_r}{\rho_r} \quad (2)$$

where, ρ_{th} is theoretical density of sample (g/cm^3), M_m is the mass fraction of matrix (g), M_r is mass fraction of reinforcement (g), $\rho_m$is the density of matrix (g/cm^3) and ρ_r is the density of reinforcement (g/cm^3).

Dry sliding wear test

Dry sliding wear experiments were conducted using Pin-on-disc apparatus following ASTM G99 standards [20]. The Pin-on-disc apparatus is as viewed in Figure 2. The cylindrical samples of length 30 mm and 8 mm diameter was used for conducting the wear test. The track diameter of 70 mm was maintained for conduction of all experiments. Before the test, the sample and the counter face were cleaned with acetone for removal of debris. The mass of sample before test and after test are measured to determine the wear volume of samples using equation (3).

$$V_w = \frac{m_i - m_f}{\rho_{ex}} \quad (3)$$

where, V_w represents wear volume (mm^3), m_i is the initial mass of sample (g), m_f is the final mass of sample (g), ρ_{ex} represents measured density of the sample (g cm^{-3}). Further, the specific wear rate (K_s) is determined using equation (4).

$$K_S = \frac{V_W}{P \times D} \quad (4)$$

where, K_S represents specific wear rate (mm^3/ N-m), V_w is the wear volume (mm^3), P is the applied normal load (N), and D is the sliding distance (m).

Figure 2. Pin on disc apparatus

In this present experimental work, the input tribo-parameters such as applied load, sliding velocity and sliding distance were modelled and optimised using Box-Behnken design via Response surface methodology (RSM) for minimizing number of trails and time in finding minimum K_S as response parameter. Minitab 19.1, a statistical software was used for creating and analysing of design matrix. Table 3 summarizes the factors and their levels considered for experimental design.

Table 3. Factors and their levels considered in Box Behnken Design

Control factors	Designation	Levels		
		Low (-1)	Medium (0)	High (+1)
Applied load (N)	A	10	20	30
Sliding velocity(m/s)	B	0.5	1.0	1.5
Sliding distance(m)	C	1000	2000	3000

3 RESULTS AND DISCUSSION

Effect of carbon fiber on density of PLA composite

The results of measured density for PLA and CF-PLA composites are listed in Table 4. The difference between experimental and theoretical densities indicate presence of voids and pores in the fabricated composite samples, that have a considerable impact on the material properties [17]. The CF-PLA samples showed increase in density with comparison to PLA samples, this is because the carbon fibers possess high density and good adhesion with PLA matrix thereby resulting in increased density as compared to PLA. The decreased void percent reported for PLA and CF-PLA samples was owing to strong interlayer bonding obtained between layers of samples with less air gaps, confirming the efficiency of the FDM process used for composite fabrication.

Table 4. Measured density of PLA and CF-PLA composites

Sl. No	Material	Theoretical Density (g cm^{-3})	Experimental Density (g cm^{-3})	Voids (%)
1.	PLA	1.23	1.20	2.44
2.	CF-PLA	1.26	1.22	3.18

Effect of carbon fibers on K_S of PLA composite

Table 5 presents the experimental design obtained by Box-Behnken design and the experimental outcomes and the Specific wear rate (K_S) obtained for PLA and CF-PLA composite. The experiments were performed as per design and the K_S values were determined. The minimum Ks value for PLA composites obtained was 0.0931 under applied load of 10 N, sliding velocity of 1 m/s and 1000 m sliding distance, whereas the Ks value for CF-PLA composites was 0.0707 at the same wear test conditions.

Table 5. Experimental design and results of composites

Sl. No.	Load (N)	Velocity (m/s)	Distance (m)	Ks (10^{-13} mm^3/N m)	
				PLA	CF-PLA
01	10	0.5	2000	0.2014	0.1017
02	30	0.5	2000	0.9652	0.9154
03	10	1.5	2000	0.2115	0.1515
04	30	1.5	2000	1.2500	1.2514
05	10	1.0	1000	0.0931	0.0707
06	30	1.0	1000	0.6525	0.6535
07	10	1.0	3000	0.3317	0.1371
08	30	1.0	3000	1.5021	1.5012
09	20	0.5	1000	0.8012	0.7501
10	20	1.5	1000	1.0100	0.9317
11	20	0.5	3000	1.0512	0.9615
12	20	1.5	3000	1.4012	1.3512
13	20	1.0	2000	1.0221	0.9514
14	20	1.0	2000	1.1125	0.9914
15	20	1.0	2000	1.1254	0.9774

The statistical approach of Analysis of variance (ANOVA) was used to generate a set of conclusions according to the experimental results and significant contributions of each factor on the response [21]. ANOVA was performed for 5% level of significance and 95% degree of confidence. Tables 6 and 7 list the results of the ANOVA test for K_S of PLA and CF-PLA composites respectively.

Table 6. ANOVA results of *Ks* for PLA composite

Source	DF	Adj SS	Adj MS	F-Value	P-Value	Contribution (%)
Linear	3	2.1044	0.7014	74.31	0	71.3766
A	1	1.6129	1.6129	170.88	0	54.7101
B	1	0.0911	0.0911	9.65	0.027	3.0901
C	1	0.4002	0.4002	42.41	0.001	13.5762
Square	3	0.6968	0.2323	24.61	0.002	23.6387
A×A	1	0.6917	0.6917	73.29	0	23.4651
B×B	1	0.000	0.0000	0	0.952	0.00135
C×C	1	0.0021	0.0021	0.22	0.655	0.0719

R-sq=98.40%, R-sq(adj)= 95.52%, R-sq (pred)= 77.34%

Table 7. ANOVA results for *Ks* of CF-PLA composite

Source	DF	Adj SS	Adj MS	F-Value	P-Value	Contribution (%)
Linear	3	2.2758	0.7586	248.09	0	74.75504
A	1	1.8629	1.8629	609.23	0	61.19263
B	1	0.1145	0.1145	37.45	0.002	3.761369
C	1	0.2983	0.2983	97.58	0	9.801043
Square	3	0.5693	0.1897	62.06	0	18.70173
A×A	1	0.5564	0.5564	181.97	0	18.27734
B\×B	1	0.0014	0.0014	0.47	0.522	0.047629
C×C	1	0.0001	0.0001	0.04	0.858	0.003613

R-sq=99.5%, R-sq(adj)= 98.59%, R-sq (pred)= 92.34%

It is clear from the ANOVA results that applied load has the highest significant contribution of 54.71% and 61.19% on *Ks* of PLA and CF-PLA composites respectively, followed by sliding distance (13.97% and 9.80%) and sliding velocity (3.09% and 3.76%). The R-square values obtained are nearer to 100% which states that the experimental models are more valid and in good correlation to experimental data. The higher values of R-square (adj) and R-square (pred) resembles good predictability of model. The normal probability plots of *Ks* are as depicted in figure 3, wherein the residuals are approaching towards straight line indicating that the experimental results obtained are normally distributed and dependable. The quadratic regression models obtained with respect to *Ks* of PLA and CF-PLA samples are shown in equations 5 and 6 respectively. The K_S values obtained by experimentation were compared with the K_S values obtained by regression models and were closer within acceptable error.

$$Ks = -1.320 + 0.176 \times A - 0.228 \times B - 0.000027 \times C - 0.004328 \times A^2 + 0.013 \times B^2 + 0.000054 \times C^2 + 0.01374 \times A \times B + 0.000014 \times A \times C + 0.000071 \times B \times C \quad (5)$$

$$Ks = -0.794 + 0.1502 \times A - 0.414 \times B - 0.000323 \times C - 0.003882 \times A^2 + 0.079 \times B^2 + 0.00032 \times C^2 + 0.01431 \times A \times B + 0.000020 \times A \times C + 0.000104 \times B \times C \quad (6)$$

IOP Conf. Series: Materials Science and Engineering 1272 (2022) 012022 doi:10.1088/1757-899X/1272/1/012022

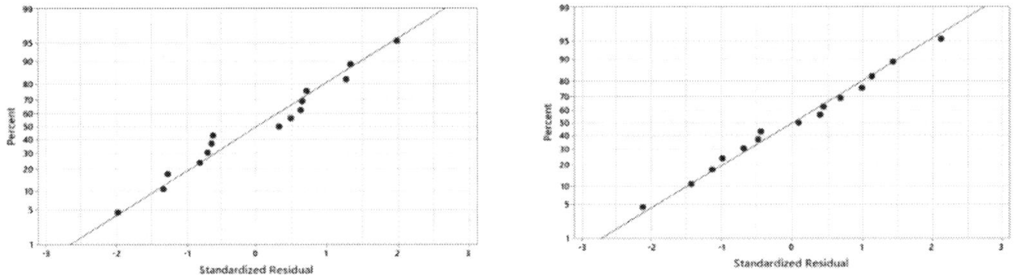

(a) (b)

Figure 3. Normal probability plot for *Ks* of (a) PLA (b) CF-PLA

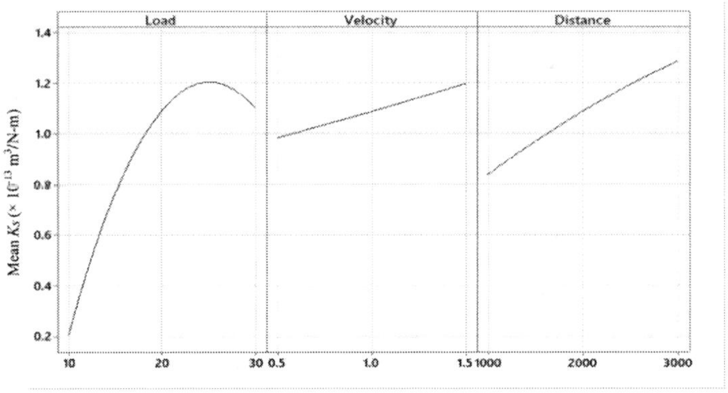

Figure 4. Effect of control factors on K_S of PLA

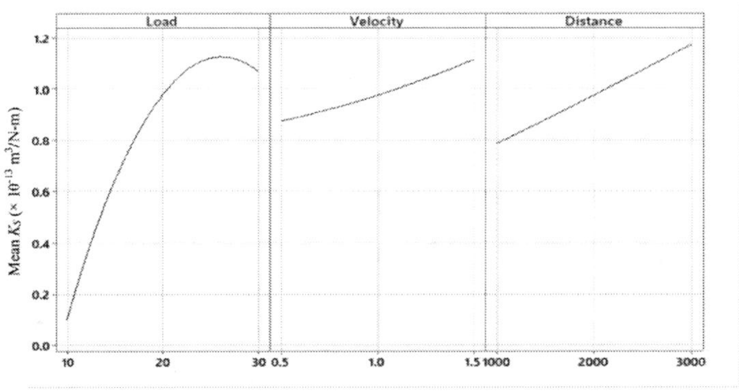

Figure 5. Effect of control factors on K_S of CF-PLA

Figures 4 and 5 represent the plots for effect of control factors on the K_S of PLA and CF-PLA composites respectively. It is noted that as the applied load increased from 10 N to 20 N, there was an increment in mean *Ks* up to 1.1 mm^3/Nm for CF-PLA composites, which was less as compared to PLA composites. PLA composites have poor wear performance due to the process of softening, as well as formation of layered film on counter face during sliding [22]. With increase in load applied, the stress developed at the interface increases, resulting plastic deformation of the sample and increase in *Ks* trends. Increase in sliding velocity from 0.5 m/s to 1.5 m/s caused the counter face to deliver more shearing action to the sample surface, resulting in increased material removal, the obtained trends are in line with previous literature works [14-16]. The *Ks* value at maximum sliding velocity for CF-PLA composite was lower as compared to PLA composites due to improvement in shear strength of PLA composite by presence of carbon fibers.

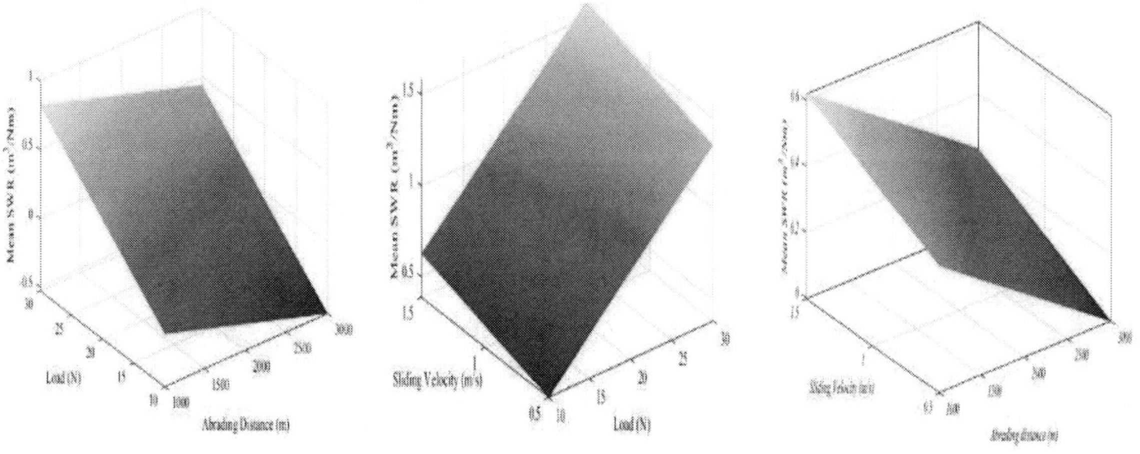

(a) (b) (c)

Figure 6. Surface plots of *Ks* for (a) load vs distance, (b) velocity vs load, (c) velocity vs distance

Figure 6 represents the surface plots obtained for different interactions between applied load, sliding distance and sliding velocity. It is noted that with increase in sliding velocity at lower load, the K_S value tend to increased significantly. The plot of applied load vs sliding distance in Figure 5c demonstrates that at lower load, with increase in the sliding distance, the K_S value increased drastically.

Optimisation of K_S

Response surface methodology desirability optimization analyses experimental results and predicts the best independent factors along with corresponding levels and responses. A number of iterations were carried out to obtain the optimize value with respect to minimum *Ks*. Figure 7 represents the optimization plot showing the optimum wear conditions for independent factors of load applied, sliding velocity and sliding distance for PLA and CF-PLA composites respectively. To verify the repeatability and reliability of the predicted optimum wear conditions, an experimental test was carried out. Table 8 represents the predicted and experimental results obtained at the optimum wear conditions with respect to minimum *Ks* for PLA and CF-PLA samples. The *Ks* value obtained by confirmation experiment are closer to the experimental *Ks* value obtained by RSM. The *Ks* value obtained experimentally for CF-PLA composite at optimised condition was lower in comparison to PLA composite. *Ks* value of PLA composite was reduced due to higher load bearing capacity along with increased strength and stiffness by reinforcement of carbon fibers, the obtained results are in line with the previous literature works [12-23].

Table 8. Confirmation experiments for *Ks*

Material	Applied load (N)	Sliding velocity (m/s)	Sliding Distance (m)	Predicted *Ks* ($10^{-13}m^3/Nm$)	Experimental *Ks* ($10^{-13}m^3/Nm$)	Error (%)
PLA	10	0.78	1000	0.0943	0.0901	4.45
CF-PLA	10	0.5	3000	0.0255	0.0270	5.55

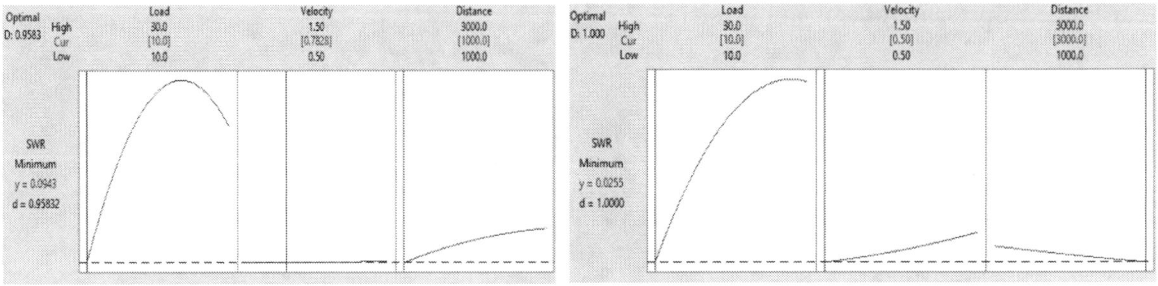

(a) (b)

Figure 7. Desirability plot for minimum K_S of PLA and CF-PLA composite

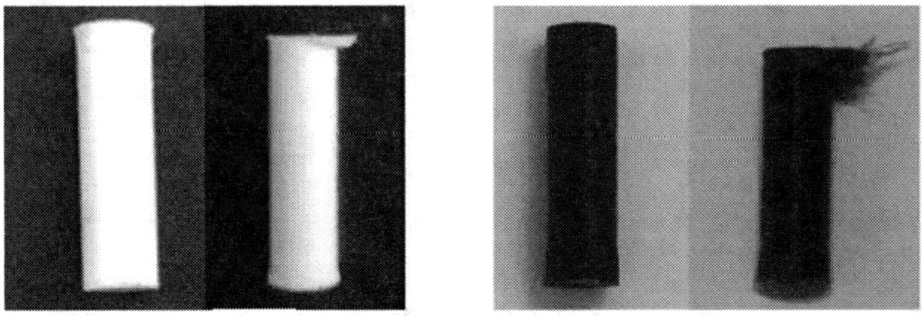

(a) (b)

Figure 8. Samples for PLA and CF-PLA composite (a) Before test, (b) After test

Worn surface morphology

The optical images as represented in figure 8 indicate the deformation characteristics of worn samples. The samples are subjected to shear stress and the material at the surface of sample is deformed along the tangential component of sliding direction. This wear type is formed due to localised bonding developed between the surface of sample and the counter disc surface as a result of increase in temperature at the interface. The wear debris from the worn-out samples was adhered to the counter face forming a thin transfer layer [14] as shown in Figure 9.

The features of worn surface of PLA and CF-PLA composites for 30 N applied load and sliding distance of 3000 m are as depicted in Figures 10 and 11. As observed from Figure 10, the depth of grooves is steeper at higher load, which indicates higher wear loss of neat PLA sample. Similar observations were obtained during experimentation at higher load and sliding distance.

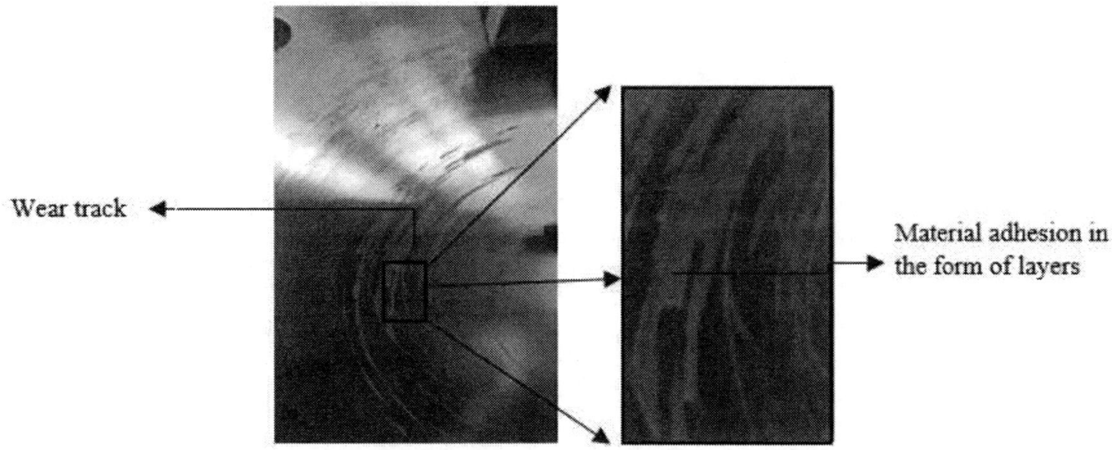

Figure 9. Surface image of counter face after wear test of composite

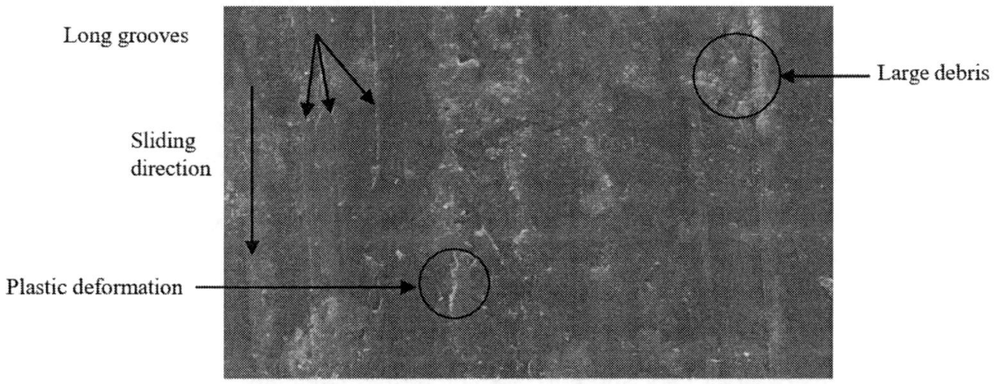

Figure 10. Worn surface features of PLA composites

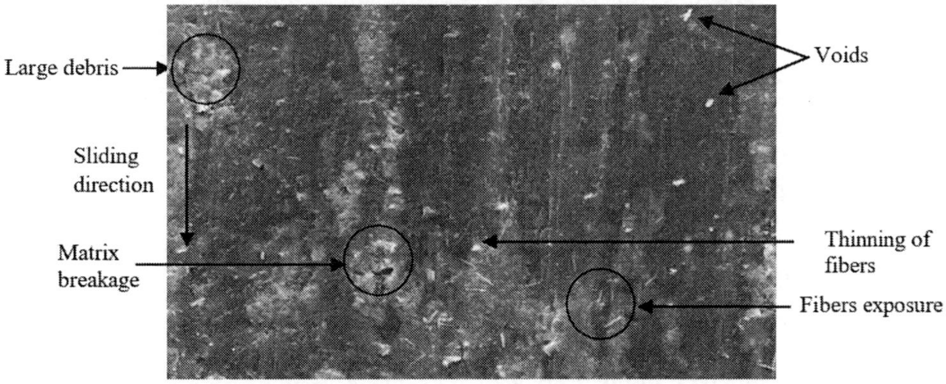

Figure 11. Worn surface features of CF-PLA composites

Figure 11 shows the exposure of carbon fibers in the matrix, thinning of fibers in the sliding direction and few fibers pull-out. The carbon fibers embedded in the matrix assist in providing support and protect the matrix from being worn during sliding, hence reducing wear loss. There was no detachment or pull out of fibers observed, replicating high strength and load bearing ability of carbon fibers. Accumulation of debris and formation of transfer film can be observed representing adhesive wear as the predominant wear mechanism.

4 CONCLUSION

The composite samples were fabricated successfully by using Fused deposition modelling technique. The effect of short carbon fibers on density and tribological behaviour on PLA-based thermoplastics were investigated. The inclusion of carbon fibers increased the density of PLA samples by 1.3 percent, owing to the higher density of carbon fibers and efficiency of FDM technique. The ANOVA findings disclosed that load applied had a substantial effect on minimization of K_S followed by sliding distance and sliding velocity. The RSM method projected the best outcomes for the combination of all independent factors, resulting in the lower specific wear rate. The optimum results obtained by statistical approach were validated by conducting a confirmation test and the obtained results were in concurrencewith experimental outcomes. By incorporation of carbon fibers, the specific wear rate was reduced by 70% in comparison to neat PLA samples. This reduction was attributed to the inclusion of carbon fibers improving the lubricating property and stiffness of PLA composite. The SEM micrographs of worn surfaces of PLA and CF-PLA samples showed adhesive wear as the primary wear mechanism.

REFERENCES

[1] Biron M., (2018). *Thermoplastics and thermoplastic composites*, William Andrew, Elsevier.

[2] Andrew J.J and Dhakal H.N., (2022). Sustainable bio based composites for advanced applications: recent trends and future opportunities–A critical review, *Composites Part C: Open Access,* Vol. 7, pp.100-220.

[3] Mueller B., (2012). Additive manufacturing technologies –Rapid prototyping to direct digital manufacturing. *Assembly Automation*, Vol. 32, Emerald Group Publishing Limited.

[4] Gibson I, Rosen D.W, Stucker B, Khorasani M, Rosen D, Stucker B, Khorasani M.,(2021). *Additive manufacturing technologies*, Vol. 17, Cham, Switzerland: Springer.

[5] Pavan Kumar Penumakala, Jose Santo, Alen Thomas., (2020). A critical review on the fused deposition modelling of thermoplastic polymer composites, *Composites: Part B,* Vol. 201, pp. 108336.

[6] Tymrak B.M, Kreiger M, Pearce J.M., (2014). Mechanical properties of components fabricated with open-source 3-D printers under realistic environmental conditions. *Materials & Design*, Vol. 58, pp. 242-246.

[7] Sajjad Mamaghani Shishavan, Taher Azdast, Samrand Rash Ahmadi., (2014). Investigation of the effect of nano clay and processing parameters on the tensile strength and hardness of injection molded Acrylonitrile Butadiene Styrene–organoclay nanocomposites. *Materials and Design,* Vol. 58, pp527-534.

[8] Lopes B.J. and Almeida D., (2019). Initial development and characterization of carbon fiber reinforced ABS for future Additive Manufacturing applications. *Materials Today: Proceedings*, Vol. 8, pp.719-730.

[9] Wang Kui, Li Shixan, Rao Yanni, Yiyun Wu, Peng Yong, Yao Song, Honghao Zhang, Said Ahzi.,(2019). Flexure behaviours of ABS-based composites containing carbon and Kevlar fibers by material extrusion 3D printing. *Polymers*, Vol. 11, pp.1878.

[10] Sonsalla T, Moore A.L, Meng W.J, Radadia A.D, Weiss L., (2018). 3-D printer settingseffects on the thermal conductivity of acrylonitrile butadiene styrene (ABS). *Polymer Testing*, Vol. 70, pp.389-395.

[11] Aqzna S.S, Yeoh C.K, Idris M.S, Teh P.L, Bin Hamzah K.A, Atiqah T.N., (2018). Effect of different filler content of ABS–zinc ferrite composites on mechanical, electrical and thermal conductivity by using 3D printing. *Journal of Vinyl and Additive Technology*, Vol. 24, pp. E217-E229.

[12] Shrinivasan R, Suresh Babu, Udhaya Rani V, Suganthi M, Dheenasagar R., (2020). Comparison of tribological behaviour for parts fabricated through fused deposition modelling process on ABS and 20% carbon fibre PLA, *Materials today: Proceedings*, Vol. 27, pp. 1780-1786.
[13] Abdollah M.F.B, Norani M.N.M, Abdullah M.I.H.C, Amiruddin H, Ramli F.R, Tamaldin N., (2020). Synergistic effect of loads and speeds on the dry sliding behaviour of fused filament fabrication 3D-printed acrylonitrile butadiene styrene pins with different internal geometries, *The International Journal of Advanced Manufacturing Technology*, Vol. 108, pp.2525- 2539.
[14] Sahar Zhiani Hervan, Atakan Altın kaynak, Zeynep Parlar., (2020), Hardness, friction and wear characteristics of 3D-printed PLA polymer, Proc. I, Part J: *Journal of Engineering Tribology*. Vol. 235, pp.1590-1598.
[15] Kumar S. and Roy B.S., (2020). Tribological properties of acrylonitrile butadiene styrene in self–mated contacts and against steel disc. *Materials Today: Proceedings*, Vol. 26, pp.2388-2394.
[16] Pawlak W, Wieleba W, Wroblewsk R., (2019). Research of tribological properties of polylactide (PLA) in the 3D printing process in comparison to the injection process. *Tribologia*. Vol. 1, pp. 25-28.
[17] Pramendra Kumar Bajpai, Inderdeep Singh, Jitendra Madaan., (2013). Tribological behaviour of natural fiber reinforced PLA composites, *Wear,* Vol. 297, pp. 829-840.
[18] Abhijeet Nayak, Dipak Kumar Jesthi, Bharat Chandra Routara, Diptikanta Das, Ramesh Kumar Nayak., (2018). Tribological properties of glass/carbon hybrid composites through inter-ply arrangement using Response Surface Methodology. *Materials Today: Proceedings,* Vol. 5, pp. 19828- 19835.
[19] ASTM D792-20., (2020). *Standard Test Methods for Density and Specific Gravity (Relative Density) of Plastics by Displacement,* West Conshohocken, ASTM International.
[20] ASTM G99-17., (2017). *Standard Test Method for Wear Testing with a Pin-on-Disk Apparatus*, West Conshohocken, ASTM International.
[21] Divya G.S, Suresha B, Somashekar M, Jamadar I.M., (2021). Dynamic Mechanical Analysis and Optimization of Hybrid Carbon-Epoxy Composites Wear Using Taguchi Method, *Tribology in industry,* Vol 43, pp. 298-309.
[22] Ziyan Man, Hongjian Wang, Qinghao He, Dae-Eun Kim, Li Chang., (2021). Friction and wear behaviour of additively manufactured continuous carbon fibre reinforced PA6 composites. *Composites: Part B*, Vol. 226. Pp. 109332.
[23] Sudin M. N, Ramli F. R, Alkahari M. R, Abdullah M. A., (2018). Comparison of wear behaviour of ABS and ABS composite parts fabricated via fused deposition modelling, *International Journal of Advanced and Applied Sciences*, Vol. 5, pp. 164-169.

IOP Conf. Series: Materials Science and Engineering 1272 (2022) 012023 doi:10.1088/1757-899X/1272/1/012023

Dual band-notched antenna design using composite right/ left handed metamaterial

V Indu Nair

Biomedical Engineering Department, Park College of Engineering and Technology, Kaniyur, Coimbatore-641659, India.

E-mail: indu.elegance@gmail.com

Abstract. A small metamaterial notch-antenna with band reject at 3.5 and 4.4 GHz is proposed by introducing reactance to balanced composite left-right unit cell, thus making it non-resonant at the desired frequency without any change in overall size. Simulations were carried on HFSS and antenna with notch band implemented on RT\Duroid substrate. The type I and type II CRLH metamaterial notch-antenna illustrate that the antenna notches at 3.5 and 4.4 GHz, with VSWR > 2. Such filters suitable for UWB application where there is interference.

1. Introduction

Federal Communications Commission (FCC) has endorsed the use of spectrum ranging from 3.1–10.6 GHz for use in UWB communication since 2002. But within this wider range of UWB, lies small bands that interfere with the UWB spectrum. The WLAN frequency band, which operates at 5–6 GHz, also falls within the UWB range and causes electromagnetic interference.

To achieve the desired notch band for UWB systems, several methods are available in open literature. Single or more band reject function achieved for band rejection performance in UWB antennas with diverse design techniques has been proposed. The outstanding power of metamaterial that controls the electromagnetic radiations is a major attraction in design of notches [1, 2] and antenna system. Further study on band reject UWB antennas include; single and multiple notched UWB antennas has been proposed [3, 4]. Notch band produced by using resonators [5] incorporate use of filters along with antennas [6]

Recent review on metamaterial as notch include Modified Triangular Split Ring Resonator (TSRR) with metamaterial property obtains desirable negative permeability bands that help in accommodating all three frequency bands [7] Printed, electrically small, capacitively-loaded loop (CLL) resonators are utilized to achieve the desired notch [8], switching notched bands of UWB antennas [9] wherein the parameters of the band-notched configurations can be controlled individually to bring alteration to the upper and lower bands. Modern notch design with matamaterials for use in future communication system are discussed in [10,11 and 12]

Brief literature survey highlighted the drawbacks of existing antennas where additional parasitic elements or extra metamaterial structures were used. Many of these designs are complex, hence a simple antenna is proposed in this research work.

A metamaterial notch-antenna to work as dual notch in 3.5 and 4.4 GHz range is designed. High Frequency Structural Simulator (HFSS) is used to obtain various parameters like reflection coefficient and VSWR. Such an antenna is preferred because a slight variation in the stub length produced band reject capabilities at different frequencies. The designed double notched UWB antenna is optimized, fabricated, and measurement made for validation and verification purposes. Two novel notch-antennas using CRLH Metamaterial is designed and prototype realized to reject dual band of frequencies at 3.5 and 4.4 GHz.

IOP Conf. Series: Materials Science and Engineering 1272 (2022) 012023 doi:10.1088/1757-899X/1272/1/012023

The remaining of the chapter is organized as follows: section 2 discusses frequency notched bands with CRLH Metamaterial antenna. Section 3 provides a design of balanced unit cell CRLH Metamaterial antenna. Antenna design for Balanced and Unbalanced Unit cell is described in section 4 and 5 respectively. Simulation results and discussions are carried out in section 6. 7 concludes the chapters

2. Frequency notched bands with CRLH metamaterial antenna

UWB systems shares band with existing systems such as WLAN, WiMAX hence it mandates to avoid interference with nearby communication systems. A conventional filter can be used in front end receiver but it is possible to design UWB antennas with a band reject characteristic that aids in narrow band signal rejection.

Set of horizontal and vertical lines with turns form the meander line structure [13]. Increase in number of turn's increases efficiency as it is directly related. Shortening of meander lines leads to reduction in reflection coefficient and hence it helps realize compact antennas. Salient features of meander line antennas are wideband characteristics and compact small antennas. Size of the antenna at a given frequency is reduced by a factor that is proportional to the number of turns, as far as meander line antennas are considered Formatting the title.

This chapter presents one of methods by removing narrow band resonant structure and introducing structures to control the notched band and to notch multiple bands. This antenna is designed based on metamaterial to work as dual notch in 3.5 and 4.4 GHz.

3. CRLH MTM Antenna

Antenna design using metamaterial is extensively used in recent years owing to their unique properties of having at a particular frequency, zero propagation constant.

Fundamentals study on Balanced CRLH TL Unit-Cell reveals that in a typical transmission line (TL), the series impedance that symbolizes the stored energy of the magnetic field is inductive (L_R) and shunt admittance which epitomizes the stored energy of the electric field, is capacitive (C_R). Such a TL is termed as pure right handed transmission line (RH-TL). The equivalent model of Homogeneous RH-TL and Homogeneous LH-TL, is shown in figure 1 a and b. However, in a simulated pure left handed transmission line (LH-TL) series impedance is a capacitive (C_L) and shunt admittance is inductive (L_L) in nature.

Physical realization of a pure LH transmission line not made possible because of the occurrence of parasitic effects of right handed (RH) geometry. General model of a LH-TL, termed as composite right left handed transmission line (CRLH TL) can be realized. The equivalent model of unit-cell CRLH TL is shown in figure 1c. The balanced equivalent CRLH TL consists of a series RH inductance (L_R), a series LH capacitance (C_L), a shunt RH capacitance (C_R), and a shunt LH inductance (L_L).

IOP Conf. Series: Materials Science and Engineering 1272 (2022) 012023 doi:10.1088/1757-899X/1272/1/012023

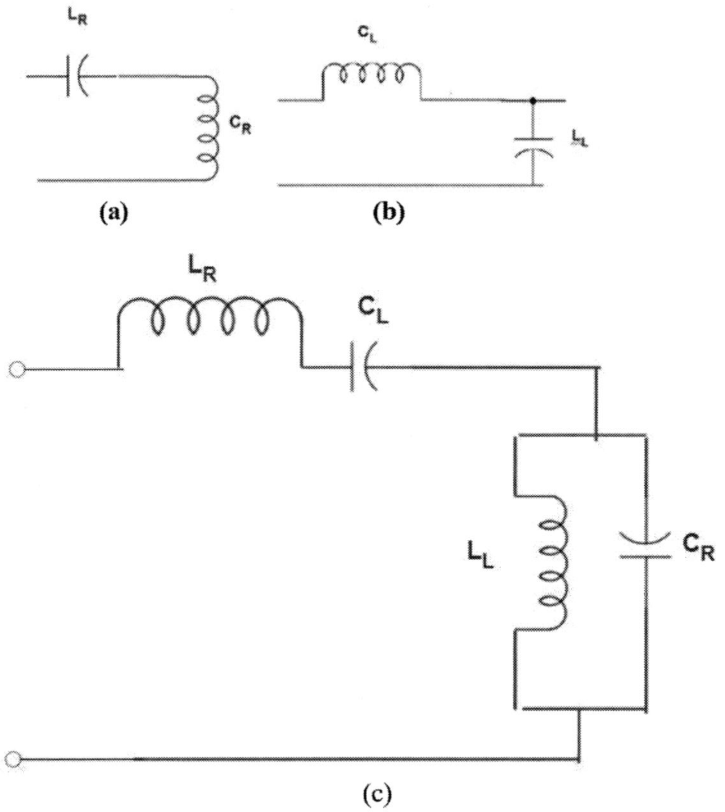

Figure 1. Equivalent circuit models for unit-cell: **(a)** Homogeneous PRH TL, **(b)** Homogeneous PLH TL, **(c)** Homogeneous CRLH TL

In most of the metamaterials the propagation of electromagnetic waves comply with the right-hand rule for the E, H and β vector field, where E is the electric field, H the magnetic field, and β is the phase constant, the propagation constant γ, defined as follows

$$\gamma=\alpha+j\,\beta=\sqrt{Z_s(\omega),Y_s(\omega)} \tag{1}$$

If Z_s and Y_p are the series impedance per unit length and shunt admittance per-unit length of CRLH TL unit cell respectively. The propagation constant for the CRLH unit-cell is given by

$$\beta(\omega)=s(\omega)\sqrt{\omega^2 L_R C_R+\frac{1}{\omega^2 L_L C_L}-(\frac{L_R}{L_L}+\frac{C_R}{C_L})} \tag{2}$$

where $S(\omega)$ is the sign function given by

$$s(\omega)=\begin{cases}-1,\text{if } \omega<\omega_{se}=\min\left(\frac{1}{\sqrt{L_R C_L}},\frac{1}{\sqrt{L_L C_R}}\right)\\ 0 \qquad \text{if } \omega_{se}<\omega<\omega_{sh}\\ +1,\text{if } \omega<\omega_{sh}=\max(\frac{1}{\sqrt{L_R C_L}},\frac{1}{\sqrt{L_L C_R}})\end{cases} \tag{3}$$

where

ω_{se}, is the series frequency and ω_{sh} ,shunt resonance frequency, the ω_{se} and ω_{sh} are given in equation (4)

$$\omega_{se}=1/\sqrt{L_R C_L} \tag{4}$$

$$\omega_{sh}=1/\sqrt{L_L C_R}\text{ , and}$$

$$Z_s(\omega)=j(\omega L_R-1/\ \omega C_L) \tag{5}$$

$$Y_p(\omega)=j(\omega C_R-1/\omega L_L) \tag{6}$$

The phase velocity, V_p in the same direction as the direction of signal energy propagation as given by the following equation (7) and (8).

$$V_p=\omega\ /\beta=\omega^2 L_L C_L \tag{7}$$

$$V_g=(\delta\beta/\delta\ \omega)^{-1}=\ \text{-}\omega^2 L_L C_L \tag{8}$$

Where V_g = Group Velocity

4. Antenna Desi

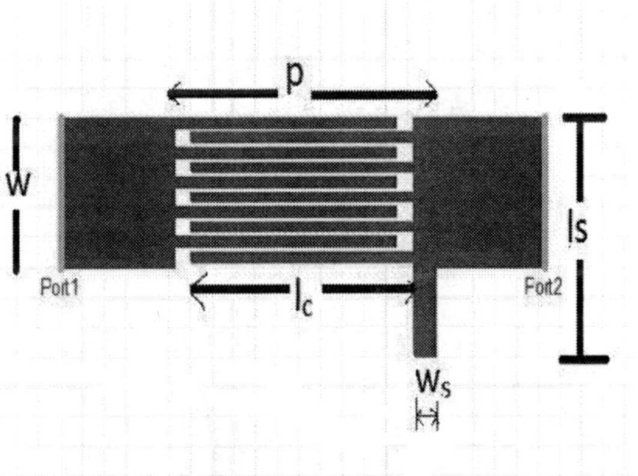

Figure 2. Basic unit cell of CRLH MTM antenna

Figure 2 shows the balanced basic unit cell designed using CRLH metamaterial. The substrate used is RT/Duroid 5880 with a relative dielectric constant, ε_r=2.2 and thickness h=1.57 mm and the inter-digital capacitor (IDC) for a unit cell includes 7 fingers each has a unit-cell period (p) 12.3 mm, stub length (l_s) of 10.9 mm, stub width (W_S) of 1.00 mm, interdigital finger length (l_c) of 10.5 mm, width of interdigital finger (W_C) of 0.30 mm, spacing between the interdigital fingers (s) are 0.20 mm.

Two edge ports namely port1 and port 2 are defined at both ends of the structure by taking micro strip line of width 5mm corresponding to 50Ω impedance at the centre frequency of 3.5 GHz for the substrate RT/Duroid 5880. Unlike conventional resonating-wave antennas, the (Leaky Wave Antenna) LWA is based on the concept of a traveling wave. When a radio-frequency signal is applied to one of the input ports, the traveling wave leaks out energy as it progressively travels toward the second port. Directivity of the radiated beam depends on the leaked energy is a function of the propagation constant along the structure.

4.1 Bandstop Filter Design

In this work, compact type I and type II UWB antenna having dual band notched characteristics has been proposed. The notched bands have been achieved by introducing omega shaped and meander shaped structure in stub length of CRLH metamaterial. A radiating antenna is considered, reactance added to obtain stop band, hence the initial design under consideration is resonant antenna. The formula for stop band and resonant antenna can be obtained from the unbalance and balance CRLH transition frequency.
For the unbalance and balance CRLH the transition frequency is given as [16]

$$\omega_o \text{ (Unbalanced} = \tfrac{1}{4}\sqrt{L_R C_R L_L C_L}\text{)} \tag{9}$$

$$\omega_o \text{ (balanced} = 1 \backslash \sqrt{LC}\text{)} \tag{10}$$

The proposed notch-antenna is designed by removing a rectangle 17mm by 17mm from the stub length of the basic unit cell and the meander-line has been introduced.

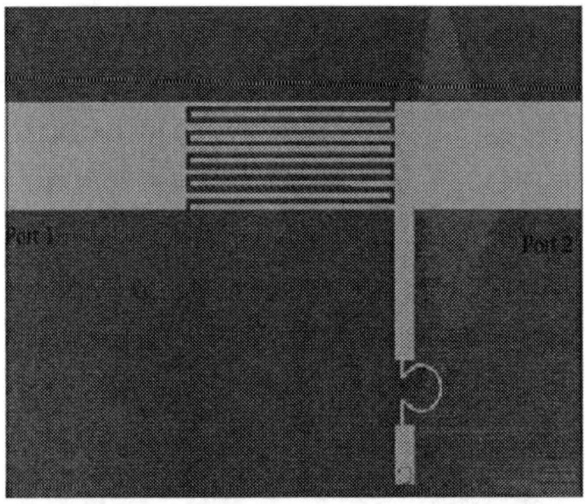

Figure 3. Omega shaped unit cell antenna (type I)

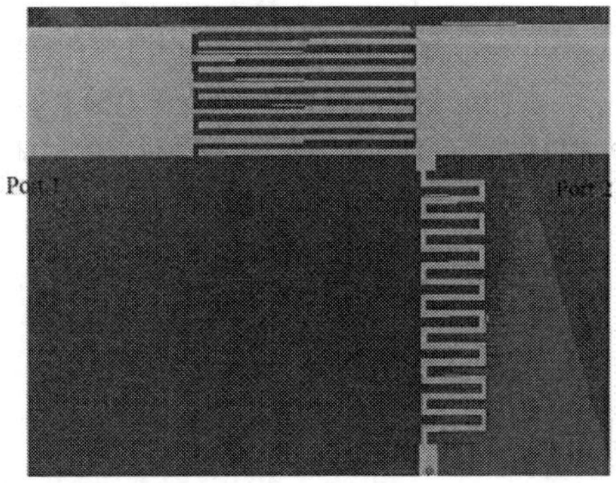

Figure 4. Meander line unit cell antenna (type II)

In figure 3, a rectangle of dimension 17mm by 17mm is removed and omega shaped inductor is introduced. The omega shape is designed with inner radius of the arc, 0.35mm and outer radius of 0.55mm. This type of antenna with omega shaped structure in unit cell is classified as type I. The parametric study has been carried out in the proposed antenna.

The design of CRLH with meander line is shown in figure 4. The design of the meander line antenna began from the single element, later on developed into the 16 elements of 2mm by 1.5mm. This type of antenna with meander line in stub length is classified as type II. Finally based on the evaluation of reflection coefficient, gain and radiation pattern characteristics; the performance of antenna has been studied.

5. Results and Discussions

The type I and type II antenna are simulated in HFSS 14.0 tool. Results of simulation, namely reflection coefficient are shown in table 1. Table 1 indicates that type I Antenna notches at 3.5 and 4.4 GHz. The reflection coefficient and VSWR were analyzed to show the antenna performance. The filters were implemented on Rogers RT/Duroid 5880 substrate with dielectric constant ε_r= 2.2 and thickness h= 1.57 mm. As phase shift is determined by $\phi = -\beta d = 0$, there is no phase shift across the resonator at $\beta = 0$, but the introduction of omega shape and meander line in the stub, introduces shunt inductance and shunt capacitance, thus β is greater than zero resulting in band stop or band reject characteristic. Prototypes of the proposed design (type I and type II antenna) were fabricated as shown in figures 5 and 6 to validate simulated performance.

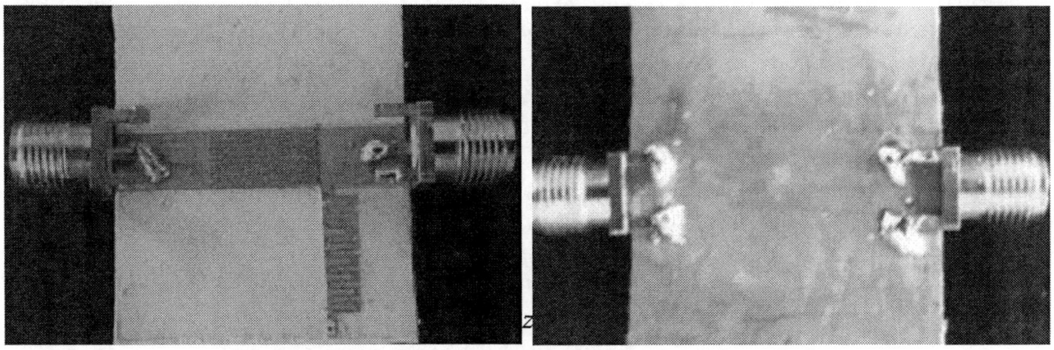

Figure 5. Fabricated antenna type I Front & rear-view

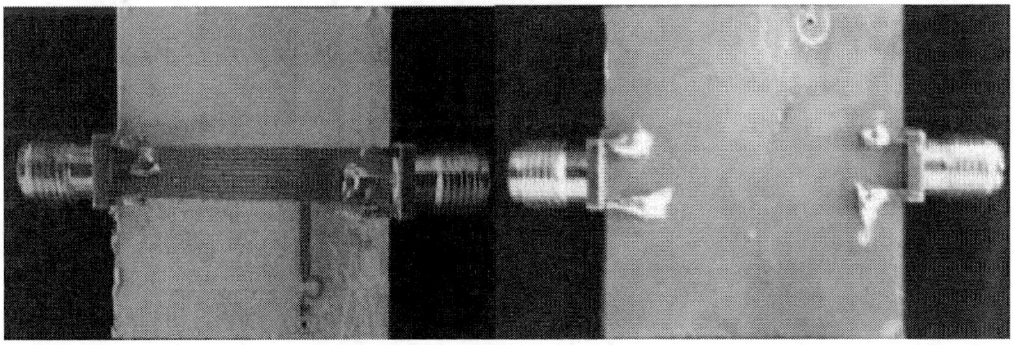

Figure 6. Fabricated antenna type II Front & rear-view

Table 1. Stop band for stub length of 15.7mm

Stub length	Antenna Type	Stop band	S11 (dB)
15.7mm	type I	3.5GHz	-0.645
		4.4GHz	-0.299
	type II	3.5GHz	-0.545
		4.4GHz	-0.199

a. Reflection coefficient.

Figure 7 depicts the simulated and measured reflection coefficient for type I and type II antenna. It is clear from figure 7 that stop band or reject band occurs at 3.5 GHz and 4.4 GHz where S_{11} = -0.645 dB and -0.299 dB and S_{11} = -0.545 dB and -0.199 dB for type I for type II respectively. The stop bands for type I and type II, antenna is tabulated in Table 1. Table 1 clearly indicates that notch or reject band can be achieved at 3.5 GHz and 4.4 GHz with meander lines and omega shaped reactance introduced with stub length of 15.7 mm for both type 1 and type 2. However, in its stop-band the CRLH has a negative permeability ($-\mu$) or a negative permittivity ($-\varepsilon$). At the resonant frequency, due to high impedance matching, the impedance of balanced CRLH metamaterial antenna is purely resistive and antenna radiates well, but in case of unbalanced CRLH structure inductive or capacitive reactance get added up and the structure no longer radiates. From the measured value of S_{11}, stop band occurs at
3.5 GHz and 4.4 GHz where all the energy is virtually reflected.

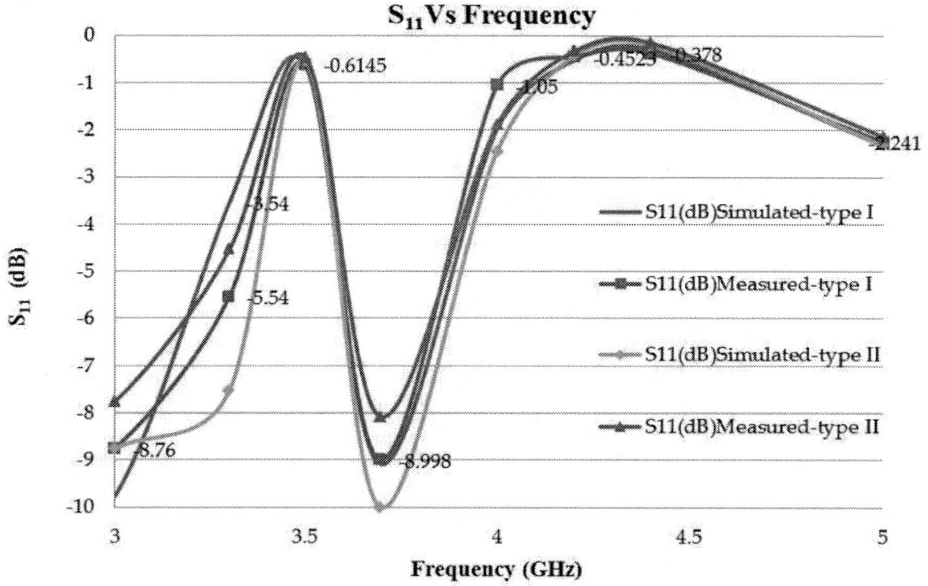

Figure 7. Reflection coefficient, Simulated and measured, type I and type II Antenna

b. *Voltage Standing Wave Ratio (VSWR)*

VSWR of the proposed filter has been measured and compared with simulated value as shown in figure 8. There is good agreement in the shapes of both the plots. The measured rejection level is not as high as simulation value due to fabrication process and SMA connectors used. The value of VSWR from figure 3.8 indicates that the antenna notches at 3.5 and 4.4 GHz, with VSWR > 2.

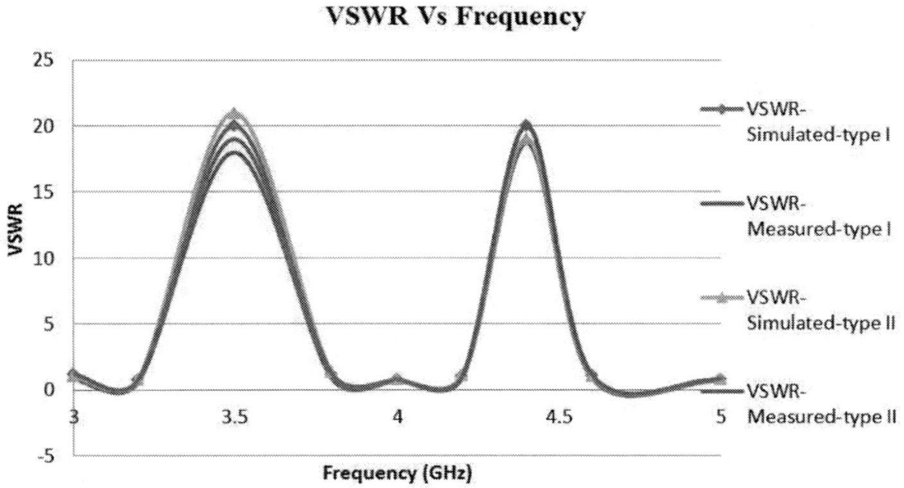

Figure 8. VSWR, Simulated and measured, type I and type II antenna

6. Comparative Study

To validate the efficiency of the proposed design, the dual band type I and type II antenna are compared with existing work. Table 2 represent the performance of proposed notch antenna with other recently reported antennas in terms of size, notch frequencies and maximum VSWR. It is evident from table 1.2 that the proposed antenna is compact and of much smaller size than existing antenna

Table 2. Comparison between different UWB notched antenna designs

	Size[mm^2]	Notch frequencies [GHz]
[2]	36 × 32	3.3,6 and 8.4
[1]	35 × 35	3.55,5.75 and 8.27
[9]	44 × 12	2.4,3.8 and 5.5
[14]	50 × 42	3.45,5.55 and 7.4
[15]	42 × 45	2.4
type I Antenna	30 × 30	3.5 and 4.4
type II Antenna	30 × 30	3.5 and 4.4

7. Conclusions

Design of compact size CRLT MTM UWB antenna with dual notch band has been presented in this chapter. CRLH MTM type I and type II new UWB antenna with dual notch bands has been proposed and discussed. The notch band was designed by selecting the appropriate length and width of stub and the length of IDC fingers. These antennas have been analysed for reflection coefficient and VSWR. The analysed result shows that the proposed antennas exactly notched at 3.5 GHz and 4.4 GHz. The proposed UWB notch-antenna is used for UWB wireless technology due to simple and easy implementation. Reflection coefficient for CRLH with meander line and omega shaped inductor in stub length yields the desired notch at the two specified frequency of 3.5 and 4.4 GHz. In future, this antenna with tuneable capability will be applied for UWB wireless technology. Thus the notch antenna is successfully simulated, fabricated and measured.

IOP Conf. Series: Materials Science and Engineering 1272 (2022) 012023 doi:10.1088/1757-899X/1272/1/012023

References

[1] Sarkar D, Srivastava K V and Saurav K 2014 A compact microstrip-fed triple band-notched UWB monopole antenna, *IEEE Antennas and Wireless Propagation Letters*, *13*, 396-399

[2] Ali W Ibrahim A A and Machac J 2017 Compact size UWB monopole antenna with tripleband-notches, *Radioengineering*, 26, 57-63

[3] Gao G, Hu B, He L, Wang S and Yang C 2017 Investigation of a reconfigurable dual notched UWB antenna by conceptual circuit model and time-domain characteristics, *Microwave and Optical Technology Letters*, *59*(6), 1326-32.

[4] Nejatijahromi M, Rahman M and Naghshvarianjahromi M 2018 Continuously tunable WiMAX band-notched UWB antenna with fixed WLAN notched band, *Progress In Electromagnetics Research Letters*, *75*, 97-103

[5] Rahman M, NaghshvarianJahromi M, Mirjavadi S S and Hamouda A M 2018 Resonator based switching technique between ultra wide band (UWB) and single/dual continuously tunable-notch behaviors in UWB radar for wireless vital signs monitoring, *Sensors*, *18*(10), 3330.

[6] Lotfi P, Azarmanesh M and Soltani S 2013 Rotatable dual band-notched UWB/triple band WLAN rcconfigurable antenna, *IEEE Antennas and Wireless Propagation Letters*, *12*,104-7

[7] Rajkumar R and Kiran K U 2016 A compact metamaterial multiband antenna for WLAN/WiMAX/ITU band applications, *AEU-International Journal of Electronics and Communications*, *70*(5), 599-604.

[8] Lin C C, Jin P and Ziolkowski R W 2011 Single, dual and tri-band-notched ultrawideband (UWB) antennas using capacitively loaded loop (CLL) resonators, *IEEE Transactions on Antennas and Propagation*, *60*(1), 102-9.

[9] Fischer T W, Kelleci B Shi K, lker Karsilayan A and Serpedin E 2007 An analog approach to suppressing in-band narrow-band interference in UWB receivers. *IEEE Transactions on Circuits and Systems I: Regular Papers*, *54*(5), 941-50.

[10] Khan, M.I., Khattak, M.I., Witjaksono, G., Barki, Z.U., Ullah, S., Khan, I. and Lee, B.M., 2019. Experimental investigation of a planar antenna with band rejection features for ultra-wide band (UWB) wireless networks. International Journal of Antennas and Propagation, 2019

[11] Jairath, K., Singh, N., Shabaz, M., Jagota, V. and Singh, B.K., 2022. Performance Analysis of Metamaterial-Inspired Structure Loaded Antennas for Narrow Range Wireless Communication. Scientific Programming, 2022.

[12] Vallappil, A.K., Khawaja, B.A., Rahim, M.K.A., Iqbal, M.N., Chattha, H.T. and Ali, M.F.B.M., 2022. A compact triple-band UWB inverted triangular antenna with dual- notch band characteristics using SSRR metamaterial structure for use in next- generation wireless systems. Fractal and Fractional, 6(8), p.422.

[13] Elsheakh D M and E Safwat A M 2012 Meander line-loaded planar monopole antennas, *Microwave and optical technology letters*, *54*(8), 1851-4.

[14] Jaglan N, Kanaujia B, Gupta S D and Srivastava S 2016 Triple band notched UWB antenna design using electromagnetic band gap structures, *Progress In Electromagnetics Research C*, *66*,139-47.

[15] Al-Yasir, YIA, M. K. Alkhafaji, MK, Alhamadani, HA, Parchin, NO, Elfergani, I Saleh, AL, Rodriguez, J and R. A. Abd-Alhameed, RA 2020 'A new and compact wide-band microstrip filter-antenna design for 2.4 GHz ISM band and 4G applications', Electronics, vol. 9, no. 7, p. 1084

[16] Lai A, Itoh T and Caloz C 2004 Composite right/left-handed transmission line metamaterials, *IEEE microwave magazine*, *5*(3), 34-50.

V. Indu Nair received the Bachelor Degree in Electronics and Communication Engineering in 1992 from Bharathiar University, Coimbatore, Tamil Nadu, India. She received M.E. degree in Communication System in the year 2004 and has 17 years of teaching experience. Currently working in Park College of engineering and technology, Tamil Nadu, India. Her research area includes Antenna design, communication, etc. Her research papers published in various journals and she presented papers in national and international conferences

IOP Conf. Series: Materials Science and Engineering 1272 (2022) 012024 doi:10.1088/1757-899X/1272/1/012024

Tensile and thermal behaviour of linear low-density polyethylene nanocomposite films

R B Jagadeesh Chandra[1], B Shivamurthy[2*], Vishwanatha H M[2], and M Sathish kumar[1]
[1]Department of Electronics & Communication Engineering, Manipal Institute of Technology, Manipal Academy of Higher Education, Manipal-576104, India.
[2]Department of Mechanical & Industrial Engineering, Manipal Institute of Technology, Manipal Academy of Higher Education, Manipal-576104, India.

*Corresponding author email: shiva.b@manipal.edu

Abstract. In this work Graphene nanoplatelets (GNPs) filled with Linear low-density polyethylene (LLDPE), nanocomposite films designated as LLDPE+1%GNP and LLDPE+5% GNP were manufactured using LLDPE pristine granules, 1 wt. % GNPs-LLDPE, and 5wt. % GNPs-LLDPE masterbatch granules by the extrusion-blow molding process. It was found that the ductile behaviour of neat LLDPE film and nanocomposite films are anisotropic in the machine direction (MD) and transverse direction (TD). The tensile strength of neat LLDPE films in the MD and TD is 25.93 & 20MPa. LLDPE+1GNPs nanocomposite films showed higher tensile strength in the MD and TD 28.38 & 23MPa, and LLDPE+5GNPs nanocomposite films showed 24.36 & 23MPa respectively. The high concentration of GNPs in LLDPE acts as plasticizers increased the elongation, reduced the strength, and increased the rate of thermal degradation. The XRD results reveal additional peeks observed in LLDPE+5GNPs nanocomposite films infer the slight change in crystallinity as compared to neat LLDPE film. This is evident in the change of tensile behaviour of nanocomposite as compared to LLDPE.

1. Introduction

The polymer nanocomposites have wide range of application due to their attractive properties even at low concentration of nanofiller [1–4]. Due to better dispersion and high surface area of nanofillers in polymer can develop a higher interface than conventional micros fillers [5, 6]. Hence, researchers are tried to improve the properties of a polymer by reinforcing nanofillers [7–9]. In particularly the carbonaceous nano fillers are used with thermoset and thermoplastic polymer to achieve good thermal and electrical properties [10-13].

The nano polymer composites with graphene nanoplatelets (GNPs) have various applications in the electronics sector. The GNPs consist of multi-layer hexagonal nanocrystals that are formed by Van der Waals force of attraction. The 2D structure of GNPs mixed with polymers develop a network structure due to different orientation of nano filler dispersed in the polymer that lead to better electrical conductivity and magnetic permeability [10]. Cabrera et.al [11] investigated multi-micro/nanolayer films and observed that polypropylene/liner low-density polyethylene/ polypropylene (PP/LLDPE/PP) multi-layer nano films exhibit maximum tensile strength at yield is 8.3MPa and break is 25.6 MPa as compared with micro layered films. The authors found an influence of recycling process on mechanical properties of the multi-layer films. Al-Saleh et.al [12] developed PP-GNPs nano composites films and found significant influence on thermal performance and increase in the tensile and flexural strength due to the addition of 5wt.% GNPs. But Jun et.al [13] found that the modulus increased with addition of GNPs loading to PP and reported not much appreciable change in tensile strength. Li et.al [14] reported increased in tensile strength and Young's modulus due to addition of up to 2 wt. %graphene oxide (GO) to LLDPE. In this work an attempt has been made to study the tensile and thermal properties of multi-layer extrusion blown LLDPE/GNPs nano composite films.

2. Materials and Methods

The LLDPE pristine granular supplied by Reliance Polymers Industries Pvt. Ltd. has a density of 0.918 g.cm-3 and melt mass flow rate of 1.0 g per min. used as matrix. The GNPs procured from M/s. Adnano Technologies Pvt. Ltd., Shivamogga, Karnataka, India. The GNPs used to have a surface area of 60-200 m2. g-1, average lateral dimension; < 1 μm, approximate thickness; 5-10 nm and bulk density 0.25 g.cm-3 used as fillers. The 1 wt.% GNPs-LLDPE masterbatch and 5 wt. % GNPs-LLDPE masterbatch was produced by compounding. Further neat LLDPE and LLDPE+1GNPs and LLDPE+5GNPs nanocomposite films were developed by an extrusion blow molding process. The tensile properties of neat LLDPE films and LLDPE+1GNPs and LLDPE+5GNPs nanocomposite films were tested using a universal testing machine according to ASTM D882. Thermal behaviour was investigated by thermogravimetric analysis. The wide-angle X-Ray Diffraction (XRD) was performed for GNPs, neat LLDPE film, and LLDPE+1GNPs and LLDPE+5GNPs nanocomposite film samples on a RIGAQU-MINI FLEX 600 powder diffractometer with Cu Kα radiation at a scanning rate of 1°/min. The XRD spectra were recorded in the 2θ range of 0°- 60°. Differential scanning calorimeter (DSC) recorded thermograms of LLDPE+1GNPs and LLDPE+5GNPs nanocomposite films were used to estimate the crystallinity

3. Results and Discussion

3.1. Morphology of GNPs

The GNPs morphology was investigated using Hitachi SU 3500 scanning electron microscope at an accelerating voltage of 10kV and the images are taken at 8k magnification as shown in Figure 1. it is observed flake structure is in the form of stacked type.

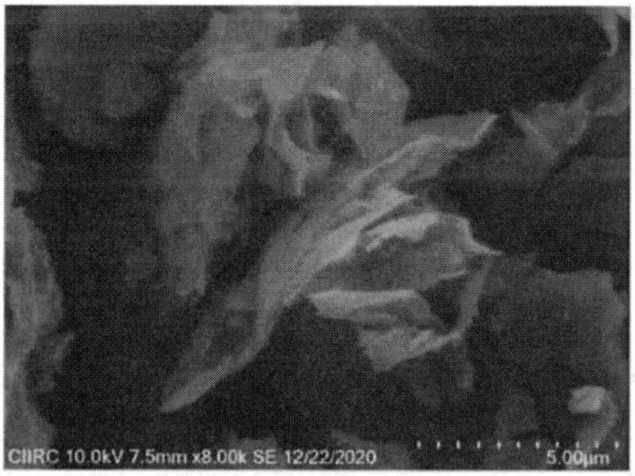

Figure 1. SEM Image of GNPs

3.2. Manufacturing of multi-layer films

The three-layer LLDPE+1GNPs and LLDPE+5GNPs nanocomposite films are manufactured using co-extrusion and blowing. The 1wt. % GNPs-LLDPE masterbatch granules with LLDPE pristine granules, 5wt.% GNPs-LLDPE masterbatch granules with LLDPE pristine granules are taken in the ratio of 1:100 by weight to produce nanocomposite films. These films are designated as LLDPE+1GNPs and LLDPE+5GNPs respectively. Three single-screw extruders with L/D ratio-2100/50 was used with a Maddock mixer to get uniform dispersion of GNPs in the LLDPE melt. The blown die has a diameter of 350 mm and was used for blowing. The screw speed of 110 rpm, die gap of 2.2 mm, and blow-up ratio of 1: 2.5 were maintained throughout the process.

3.3. *X-ray diffraction analysis*
The XRD patterns of GNPs used as filler in this study is shown in Figure 2. The XRD pattern of the GNPs, neat LLDPE film, LLDPE+1GNPs and LLDPE+5GNPs nanocomposite films in the range 5° to 65° is shown in Figure 2. The XRD spectrum of graphene corresponding to the (001) plane in LLDPE+5GNPs nanocomposite film confirmed the presence of graphene. The XRD patterns of LLDPE exhibits reflection peaks at 21.68°, 24.02°, and 36.48° with the strongest peak at 21.68°. The LLDPE+1GNPs nanocomposite film consists of reflection peaks at 21.46°, 23.8°and 36.22° with the strongest peak at 21.46° as like neat LLDPE [14]. In addition, observed a slight peak at 26.5° (002) plane in LLDPE+1GNPs and LLDPE+5GNPs nanocomposite film indicates the content of graphene [15] as shown in the magnified image in Figure 2. However, it is noted similar XRD patterns of neat LLDPE film and LLDPE+1GNPs nanocomposite films. It suggests the crystalline structure of neat LLDPE remains unchanged upon compounding using 1wt.% GNPS-LLDPE master batch. Observed there are extra peaks in LLDPE+5GNPs nanocomposite film, a peak observed at $2\theta = 10°$ corresponding to (001) plane due to oxidation of graphene [16]. Further, a more intensity and wider visible peak at 26.5° (002) plane due to a higher concentration of graphene as compared to LLDPE+1GNPs nanocomposite film. However, it is noticed that not much changes in crystalline structure due to the addition of 5wt.% GNPs-LLDPE masterbatch in neat LLDPE.

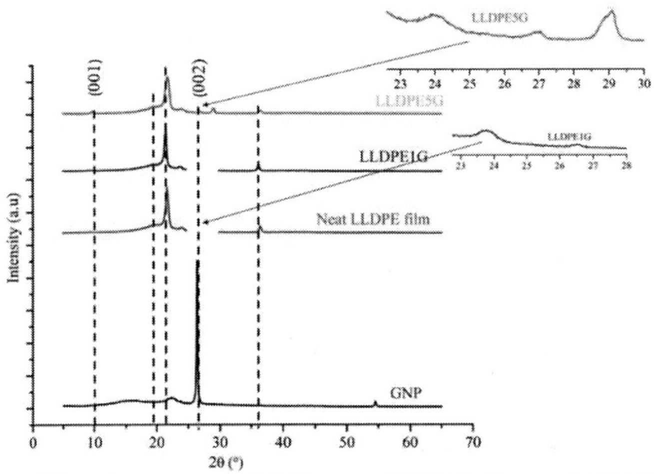

Figure 2. X-ray diffraction spectra of GNP, neat LLDPE film, and LLDPE+1GNPs and LLDPE+5GNPs nanocomposite films

3.4. *Tensile behavior*
The tensile properties of the neat LLDPE and LLDPE+1GNPs and LLDPE+5GNPs nanocomposite films are studied using a universal testing machine subjected to the machine direction (MD) and transverse direction (TD) in the strain rate of 0.02 S-1 as per ASTM D 638. The stress-strain behavior of neat LLDPE and nanocomposite films are plotted in MD and TD as shown in Figure 3(a) and 4(b) respectively. It is observed from the stress-strain plot that the blown extrusion process leads to anisotropy in the ductile behavior of neat LLDPE films in MD and TD. This is due to the stretched polymer chain and orientation of polymer molecules in neat LLDPE in the MD leads to decreased ductility as compared to TD. The average tensile strength of neat LLDPE films in the MD and TD is 25.93 & 20MPa respectively. Whereas the LLDPE+1GNPs nanocomposite films showed the highest tensile strength of 28.4 & 23MPa in MD and TD directions respectively with decreased ductility. The contribution of the inherent strong lateral direction strength of GNPs as reinforcement leads to higher tensile strength with a lower strain rate in the LLDPE+1GNPs nanocomposite films. This is attributed to the GNPs oriented in MD of the LLDPE film acting as reinforcement and sharing the tensile load in its lateral direction. The LLDPE+5GNPs nanocomposite films showed 24.4, 23MPa in MD and TD

direction respectively. Although the strength in the TD remains the same, the strength in the MD is lower as compared to both neat and LLDPE+1GNPs nanocomposite films. There exists an increased concentration of GNPs that leads to stacking up of GNPs in multiple layers in MD. These layers during tensile loading undergo slippage due to weak Van der Wall's force of attraction between the layers (as observed slippage in the stress-strain curve Figure 3(a) resulting in decreased strength in MD as compared to lower concentration GNPs nanocomposites.

The dispersion status of GNPs in LLDPE is shown in schematic Figure 3(c-d). At lower concentration due to shear action in the extruder, the stacked GNPs breaks into less thicker stacking with few layers separated and the LLDPE moved in between layers and structure formed as shown in Figure 3(c). Whereas at higher concentration, the degree of delamination of stacking is reduced, and stacking layers of GNPs remain in the resultant nanocomposite as shown in Figure 3(d). Thus, in the case of 5wt.% GNPs master batch nanocomposite films the GNPs act as plasticizers, whereas in the case of LLDPE+1GNPs nanocomposite films behave as reinforcement with LLDPE matrix. Nanocomposite film subjected to tensile loading in TD direction showed slightly increased tensile strength than neat LLDPE but not much change was observed in ductility

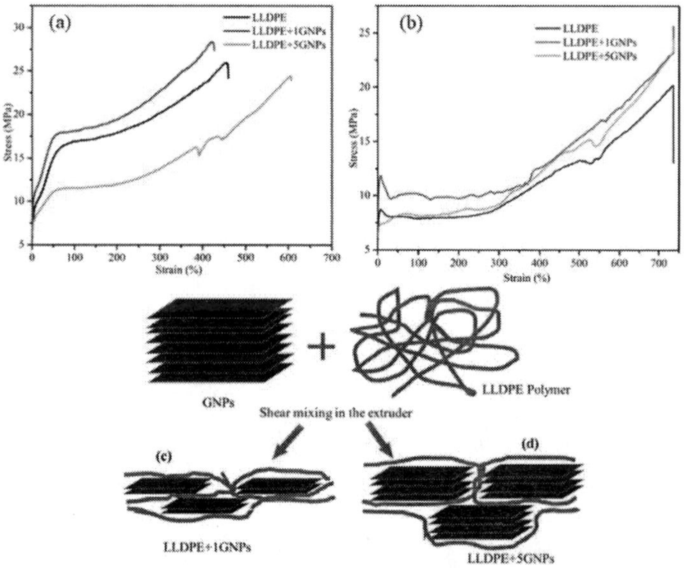

Figure 3. Stress-Strain plot of LLDPE film, LLDPE+1GNPs and LLDPE+5GNPs nanocomposite films (a) MD-machine direction (b) TD-transverse direction, Schematic diagram of GNPs dispersion in nanocomposites (c) at lower concentration and (d) at higher concentration.

3.5. Thermal behavior

The thermal behavior of the LLDPE film and LLDPE+1GNPs and LLDPE+5GNPs nanocomposite films are represented by TGA curves shown in Figure 4(a) with the onset, end set, and peak temperatures. It is noticed that the samples undergo decomposition at 460 ±9 °C. It is observed that the LLDPE+1GNPs nanocomposite film sample is decomposing at a little higher temperature than the pristine LLDPE film, which infers that the GNPs reinforcing the LLDPE. In contrary to this, the LLDPE+5GNPs nanocomposite film sample is decomposing at a much lower temperature than the neat LLDPE. This infers that the GNPs are acting as plasticizers, which decreased the interactions between the LLDPE chains and enhanced the decomposition process. This may be argued due to the total surface area of GNPs, which favored the decomposition. This observation is also reflected in the appreciable lower onset decomposition temperature of LLDPE+5GNPs nanocomposite film samples than the pristine LLDPE.

The mechanical properties of the polymer-based nanocomposites are influenced by the degree of crystallinity of the composite. To realize the degree of crystallinity of the composites, the DSC thermograms are recorded for the neat LLDPE and its nanocomposites with GNPs are shown in Figure 4(b) with the melting temperature along with the enthalpy change (ΔH). The degree of crystallinity of the samples is calculated using the equation, $X_C = \frac{\Delta H_m}{\Delta H_f}$, where the ΔHm is estimated from the DSC thermograms of each sample and ΔHf is the reference material's (polyethylene) melting enthalpy change with 100% crystallinity. It is noticed that the crystallinity is decreased with increasing the content of the GNPs in the present studies. This supports the decreased tensile strength and thermal stability at higher concentrations of GNPs fillers. The GNPs fillers at higher concentrations act as the plasticizers in these composites and hence reduce the intermolecular forces and the crystallinity. Due to this, the melting temperatures of the composites are decreasing with the content of GNPs fillers.

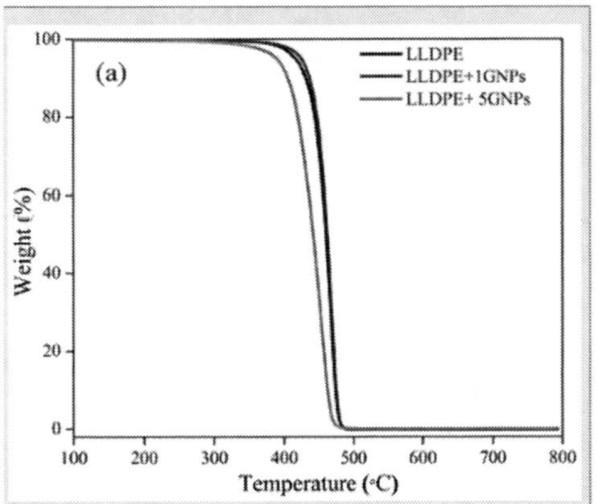

Sample	Temperature (°C)			
	T_{onset}	$T_{50\%}$	T_{Peak}	T_{end}
LLDPE	338.4	459.6	467.6	495.9
LLDPE+1GNPs	338.4	461.3	468.8	501.1
LLDPE+5GNPs	255.0	441.5	454.0	517.4

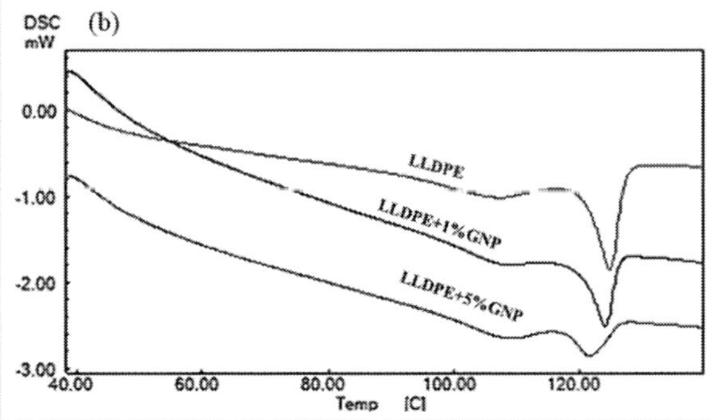

Composition	T_m (°C)	ΔH_m (Jg^{-1})	X_c (%)
LLDPE	124.9	29.9	10.2
LLDPE+1GNPs	124.1	26.1	8.9
LLDPE+5GNPs	121.7	15.5	5.3

Figure 4 (a) TGA and (b) DSC curves of neat LLDPE and LLDPE+1GNPs and LLDPE+5GNPs nanocomposite films.

4. Conclusions

- The blown extrusion process leads to anisotropy in the ductile behavior of films subjected to the machine direction (MD) and transverse direction (TD).
- Due to addition of the 1 wt. % GNPs+LLDPE masterbatch, LLDPE becomes brittle in MD, due to the 5 wt. % GNPs+LLDPE master batch LLDPE becomes ductile due to plasticizing effect. This is correlated with TGA results, due to a higher surface area of 5 wt.% GNPs+ LLDPE master batch decomposes at a lower temperature.
- XRD spectra reveal additional peeks observed in LLDPE+5GNPs nanocomposite film infer a slight change in crystallinity as compared to neat LLDPE film. This is evident in the change of tensile behavior nanocomposite as compared to neat LLDPE.

Declarations

The authors declare that they have no conflict of interest.

References

1. Tjong SC (2014) Polymer composites with graphene nanofillers: electrical properties and applications. J Nanosci Nanotechnol 14:1154–1168. https://doi.org/10.1166/jnn.2014.9117
2. Sengupta R, Bhattacharya M, Bandyopadhyay S, Bhowmick AK (2011) A review on the mechanical and electrical properties of graphite and modified graphite reinforced polymer composites. Prog Polym Sci 36:638–670. https://doi.org/https://doi.org/10.1016/j.progpolymsci.2010.11.003
3. Papageorgiou DG, Kinloch IA, Young RJ (2017) Mechanical properties of graphene and graphene-based nanocomposites. Prog Mater Sci 90:75–127. https://doi.org/https://doi.org/10.1016/j.pmatsci.2017.07.004
4. Huang X, Qi X, Boey F, Zhang H (2012) Graphene-based composites. Chem Soc Rev 41:666–686. https://doi.org/10.1039/C1CS15078B
5. Ray SS (2013) Clay-Containing Polymer Nanocomposites From Fundamentals to Real Applications. Elsevier, Amsterdam
6. Bumbudsanpharoke N, Lee W, Choi JC, et al (2017) Influence of Montmorillonite Nanoclay Content on the Optical, Thermal, Mechanical, and Barrier Properties of Low-Density Polyethylene. Clays Clay Miner 65:387–397. https://doi.org/10.1346/CCMN.2017.064071
7. Yang K, Endoh M, Trojanowski R, et al (2015) The thermo-mechanical response of PP nanocomposites at high graphene loading. Nanocomposites 1:126–137. https://doi.org/10.1179/2055033215Y.0000000008
8. Liang J-Z (2019) Effects of tension rates and filler size on tensile properties of polypropylene/graphene nano-platelets composites. Compos Part B Eng 167:241–249. https://doi.org/https://doi.org/10.1016/j.compositesb.2018.12.035
9. El Achaby M, Arrakhiz F-E, Vaudreuil S, et al (2012) Mechanical, thermal, and rheological properties of graphene-based polypropylene nanocomposites prepared by melt mixing. Polym Compos 33:733–744. https://doi.org/https://doi.org/10.1002/pc.22198
10. Kalaitzidou K, Fukushima H, Drzal L (2007) A new compounding method for exfoliated graphite-polypropylene nanocomposites with enhanced flexural properties and lower percolation threshold. Compos Sci Technol 67:2045–2051. https://doi.org/10.1016/j.compscitech.2006.11.014
11. Cabrera G, Touil I, Masghouni E, et al (2021) Multi-Micro/Nanolayer Films Based on Polyolefins: New Approaches from Eco-Design to Recycling. Polymers (Basel) 13:. https://doi.org/10.3390/polym13030413

12. Al-Saleh MA, Yussuf AA, Al-Enezi S, et al (2019) Polypropylene/Graphene Nanocomposites: Effects of GNP Loading and Compatibilizers on the Mechanical and Thermal Properties. Mater (Basel, Switzerland) 12:3924. https://doi.org/10.3390/ma12233924
13. Jun Y-S, Um JG, Jiang G, et al (2018) Ultra-large sized graphene nano-platelets (GnPs) incorporated polypropylene (PP)/GnPs composites engineered by melt compounding and its thermal, mechanical, and electrical properties. Compos Part B Eng 133:218–225. https://doi.org/https://doi.org/10.1016/j.compositesb.2017.09.028
14. Li J, Gunister E, Barsoum I (2019) Effect of graphene oxide as a filler material on the mechanical properties of LLDPE nanocomposites. J Compos Mater 53:2761–2773. https://doi.org/10.1177/0021998319839215
15. La Mantia FP, Ceraulo M, Mistretta MC, Morreale M (2018) Rheological Behaviour, Mechanical Properties and Processability of Biodegradable Polymer Systems for Film Blowing. J Polym Environ 26:749–755. https://doi.org/10.1007/s10924-017-0995-4
16. Rashad M, Pan F, Tang A, et al (2014) Development of magnesium-graphene nanoplatelets composite. J Compos Mater 49:285–293. https://doi.org/10.1177/0021998313518360

Investigation of Dielectric Spectroscopy of Tellurium (Te) doped Bismuth Antimony (BiSb) pellet

Reena Panchal [1], Maunik Jani [2], S M Vyas [3] and Himanshu Pavagadhi [4]

[1]*Shri Govind Guru University, Godhra, Gujarat*
[2]*Sigma Institute of Science, Shri Govind Guru University, Godhra, Gujarat*
[3,4]*Department of Physics, School of Science, Gujarat University, Ahmedabad, Gujarat*

Email: *[*1]reena.jp05@gmail.com, [2]drmaunikjani@gmail.com, [3]s_m_vyas_msu@yahoo.com, [4]Hbpavagadhi@gmail.com*

Abstract: Tellurium (Te) doped Bismuth Antimony (BiSb) sample has been prepared in the form of the pellet (Bulk) with the help of a compression method using a Pelletizer. Using Tellurium (Te) doped Bismuth antimony (BiSb) pellet was used for the study of dielectric properties through VNA (Vector Network Analysis) in the high range of frequency 200 MHz to 20 GHz. From initial data obtained for the study, DC conductivity, Dielectric loss, Dielectric constant, AC conductivity, and Real & Imaginary Electric modulus have been figured. All these properties have been used for exploring different mechanisms involved in the examination of dielectric spectroscopy.

Keywords: Dielectric constant, AC conductivity, VNA meter, Dielectric Loss, bismuth antimony, Electric modulus.

1. Introduction

Bismuth and antimony both are semi-metal in their bulk form [1-2]. There are various applications in various fields of bismuth antimony (BiSb) in the modern era like the fabrication of ionizing radiation detectors, LED [3-6], solid-state electrodes, gas sensors, optoelectronic devices, photosensitive heterostructure, solar cells, ionic batteries, etc. [7-11] The Implementation of the dielectric characterization method is more constructive to gain knowledge about the properties of electrical of a bulk sample (pellet). Himanshu Pavaghadhi et.al. [12] has Sn (tin) doped CdS (Cadmium Sulphide) pellets used and worked on the dielectric measurement. They concluded with the results that the use of the LCR meter for the span of frequency from 20Hz to 2MHz is an effective technique for electrical properties. Santosh Mani et.al. [20] used Polymer Dispersed Liquid Crystals (PDLC) and studied Electrical and Optical properties of material. Also they have determined dielectric and conductivity by impedance analyser. Santosh Mani et.al. [21] prepared Polymer Dispersed Liquid Crystals (PDLC) and studied thermal and electrical properties by using differential scanning calorimetry and fabry perot scattering technique. They determined dielectric constant and conductivity of the material. Santosh Mani et.al. [22] prepared Polymer Dispersed Liquid Crystals (PDLC) by using thermotropic CNLC (Chiral Nematic Liquid Crystal) and Conducting Polymer. They examined molecular structure by FTIR Spectroscopy, morphological and Phase transition by Polarizing Optical Microscopy. Also found out

IOP Conf. Series: Materials Science and Engineering 1272 (2022) 012025 doi:10.1088/1757-899X/1272/1/012025

Refractive index of the material. In the present study, VNA data is used to examine the dielectric spectroscopy for the frequency interval from 200 MHz to 20 GHz.[14]. From this study, Electric modulus (Real and imaginary), Dielectric constant, conductivity, and Dielectric loss have been calculated at room temperature [12-14].

2. Experimental procedure

The most significant procedure for pellet preparation (BULK) is pelletizing. In this process, a material is compressed or molded to get the shape of a pellet [12-13]. There are various methods have been adopted by researchers for the development of pellet-like X-ray fluorescence (XRF) analysis [13], hydraulic press technique, potassium bromide (KBR) method [13], etc. In the present study, the hydraulic pressure technique is used for making the pellets. Stoichio metric proportion of materials (Bi, Sb & Te) were calculated and are mixed then the composition of materials converted into a fine powder with the use of a Mortar and Pestle. This powder was placed between two core dies in a Dry pressing Die set (Pelletizer) and pressed by a Hydraulic press machine under the suitable pressure of 10 tones. The pressure on the die set was kept for 10 min and then released. By this procedure, prepared a pellet $Bi_2Sb_{2.9}:Te_{0.1}$ (BULK) of 1 mm thickness. This pellet is used for dielectric study for 200 MHz to 20 GHz frequency span using the VNA meter.

3. Results and Discussion

From the figures shown below, the dielectric parameters of the pellet (Bulk) could be observed, and from that, we can find out the properties of the pellet-like AC conductivity, DC conductivity, Electric modulus (Real and imaginary), Dielectric constant, and Dielectric loss.

3.1 Dielectric Constant

By using a precision VNA meter, it is observed that the Dielectric constant is one of the essential property of the dielectric material for the frequency range from 200 MHz to 20 GHz [13]. The value of the dielectric constant rely on the frequency [13, 15]. Under an AC electric field, The quality of energy stored in the dielectric material is a very crucial part of the dielectric [13, 16]. Estimation of The dielectric constant (ε') of the pellet has been done using Equation 1 and the estimated results are shown in the following Figure 1.

$$\varepsilon' = \frac{C_p\, t}{\varepsilon_0 A} \tag{1}$$

Where t = thickness of the pellet (meter), C_P = parallel capacitance with sample (farad), A = Area of the surface of the pellet (meter^2), ε_0 = permittivity of free space (8.854×10^{-12} F/m) [13]

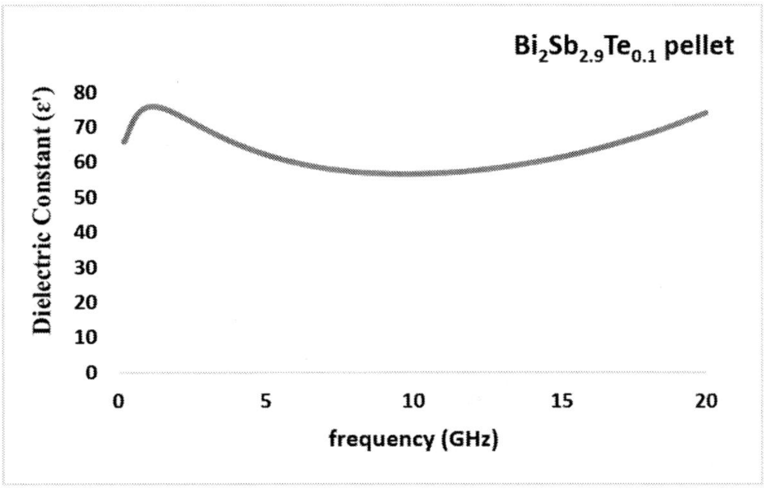

Figure 1. Dielectric Constant verses Frequency.

Figure 1 describes the change in the dielectric constant in relation to the variation in the frequency of the ac signal [19]. Initially, It is gradually increasing as frequency is increasing up to 1 GHz, due to the movement of charge carriers and polarization. After the initial increment, the dielectric constant is dropped down with frequency increment. Because the mobile carriers started to move at a high frequency, depletion of dielectric constants is stopped and it is almost steady with this high frequency.

3.2 Dielectric loss

The following equation 2 is used to evaluate The dielectric loss (ε”) of the pellet and the results of the observation are shown in Figure 2.

$$\varepsilon'' = \frac{1}{\omega C_1 \, R_p} \tag{2}$$

where $C_1 = \frac{t}{AX\varepsilon_0}$,

Rp = Air capacitance with a sample, ω= Angular frequency (2πf), t = thickness of the pellet (meter), ε_0 = permittivity of free space (8.854×10^{-12} F/m), A = Area of the surface of the pellet (meter^2) [13].

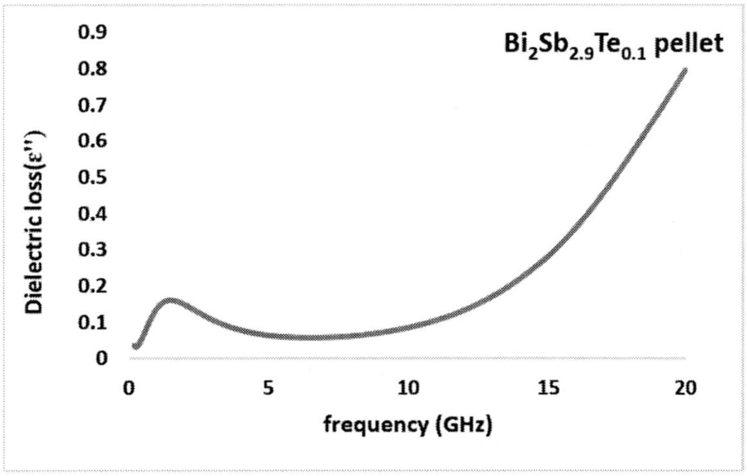

Figure 2. Dielectric loss verses Frequency.

From the figure 2, it can be seen that the dielectric loss is the loss of energy when the material is heated at a varying electric field. Initially, the dielectric loss (ε”) value is gradually increasing as the frequency is increasing by nearly 1 GHz because of polarization in the material. After the initial increment, dielectric loss is constant up to 9 GHz frequency. Due to dipole rotation at the high-frequency region occurred maximum dielectric loss which helps to discriminate the bulk material properties.

3.3 AC Conductivity

Equation 3 is used to calculate The AC Conductivity of the pellet and the results of observations are shown in the following figure -3.

$$\sigma'_{ac} = \omega\, \varepsilon_0\, \varepsilon'' \tag{3}$$

where, ε" = Dielectric Loss, ε_0 = permittivity of free space (8.854×10^{-12} F/m), ω = Angular frequency (2πf) [12].

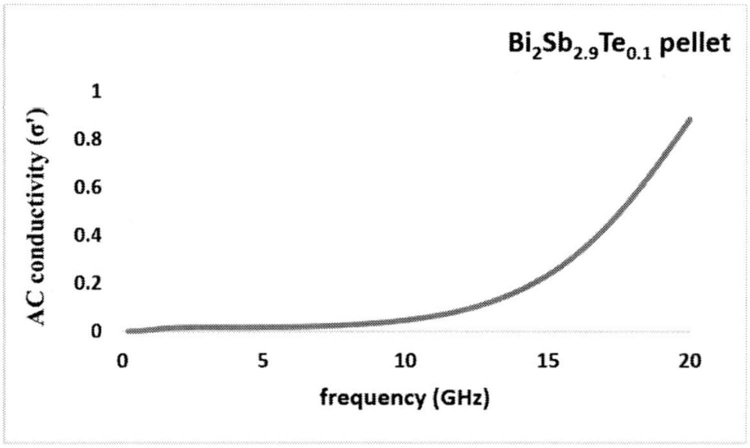

Figure 3. AC Conductivity verses Frequency.

Figure 3 describes the study of frequency-dependent on conductivity (AC Conductivity). Initially, at a low-frequency range up to 8 GHz, the conductivity value is almost zero. Then the conductivity is linearly increasing as frequency increases. It gives the description of the material behaviour in view of the effect of tunneling of charge particles and thermionic emission across the Potential barrier [12, 13].

3.4 Electric Modulus

The following equation 4 is employed to estimate The Electric Modulus (imaginary and real part) of the pellet [12, 13, 17], and the results of observations are shown in the following Figure 4 & Figure 5.

$$M^*(\omega) = M' - j\,M'' = \frac{\varepsilon'}{\varepsilon'^2+\varepsilon''^2} - j\,\frac{\varepsilon'}{\varepsilon'^2+\varepsilon''^2} \tag{4}$$

Figure 4. - Figure 5. Indicate the change of the Real & Imaginary parts of electric modulus with function of frequency [13].

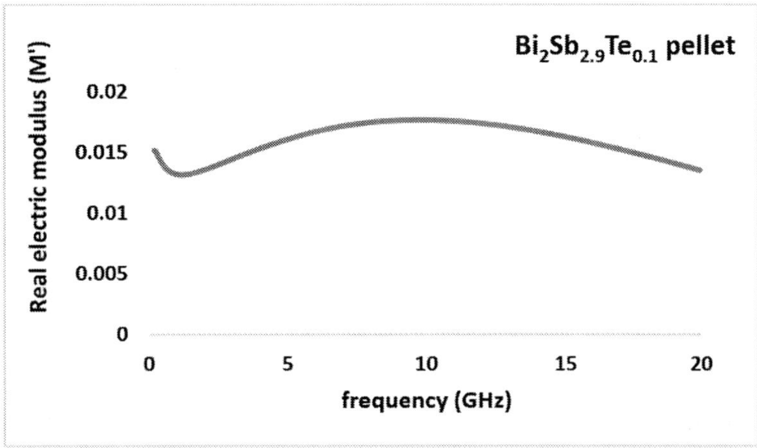

Figure 4. Electric Modulus (Real part) verses Frequency.

The relaxation phenomena, happening in ionic and electrically conducting materials can be determined [13], analyzed and interpreted with the help of Modulus spectroscopy, a well–developed tool [12, 18]. From figure 4, at the initial level, the value of Real Modulus (M') starts to fall down to 1 GHz frequency. With the increment of the frequency the value of M' is start to increase. It can be inferred that electrode polarization occurs at very high frequencies (more than 1 GHz).

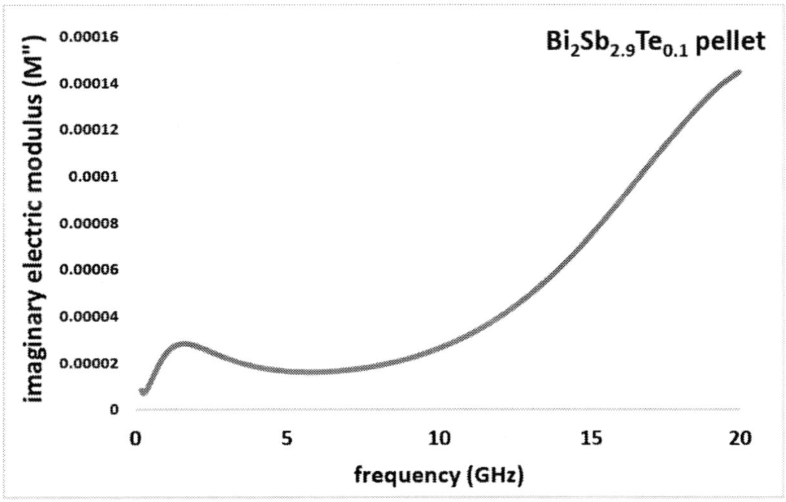

Figure 5. Electric Modulus (Imaginary part) verses Frequency.

From Figure 5 the behaviour of the of Electrical Modulus for Imaginary part (M") can be observed [13]. At the initial level, the value of the M' 'has started to increase with an increase in the value of frequency. Getting local peak at 2 GHz, it is static up to 7 GHz and then it is exponentially increased. Thus the behaviour of M" provides the information about the charge carrier flow through a material such as electrical transport and conductivity relaxation [12,13].

4. Conclusions

Pellet (Bismuth Antimony with doped tellurium) has been prepared under the pressure of 10 tons by using a pelletizer and hydraulic press machine. Different frequency values were studied AC conductivity, Dielectric constant (ε'), Dielectric loss (ε"), and Electric modulus (imaginary and real) of the pellet. The dielectric constant (ε') has a depletion followed by a peak and then remains constant for high frequency. Other properties like Dielectric loss (ε") and AC Conductivity (σ') values started to increase with the increase in frequency. This indicates maximum dielectric loss is occur at a higher frequency and the behaviour of a material. And the Real Modulus (M') declined at the low-frequency value while it is smoothly increasing for the high-frequency while the Imaginary Modulus (M'') is exponentially increasing at the high frequency.

5. References

[1] Friedrich K, Schafer K 1979 *J. Mater. Sci* **14** 483-486

[2] Tanuma S 1961 *J. Phy. Soc. Japan* **16** 2349

[3] Colvin V, Schlamp M and Alivisatos A 1994 *Nature* **370** 354 – 357

[4] Schlamp M C, Peng X and Alivisatos A P 1997 *J appl. Phy.* **82** 5837

[5] Nirmal M, Dabbousi B O, Bawendi M G, Macklin J J, Trautman J K, Harris T D and Brus L E 1996 *Nature* **383** 802 – 804

[6] Schreuder M A, Xiao K, Ivanov I N, Weiss S M and Rosenthal S J 2010 *Nano Lett* **10** 573 – 576

[7] Jie J S, Jhang W J, Jiang Y and Leea S T 2006 *Appl. Phy. Lett* **89** 113-118

[8] Karl W and Frese Jr 1982 *App. Phy. Lett* **40** 275

[9] Hamilton C E, Flood D J and Barron A R 2013 *Physical Chemical Phy.* **15** 3930 – 3938

[10] Amin S A, Salim S T and Salam K M 2013 *Int. J. Electrical Energy* **1** 1 – 5

[11] Sun B, Marx E and Greenham N C 2003 *Nano Lett.* **3** 961 – 963

[12] Pavagadhi H, Patel V, Vyas S M, Patel P, Jani M P and Shah N 2020 *AIP Conference Proceedings* **2220** 040027-1 - 040027-6

[13] Shah N, Vyas S M, Patel V, Patel P, Pavagadhi H and Jani M P 2020 *AIP Conference Proceedings* **2220** 040003-1 – 040003-6

[14] Patel V, Vyas S M, Patel P, Pavagadhi H and Jani M P 2020 *AIP Conference Proceeding* **2220** 040016-1 – 040016-4

[15] Kao K C 2004 *Elsevier Academic Press* London

[16] Deshmukh L S, Kumar K K, Laxman S B, Krishna A R and Sathaiah G 1998 *Bullet in Mater. Sci.* **21** 219 – 226

[17] Pradhan D K, Choudhary R N P and Samantray B K 2008 *Express Polymer Lett.* **2** 630 – 638

[18] Pandit R, SharmaK K, Kaur P, Kotnala R K, Shah J and Kumar R 2014 *J. Phy. And Chem. Solid* **75** 558 – 569

[19] Bamzai K K,Nidhi, Vishal singh, Kotru P N and Wanklyn 2009 *Ferroelectrics*

[20] Mani S, Patwardhan S, Hadkar S, Mishra K and Sarawade P 2022 *Mater. Today: Proceedings* **62** 7035 – 7039

[21] Mani S, Khosla S and Sarawade P 2022 *Mater. Sci. Forum* **1070** 33 – 43

[22] Mani S, Patwardhan S, Khosla S and Sarawade P 2022 *Mater. Today: Proceedings* **65** 3453 – 3460

Structural, Dielectric and Ferroelectric Characterizations of Sol-gel Prepared $Ba_{1-x}Sr_xTiO_3$ Nanoceramics

Devendra Singh[1], Anju Dixit[1*] and Pramod S Dobal[2]

[1]Department of Physics, School of Basic Sciences UIET, CSJM University, Kanpur-208024, India

[2]Department of Physics, VSSD College, Kanpur-208002, India

[*]Corresponding author E-mail: anjudixit@csjmu.ac.in

Abstract. $Ba_{1-x}Sr_xTiO_3$ ($0.0 \leq x \leq 0.5$) nanoceramics were synthesized via sol-gel route and the structural and electric properties of the resulting compositions were investigated. revealed A tetragonal crystal structures for x = 0.0 - 0.3 and a cubic structure for x = 0.5 compositions was revealed by X-ray diffraction. The smaller ionic radii of Sr ions resulted in the lowering of c/a ratio with increasing x and a tetragonal to cubic structural transformation was observed at around x = 0.4. The average crystallite size gradually decreased with increasing x from 0 to 0.5. This trend was visible in electron micrographs. The room temperature dielectric constant in these nanoceramics increases with increasing Sr and a maximum value of 1553 was observed for x=0.3. Curie temperature (T_C) of 125 °C was obtained for x=0.0, which shifted toward lower temperature with increasing x value. The value of T_c was observed as 100 and 75 °C for x = 0.1 and 0.3, respectively. The remnant polarization (P_r), saturation polarization (P_S), and coercive field (E_C) decreased with increasing Sr concentration in BT due to the structural modifications. Value of P_r decreases from 0.637 to 0.229 μC/cm^2, P_S decreases from 8.910 to 3.238 μC/cm^2 and E_C decreases from 0.631 to 0.255 kV/cm with increasing x from 0 to 0.5.

Keywords: XRD, Ferroelectrics, SEM, Dielectrics, Sol-gel

1. Introduction

Barium titanate $BaTiO_3$ (BT) has been one of the most extensively studied perovskite material. It is considered to be one of the best alternatives of lead-based toxic perovskite materials. BT has been known for its excellent ferroelectric properties, usable dielectric constant with low losses [1-7]. BT has a tetragonal crystal structure at normal temperature. The crystal structure changes with temperature and undergoes three structural transformations; a) rhombohedral ($R3m$) - orthorhombic ($Amm2$) transition around -90 °C, b) orthorhombic - tetragonal ($P4mm$) transition around 5 °C, and c) tetragonal - cubic ($Pm\bar{3}m$) transition around 130 °C [8-11], which known as Curie temperature (T_c). The structure and properties of BT can further be influenced via doping at A- and B-sites with suitable elements [12, 13]. A-site doping elements such as Sr [14], Ca [15], La [16], Ce [17], Ag [18] and B-site doping elements such as Fe [19], Zr [20], Li [21], Sm [22], Mn [23] are some of the dopants of BT. These dopants cause structural distortions which strongly affect the Curie temperature (T_c) and phase transition of the resulting material [24-27]. Since radius of Sr^{+2} ions is smaller compared to that of the Ba^{+2} ions, Sr doping in BT affects the tolerance factor [28] and modifies the lattice parameters. With Sr doping in BT, the optical band gap reduces while the dielectric properties improve [9, 29]. BST also shows good frequency [30] and temperature dependent electrical as well as dielectric properties. As a result, $Ba_{1-x}Sr_xTiO_3$ (BST) ceramics are widely used in telecommunication, multilayer capacitors, microwave phase shifters, sensors, transducers, dynamic access memory, and others [9, 24, 31, 32].

Sol-gel has been one of the most useful route to synthesize BT, BT based ferroelectrics and multiferroics [31-34]. In this work, $Ba_{1-x}Sr_xTiO_3$ nanoceramics were successfully synthesized using sol-gel method for x= 0.0 - 0.5 and the resulting compositions were investigated for their structural, dielectric and ferroelectric properties. Structural modifications along with the change in average crystallite size, dielectric properties, Curie temperature, and ferroelectric properties with Sr doping are discussed.

2. Experimental Details

$Ba_{1-x}Sr_xTiO_3$ compositions for x = 0, 0.1, 0.2, 0.3, 0.4, and 0.5 were prepared using sol-gel route. Precursor materials barium acetate (99%), strontium acetate (99.5%), and titanium (IV) isopropoxide (98%) were used according to the stoichiometric ratio. Barium and strontium acetates were mixed in acetic acid (99.5%) solvent and ethylene glycol (99%) was used for the complete dissolution of precursors. This mixture was stirred continuously at 60 °C for 1 h to form a homogeneous solution. In order to avoid precipitation, 2-Methoxyethanol (99.8%) was mixed with titanium (IV) isopropoxide. This mixture was added drop-by-drop in continuously stirred precursor solution to form a transparent sol. The resulting sol was refluxed to obtain clear gel. This gel was continuously stirred at 100 °C to obtain the powder. The powders thus obtained were dried in an oven and ground before their final calcination at 800 °C for 2 h. To prepare the disk shaped pallets of 13 mm diameter 5 ton hydraulic pressure was used and polyvinyl alcohol (2 wt%) was mixed with calcined powders. Sintering of pallets was done at 1100 °C for 3 hours. The flow chart indicating the steps involved in the synthesis of Strontium doped Barium titanate nanoceramics is shown in Fig.1.

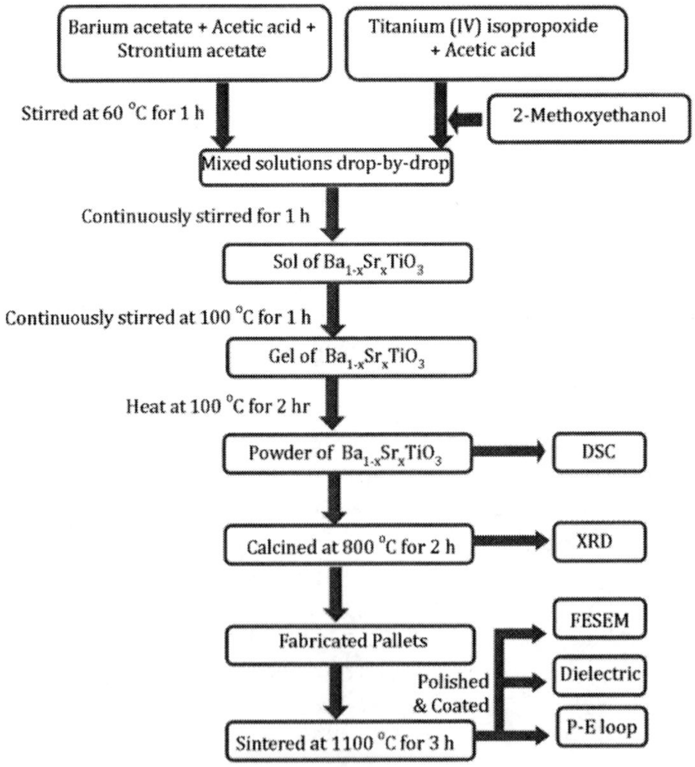

Figure 1. Flow chart indicating the steps involved in the synthesis of powders and pallets of BST nanoceramics.

The powder X-Ray diffraction (XRD) patterns of samples calcined at 800 °C were recorded in the 2θ range of $20^0 \leq 2\theta \leq 80^0$ by using X-ray powder PANalytical diffraction spectroscopy with Cu K_α radiation (1.540598 Å) at 45 kV and 30 mA. A scanning electron microscope (FESEM: Jeol JSM-7600F) was used to image the surface morphology of the samples. Silver paste was used for electrical contacts while measuring the dielectric and ferroelectric data from the pallets. Dielectric data was recorded using a Frequency Analyzer from Novocontrol Technologies Germany in the frequency range 10 Hz to 10^6 Hz and in the temperature range 30°C to 200°C. The ferroelectric data of nanoceramics was recorded using TF 2000 from Radiant Technologies.

3. Results and Discussion

3.1. Thermal Analysis

The DSC characterization was performed to understand the thermal behavior of samples. Figure 2 represents the representative DSC curve of $Ba_{1-x}Sr_xTiO_3$ sample for x = 0. The visible endothermic and exothermic peaks are indicative of the change among different phases. A small kink at around 91 °C is due to residual water vapour. The solvent decomposition and the intermediate phases reflect as peaks at three temperatures 309 °C, 477 °C, and 645 °C. The perovskite formation and exothermic peak occur up to 825 °C. Similar DSC curves with slight variations were observed for all $Ba_{1-x}Sr_xTiO_3$ compositions.

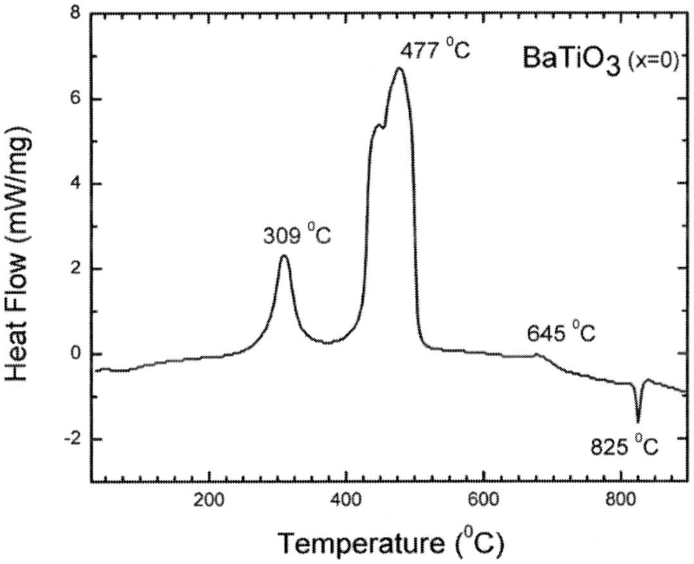

Figure 2. DSC thermograph of $Ba_{1-x}Sr_xTiO_3$ sample for x = 0

3.2. Structural Analysis

Figure 3 shows the X-ray diffraction (XRD) patterns of $Ba_{1-x}Sr_xTiO_3$ nanopowders calcined at 800 °C for x = 0-0.5 compositions. The bottom curve represents the XRD curve of x=0.0 sample sintered at 1100 °C. XRD patterns show the major peaks with Miller indices (1 0 0), (1 1 0), (1 1 1), (2 0 0), (2 1 0), (2 1 1) and (2 2 0). For x = 0 to 0.3, all XRD peaks correspond to the space group P*4mm* and confirmed the tetragonal crystal structure (JCPDS card no. 81-2203). For x = 0.4 to 0.5, all peaks (1 0 0), (1 1 0), (1 1 1), (2 0 0), (2 1 0), (2 1 1) and (2 2 0) confirmed cubic crystal structure (space group $Pm\bar{3}m$) as per the JCPDS card no. 39-1395. This shows that Sr-induces a structural transition in BT from tetragonal to a cubic structure at room temperature. For all samples, the peak (1 1 0) shifts toward a higher 2θ from 31.63° to 32.07° and shows a single phase [35]. No additional peak was observed corresponding to the secondary or impurity phase in XRD.

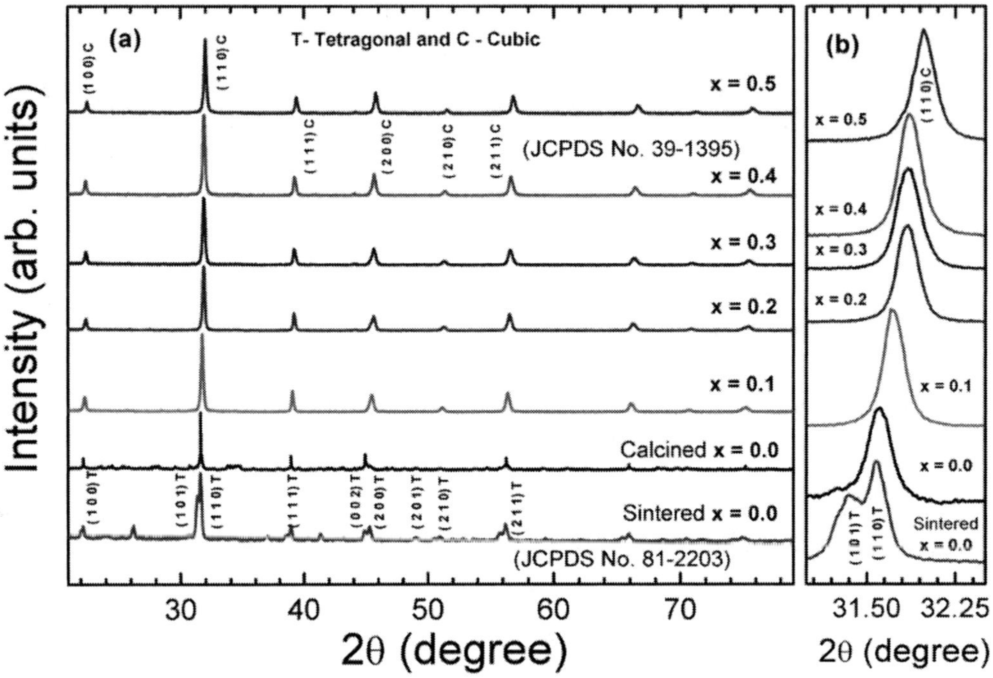

Figure 3. The XRD patterns of $Ba_{1-x}Sr_xTiO_3$ for x = 0- 0.5, calcined at 800 °C for 2 hr. The bottom curve represents the XRD curve of x=0.0 sample sintered at 1100 °C.

The XRD peak positions after the Rietveld fit were used to determine a and c parameters, volume (V) of the unit cell, the average crystallite size (D), dislocation density (δ), and lattice strain (ε) of $Ba_{1-x}Sr_xTiO_3$ compositions. The lattice parameters calculated using XRD data are tabulated in Table 1. With increasing x value, the contraction of both 'a' and 'c' parameters reduces the unit cell tetragonality, which transforms the crystal structure from tetragonal to cubic one. The ionic radii difference between Ba and Sr ions is considered to reduce the *c/a* ratio as well as to change the tetragonal symmetry to cubic structure. The lattice parameters were found to decrease with increasing of x values, due to the replacement of Ba^{+2} (1.61 Å) by smaller Sr^{+2}(1.44 Å) ions. The calculated unit cell volume of BT (x = 0) 64.65 $Å^3$decreases to 62.95 $Å^3$ for x=0.5 composition. The average crystallite size (D) was estimated from the XRD data using [36]:

$$D = \frac{k\lambda}{\beta_{2\theta} cos\theta}$$

Where, k (constant) = 0.89, $\beta_{2\theta}$is the FWHM (in radian), λ is the wavelength of X-ray radiation (1.5406 Å) and θ being the Bragg's angle. The calculated D values for studies compositions of BST are given in Table 1. The average crystallite sizes gradually decrease from 28.85 nm to 22.80 nm as Sr content increases from x = 0 to 0.5. The following equation was utilized for the calculation of the dislocation density (δ):

$$\delta = \frac{1}{D^2}$$

The dislocation density was found increasing from 12.00 x 10^{-4} to 19.23 x 10^{-4} nm^{-2} with an increase of Sr concentration in BST. The lattice strain in nanoceramics was calculated by [37]:

$$\varepsilon = \frac{D}{4tan\theta}$$

The value of lattice strain is decreasing from 5.02 x 10^{-3} lin^{-2} to 3.65x10^{-3} lin^{-2} with increasing x value. This increase in defect density with increasing x is one of the major reasons behind decreasing average crystallite sizes in BST compositions.

Table-1: *a* and *c* parameters, tetragonality (*c/a*) ratio, cell volume (V), average crystallite size (*D*), dislocation density (δ), lattice strain (ε) of $Ba_{1-x}Sr_xTiO_3$ compositions.

x	*a* (Å)	*c* (Å)	*c/a*	*Volume* ($Å^3$)	*D* (nm)	δ x 10^{-4} (nm^{-2})	ε x 10^{-3} (lin^{-2} nm)
0.0	4.006	4.029	1.006	64.65	28.87	12.00	5.02
0.1	3.998	4.017	1.005	64.22	29.08	11.82	4.67
0.2	3.993	4.013	1.005	63.99	26.08	14.70	4.57
0.3	3.900	4.008	1.005	63.82	24.67	16.43	4.37
0.4	3.988	3.988	1.000	63.41	23.45	18.18	4.13
0.5	3.978	3.978	1.000	62.95	22.80	19.23	3.65

3.3. Micro structural Analysis

SEM micrographs of $Ba_{1-x}Sr_xTiO_3$ samples for x = 0, 0.1, 0.3 and 0.5 are shown in Fig. 4. The grains in all micrographs are uniformly distributed. The grains appear more rectangular in shape for x = 0-0.3 compared to x = 0.5 cubic structure. The average grain sizes measured from these SEM images are found to decrease from 0.46 μm to 0.28 μm with increasing Sr content in BT and verify the XRD findings of grain sizes given in Table 1.

3.4. Dielectric Properties

The dielectric properties of $Ba_{1-x}Sr_xTiO_3$ compositions for x = 0, 0.1, 0.3, and 0.5 were studied at room temperature as well as in the range 27 °C to 200 °C at 1 kHz. Figures 5 (a) and (b) show the dielectric constant (ε) and dielectric tangent loss (tan δ) measured at room temperature with varying frequency (in the range 10 - 10^6 Hz), respectively. The dielectric constant values for x = 0, 0.1, 0.3 and 0.5 compositions are 1306, 1387, 1553 and 1182 respectively at 1 kHz. As the room temperature structure of BST undergoes a transition from tetragonal (polarized) to cubic (non-polarized) phase with Sr doping in BT above x=0.3, the highest dielectric constant was observed for x=0.3 composition. The dielectric constant was found higher at low frequency which decreases with increasing frequency up to 10^4 Hz and then stays almost constant in the frequency range 10^4 Hz – 10^6 Hz, due to the existence of different types of polarization [38]. The dielectric tangent loss reduced at a higher frequency and also decreased with increasing in Sr-composition. Temperature-dependent dielectric measurements at 1 kHz are shown in Fig. 5 (c) and (d), respectively for x = 0, 0.1, 0.3 and 0.5 compositions. The diffused maximum of the dielectric peak, generally associated with Curie temperature (T_C), broadens and shifts to lower temperatures with increasing Sr in BST. The value of T_C was observed 125, 100 and 75 °C for x = 0, 0.1 and 0.3, respectively. The highest dielectric constant is 2107 with tangent loss 1.31 for x = 0.3 at 75 °C and 1 kHz. This highest dielectric constant can be attributed to the thermal activation of electrons near T_C [14]. The room temperature values of dielectric constant (ε) and tangent loss (tan δ) as well as the Curie temperature (T_C), and dielectric maxima ε_{max} and tan δ at T_C are tabulated in Table 2.

IOP Conf. Series: Materials Science and Engineering 1272 (2022) 012026 doi:10.1088/1757-899X/1272/1/012026

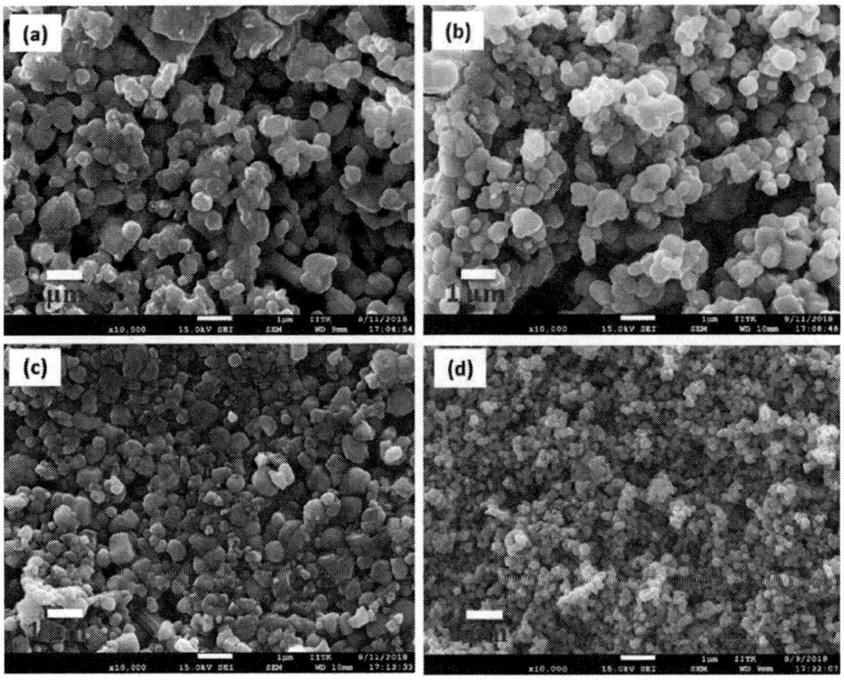

Figure 4. SEM micrographs of BST ceramics for; x = 0 (a), x = 0.1 (b), x = 0.3 (c), and x = 0.5 (d).

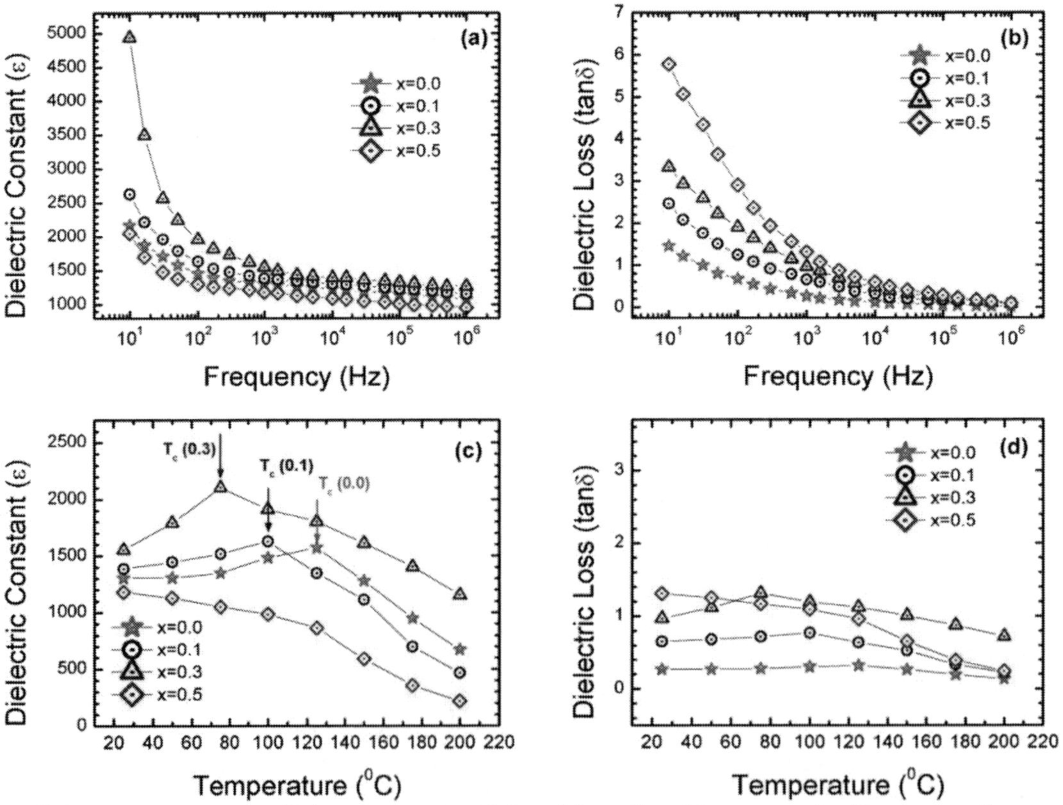

Figure 5. Room temperature dielectric constant (ε) and loss ($\tan\delta$) measured with varying frequency in $Ba_{1-x}Sr_xTiO_3$ for x = 0, 0.1, 0.2, and 0.5. (c) and (d) show the temperature variation of ε and tan δ, respectively.

3.5. Ferroelectric Properties

The ferroelectric nature in BST nanoceramics was studied. The polarization hysteresis (P-E) loops obtained from x = 0, 0.1, 0.3 and 0.5 compositions are given in Fig 6, in that order. The ferroelectric nature [9] in all compositions is clearly evident from the P-E hysteresis loops.

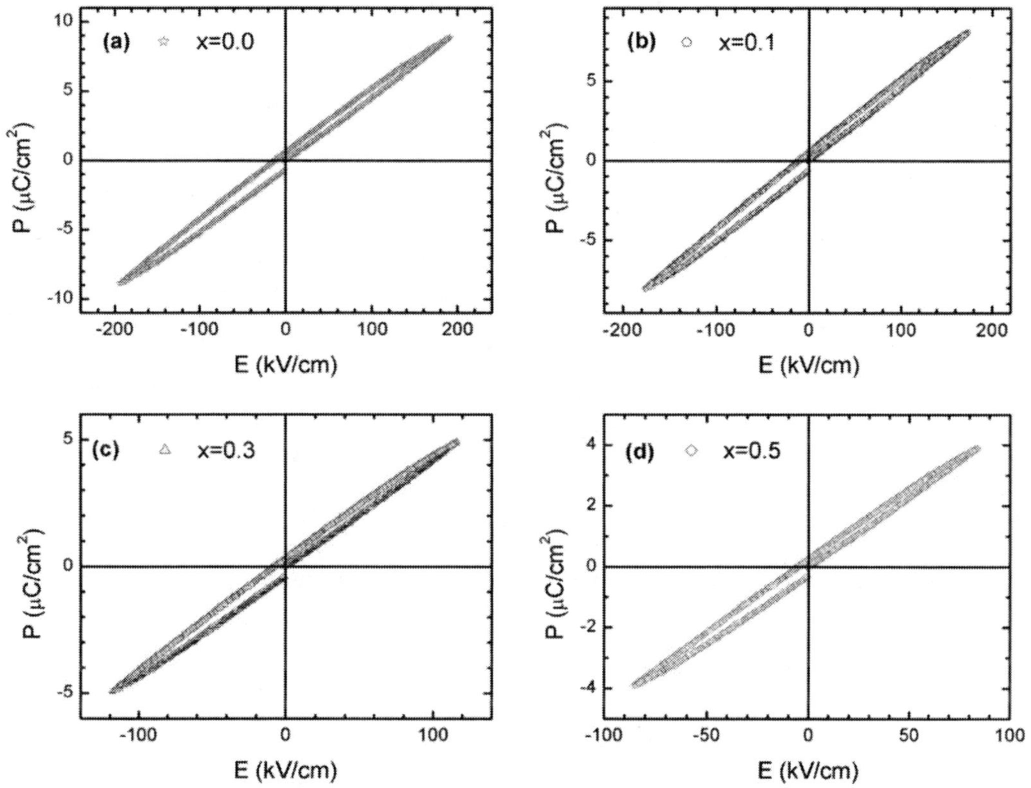

Figure 6. The polarization hysteresis (P-E) loops of $Ba_{1-x}Sr_xTiO_3$ (x = 0, 0.1, 0.3 and 0.5.)

Table 2. Dielectric [ε, tan δ, Curie temperature (T_c), dielectric maxima ε_{max} at T_c, tan δ at T_c] and Ferroelectric [P_r, P_S, and coercive field (E_C)] parameters of $Ba_{1-x}Sr_xTiO_3$ nanoceramics.

Sr Content (x)	Grain Size (μm)	Dielectric parameters					Ferroelectric parameters		
		ε (1kHz)	tanδ (1kHz)	T_C (°C)	ε_{max} (T_c)	tanδ (T_c)	P_r (μC/cm²)	P_S (μC/cm²)	E_C (kV/cm)
0	0.46	1306	0.27	125	1572	0.32	0.637	8.910	0.631
0.1	0.44	1387	0.66	100	1626	0.77	0.570	8.110	0.599
0.3	0.41	1553	0.97	75	2107	1.31	0.351	4.949	0.349
0.5	0.28	1182	1.31	---	---	---	0.229	3.238	0.255

The values of remnant polarization (P_r) and saturation polarization (P_S), and coercive field (E_C) obtained from the ferroelectric measurements of BST compositions are given in Table 2. The values of P_r and P_S are found decreasing with increasing Sr content in BST. The remnant polarization decreases from 0.637 to 0.229 μC/cm^2 while the P_S value decreases from 8.910 to 3.238 μC/cm^2 and coercive electric field (E_C) decreases from 0.631 to 0.255 kV/cm with increasing Sr composition from x = 0 to 0.5. The reduction of P_r and E_C field can be attributed to the structural modification from ferroelectric to paraelectric perovskite structure with increasing of Sr concentration. However, the formation of defects or oxygen vacancies and their contribution in electric properties cannot be completely ruled out.

4. Conclusion

$Ba_{1-x}Sr_xTiO_3$ ceramics were synthesized via sol-gel technique. Structural, dielectric and ferroelectric characterization were employed. X-ray diffraction study at room temperature suggested that the crystal structures in compositions up to x=0.3 were tetragonal. However, cubic structure was obtained in x = 0.4 and 0.5 compositions. The *c* parameter decreased more with increasing x value and hence the c/a tetragonality ratio which lead to the structural transformation towards cubic in compositions x = 0.4 and 0.5. The grain sizes gradually decreased with increasing concentration of Sr. Similar variation was observed from SEM micrographs. The room temperature dielectric constant value was found maximum in the composition x=0.3. Temperature dependent dielectric measurements revealed a dielectric maxima for x = 0 at 125 °C. The downward shift of dielectric maxima (Curie temperature) was observed till x = 0.3 (75 °C). The polarization values P_r and P_S, and coercive electric field values were also found decreasing with increasing Sr content in BT. All the observations of dielectric and ferroelectric measurements show a strong correlation with the structural modifications due to Sr in BT lattice.

Acknowledgments

The authors are indebted to Prof Vinay Gupta, DU for his valuable support during this work.

References

[1] Liu X, Zhuang Y, Xu Z, Li F, Li J, Tian Y and Dong G 2018 *J. Elec. Mater.* **47** 1099-1106 https://doi.org/10.1007/s11664-017-5862-5
[2] Aghayan M, Zak AK, Behdani M and Hashim AM 2014 *Ceram. Int.* **40** 16141-16146 https://doi.org/10.1016/j.ceramint.2014.07.045
[3] Samuvel K and Ramachandran K 2015 *Spectro. Acta Part A,* **146** 113-118 https://doi.org/10.1016/j.saa.2015.02.080
[4] Kwei GH, Lawson A, Billinge B, Cheong S 1993 *J. Phys. Chem.* **97** 2368–2377 https://doi.org/10.1021/j100112a043
[5] Everhardt AS, Denneulin T, Grunebohm A, Shao YT, Ondrejkovic P, *et al.* 2020 Appl. Phys. Rev. **7** 011402, https://doi.org/10.1063/1.5122954
[6] Islam MS and Ghosh J 2020 *Phas. Trans.* **93** 351-360 https://doi.org/10.1080/01411594.2020.1731509
[7] Madhan K and Murugaraj R 2020 *J. Sol. Sci. Tech.* **95** 11-21 https://doi.org/10.1007/s10971-020-05311-1
[8] Ren P, Wang Q, Wang X, Wang L, Wang J and Fan H 2016 *Mater. Lett.* **174** 197-200 https://doi.org/10.1016/j.matlet.2016.03.110
[9] Sindhu M, Ahlawat N, Sanghi S, Kumari R and Agarwal A 2013 *J. Alloys Compd.* **575** 109-114 https://doi.org/10.1016/j.jallcom.2013.04.026
[10] Dang NY, Kozlenko DP, Phan TL, *et. al.* 2016 *J. Elec. Mate.* **45** 2477-2483 https://doi.org/10.1007/s11664-016-4382-z
[11] Jaffari GH, Zafar A, Ali A 2021 *Physica B: Condensed matter.* **604** 412750 https://doi.org/10.1016/j.physb.2020.412750
[12] Arshad M, Huiling D, Sufyan JM, Maqsood A, Ashraf I, Hussain S, Maand W, Ran H 2020 *Ceram. Int.* **46**, 2238-2246 https://doi.org/10.1016/j.ceramint.2019.09.208
[13] Sharma H, Shilpi J and Aggarwal N 2020 *Materials Today: Proceedings* **33** 1632-1636 https://doi.org/10.1016/j.matpr.2020.06.147

[14] Kumari A and Ghosh BD 2018 *Adv. Appl. Ceram.* **117** 427-435 https://doi.org/10.1080/17436753.2018.1491166
[15] Khedhri MH, Abdelmoula N, Khemakhem H *et al.* 2019 *Appl. Phys. A* **125** 193 https://doi.org/10.1007/s00339-019-2487-y
[16] Vasilescu CA, Cernea M, Corina SE, Aldica G, Ganea P and Trusca R 2017 *J. Alloys Compd.* **706** 538-545 https://doi.org/10.1016/j.jallcom.2017.02.258
[17] Krimech FZ and Sayouri S 2020 *Materials today: Proceedings* **30** 909-917 https://doi.org/10.1016/j.matpr.2020.04.349
[18] Saleem M, Butt MS, Maqbool A, Umer MA, et al. 2020 *J. Alloys Compd.* **822** 153525 https://doi.org/10.1016/j.jallcom.2019.153525
[19] Khirade PP, Birajdar SD, Raut AV and Jadhav KM 2016 *Ceramics Int.* **42** 12441-12451 https://doi.org/10.1016/j.ceramint.2016.05.021
[20] Dixit A, Majumder SB, Katiyar RS *et al.* 2006 *J. Mater Sci.* **41** 87–96 https://doi.org/10.1007/s10853-005-5929-1
[21] Yan G, Ma M, Li C, Li Z, Zhong X, Yang J, Wu F and Chen Z 2021 *J. Alloys Compd.* **857** 158021 https://doi.org/10.1016/j.jallcom.2020.158021
[22] Zhang B, Lou X, Zheng K, Xie X et al. 2021 *Ceram. Int.* **47** 1101-1108 https://doi.org/10.1016/j.ceramint.2020.08.226
[23] Zhang M, Zhai J, Xin L and and Yao X 2017 *Mater. Chem. Phys.* **197** 36-46 https://doi.org/10.1016/j.matchemphys.2017.05.025
[24] Berenov A, Goupil FL and Alford N 2016 *Sci. Rep.* **6** 28055 https://doi.org/10.1038/srep28055
[25] Dixit A, Majumder SB, Dobal PS *et al.* 2004 *Thin Solid Films* **447-448** 284-288 https://doi.org/10.1016/S0040-6090(03)01065-4
[26] Dobal PS, Dixit A, Yu Z *et al.* 2001 *J. Appl. Phys.* **89** 8085-8091 https://doi.org/10.1063/1.1369399
[27] Dobal PS and Katiyar RS 2002 *J. Raman Spectrosc.* **33** 405-423 https://doi.org/10.1002/jrs.876
[28] Zhou L, Vilarinho PM and Baptista JL 1999 *J. Eur. Ceram. Soc.* **19** 2015-2020 https://doi.org/10.1016/S0955-2219(99)00010-2
[29] Zaman T, Islam MK, Rahman MA, Hussain A, Matin MA and Rahman MS 2019 *Ceram. Int.* **45** 10154-10162 https://doi.org/10.1016/j.ceramint.2019.02.064
[30] Arshad M, Huiling D, Sufyan JM, Maqsood A et al. 2020 *Ceram. Int.* **46** 2238-2246 https://doi.org/10.1016/j.ceramint.2019.09.208
[31] Ianculescu A, Pintilie I, Vasilescu CA, Botea M et al. 2016 *Ceram. Int.* **42** 10338-10348 https://doi.org/10.1016/j.ceramint.2016.03.152
[32] Ge PZ, Tang XG, Liu QX, Jiang YP, Li WH and Li B 2018 *J. Alloys Compd.* **731**, 70-77 https://doi.org/10.1016/j.jallcom.2017.09.330
[33] Mahani RM, Battisha IM, Aly M and Abou-Hamad AB 2010 *J. Alloys Compd.* **508** 354 - 358 https://doi.org/10.1016/j.jallcom.2010.05.060
[34] Singh D, Dixit A and Dobal PS 2021 *Ferroelectrics Journal* **573** 63-75 https://doi.org/10.1080/00150193.2021.1890464
[35] Dobal PS, Dixit A, Katiyar RS, Garcia D, Guo R and Bhalla AS 2001 *J. Raman Spectrosc.* **32** 147-149 https://doi.org/10.1002/jrs.681
[36] Holzwarth U and Gibson N 2011 *Nature Nanotech.* **6** 534 https://doi.org/10.1038/nnano.2011.145
[37] Mote V, Purushotham Y, Dole B 2012 *J Theor. Appl. Phys.* **6** 6 https://doi.org/10.1186/2251-7235-6-6
[38] Patil DR, Lokare SA, Devan RS, Chougule SS, Kanamadi CM, Kolekar YD and Chougule BK 2007 *Mater. Chem. Phys.* **104** 254-257 https://doi.org/10.1016/j.matchemphys.2007.02.027t

Study of LCF behaviour and life estimation based on modified SWT model and enhanced strain energy density of Ti–6Al–4V ELI

T D Ikhar[1] **and U Chavan**[1]

[1] Mechanical Engineering Department, Vishwakarma Institute of Technology, Bibwewadi, Pune.

tejal.ikhar20@vit.edu

Abstract. Among various types of Titanium alloy, Ti–6Al–4V ELI is the most extensively used alloy. This particular type of alloy is officially termed as Bio-material owing to its high strength, bio-compatibility, low density and good formability characteristics of titanium. Hence, it becomes necessary to evaluate its fatigue properties and low cycle fatigue behaviour for the bio-medical applications such as implants which is subjected to cyclic loading because of various physical movements performed by individual like walking, jumping or running. In this study, completely reversible strain controlled fatigue tests were carried out for various strain amplitudes at an ambient environment. The experimental data was evaluated to describe the cyclic deformation behaviour, Strain-Life plot, maximum stress variation, cyclic softening behaviour and massing behaviour. Finally, two different methods are proposed to estimate the fatigue life of Ti–6Al–4V ELI on the basis of experimental data. First is SWT model which was originally proposed by Swift, Watson and Topper to account the mean stress effect on Fatigue life is further modified in this study by replacing stress amplitude determined at the half-life with the stress amplitude determined at the first cycle. And second is estimation of Fatigue life on the basis of the Accumulated Plastic Strain Energy Density (PSED). The fatigue life estimation carried out based on above two approaches demonstrated very well agreement with experimental life.

Keywords: Low cycle fatigue; PSED; Biomedical Material; Cyclic deformation behavior; Ti–6Al–4V ELI; Titanium Alloy

1. Introduction

Titanium Alloy has wide range of applications in various industries especially in Aerospace, Biomedical, Chemical, etc. which is explained by their remarkable properties such as high strength, toughness, good fatigue behaviour, easy formability, micro-structural stability, remarkable corrosion resistance, low weight and biocompatibility [1,2,6]. Properties like high strength, low density, bio-compatibility, and good formability which are the characteristics of titanium and its alloys suits well for biomedical applications [1,3,5,6]. After the extensive research work, titanium alloy (Ti-6Al-4V ELI specifically) has been found a great implementation in medical field. It is used as an substitute metallic material in various medical, orthopaedics, cardiovascular and dental implants devices to replace the failed hard tissues. Spinal fixator, Bone fixator, artificial joints and head, implantable artificial heart, hip and knee prostheses, artificial valves, denture base, various surgical instruments and many more such medical devices where this titanium alloy is used. Owing to its wide range of biomedical applications, "Ti–6Al–4V ELI is officially regarded as an implant material in the ASTM standard (F-136)" [1,3].

Failure because of fatigue has been well-known as one of the foremost reason linked with implant failure of medical devices. Such failures are caused by repetitive loading and unloading cycles primarily in dynamic environments which leads to the progressive plastic straining of material in the stress concentrated region which further nucleates into fatigue crack and resulting into final fracture of the material [1,3,7,8].

While doing literature review, it was noticed that regarding material Ti–6Al–4V various researches and investigations were performed in the areas of fatigue behaviour mostly related to high cycle fatigue (HCF) (i.e. elastic deformation only) and the fatigue crack growth analysis as reported in [1,9]. Hence, there exists noteworthy research gap in the field of low cycle fatigue (LCF) behaviour of Titanium alloys, although few studies are still available as referred in [8-11]; "however, these available researches mostly emphasized on to study the effect of other materials, and environmental factors such as temperature, corrosion, etc. Here Strain life approach is employed since in various biomedical applications localized plastic deformation is found and also it can characterize fatigue behaviour in a better way than the stress controlled approach.

Since the failure is very certain and it is bound to happen at any condition, there is no way to evade the failure. Hence the recent researches are mainly revolved around various fatigue life prediction methods to avoid disastrous consequences of fracture in implants. Since over the last few decades, various LCF life prediction models established on strain [12-14], strain energy density (SED) [15-17], elastic-plastic fracture mechanics (EPFM) [19,41] and continuum damage mechanics (CDM) [20-22] have developed. Our study focuses on prediction tools based on strain and PSED as they are suitable for various engineering applications.

2. Experimental Program

The material investigated in this study is Ti-6Al-4V ELI. Table 1 displays the chemical composition of this titanium alloy. The hot rolled bars of present alloy having original diameter of 12.70 mm were used to manufacture the specimens required for tensile and fatigue tests. The round bars were later sectioned into long rods of length 165 mm each. Lastly, machining operation was performed to each round bar to get the required specimens with the dimensions and form as mentioned in figure 1. Now to eliminate the machining marks before conducting strain controlled fatigue tests, each specimens were polished to a 0.49 µm surface finish, thus creating finishing marks in the longitudinal direction of the specimens. For safeguarding the polished surface from the sharp edges of the extensometer, all specimens were layered with acrylic on the gage section. Acrylic was applied only at the two respective points of extensometer edges.

The use of ASTM standard E606 [28] is recommended to follow as a guideline for the strain controlled axial fatigue tests. Experiments were performed to generate experimental data for each strain amplitude in symmetrical strain controlled settings i.e. fully reversed tests with strain ratio of -1. The strain ratio, R_ε, is expressed by equation (1),

$$R_\varepsilon = \frac{\varepsilon_{min}}{\varepsilon,_{max}} \tag{1}$$

Where ε_{min} = minimum strain and $\varepsilon,_{max}$ = maximum strain.

Table 2 displays a collection of the various fatigue properties and parameters which includes the fatigue life as well of each test. The entire experimental methods which include specimen preparation and experimental procedure and experimental data of each test used in this study can be retrieved from [1,30]. The frequency were altered between 0.5 and 5 Hz with respect to the strain amplitude of the respective tests. Tests with higher strain amplitude were set in lower frequency and vice versa. These adjustment were done so that every test could be subjected to constant strain rate.

All fatigue tests were carried out at room temperature. During tests, specimens were subjected to sinusoidal strain waveform. Although testing's were performed for various strain amplitudes which are 0.012, 0.010, 0.008, 0.007, 0.006, 0.005 and 0.004 mm/mm but the entire study was focused on strain amplitudes varied from 0.012 to 0.007 mm/mm except for deriving cyclic Strain-Life Curve and Strain Life plot.

Table 1. Chemical composition (Wt%) of Ti-6Al-4V ELI used in the present research work.

Element	Aluminium	Carbon	Iron	Nitrogen	Oxygen	Vanadium	Yttrium	Titanium
Wt%	6.113	0.012	0.189	0.0039	0.123	3.87	<0.004	BAL

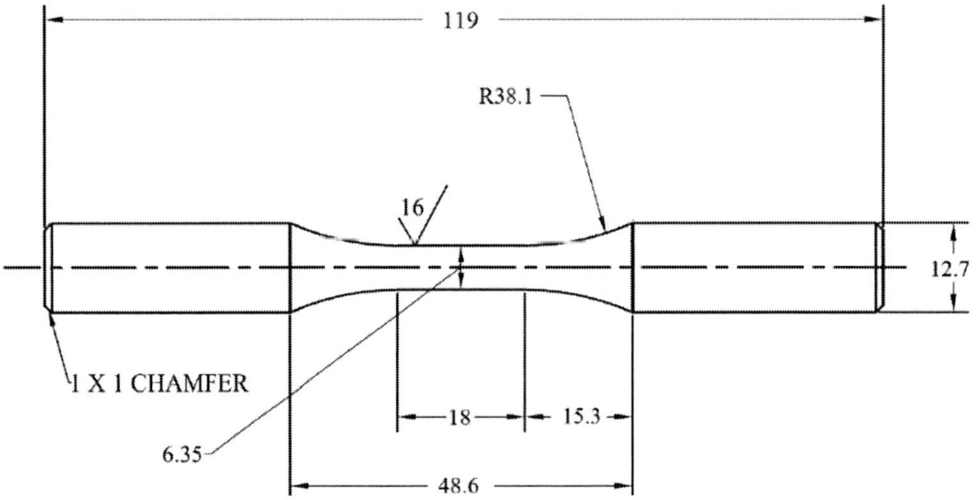

Figure 1. Round fatigue specimen with uniform gage section per ASTM standard E606/ E606M-12 used in this study [28]. All dimensions are in mm.

Table 2. Collection of the various fatigue properties and parameters of each test.

R_ε	E GPa	ε_a mm/mm	$\Delta\varepsilon_p/2$ mm/mm	$\Delta\varepsilon_e/2$ mm/mm	σ_a MPa	N_f
-1	107	0.012	0.00482	0.00717	686.9	1101
	105	0.01	0.00293	0.00706	693.6	1993
	106	0.008	0.00107	0.00692	693	6457
	106	0.007	0	0.007	745.5	24916
	105	0.006	0	0.006	648	64025
	107	0.005	0	0.005	551	1356578
	107	0.004	0	0.004	433	2215847

3. Experimental Results

3.1. Cyclic deformation behavior

The half-life hysteresis loops of four different levels of ε_a which are 0.0070, 0.0080, 0.0100 and 0.0120 mm/mm for completely reversed strain controlled loading has been plotted as shown in figure 2. It can be noticed from this figure and table 2, the strain is principally elastic, and plastic strain is insignificant for strain amplitudes of 0.0040, 0.0050, 0.0060 and 0.0070 mm/mm. The occurrence of plastic strain for the higher strain amplitudes of 0.0080, 0.0100 and 0.0120 mm/mm was observed, as indicated in figure

2. It is observed that as strain amplitude increases from 0.0080 to 0.0120 mm/mm, stress is not increasing, but rather a slight decrement as elucidated in figure 2. This decrement in stress is mainly attributed to cyclic softening.

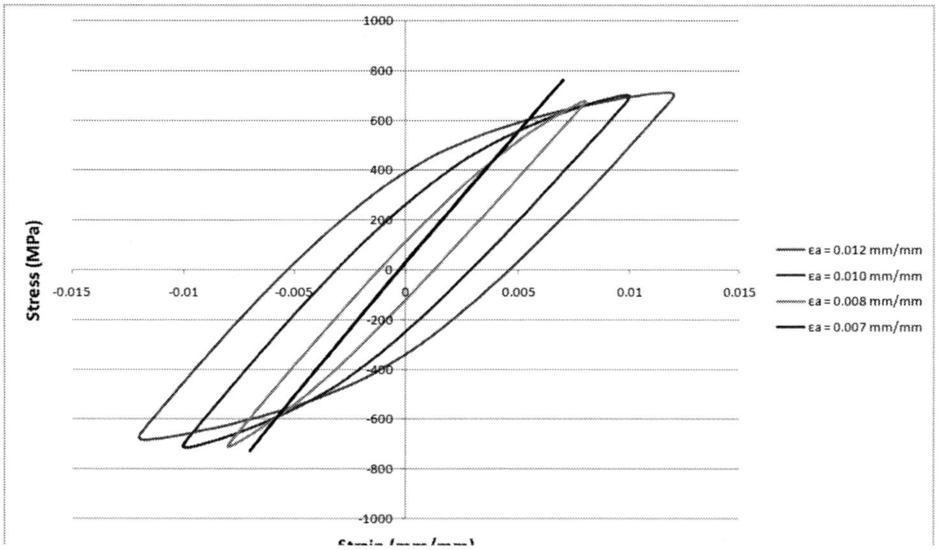

Figure 2. Half cycle loops of strain amplitude tests.

3.2. Maximum stress variation

Variation of maximum stress with each number of cycles under fully reversed condition is shown in figure 3. It can be noticed that this particular alloy displays clear cyclic softening for the strain amplitude of 0.012 and 0.010. But for strain amplitude 0.008, material behaves oddly i.e. for few initial cycles it exhibits cyclic hardening and then after that it shows cyclic softening behaviour. Again at the small strain amplitude of 0.007 mm/mm, material remains stable throughout the cycles until the failure occurs as illustrated in figure 3. Hence the above trends show that the cyclic softening/hardening behaviour of material is majorly dependent on strain amplitude. Similar behaviour is also reported in several literatures [1, 2, 3].

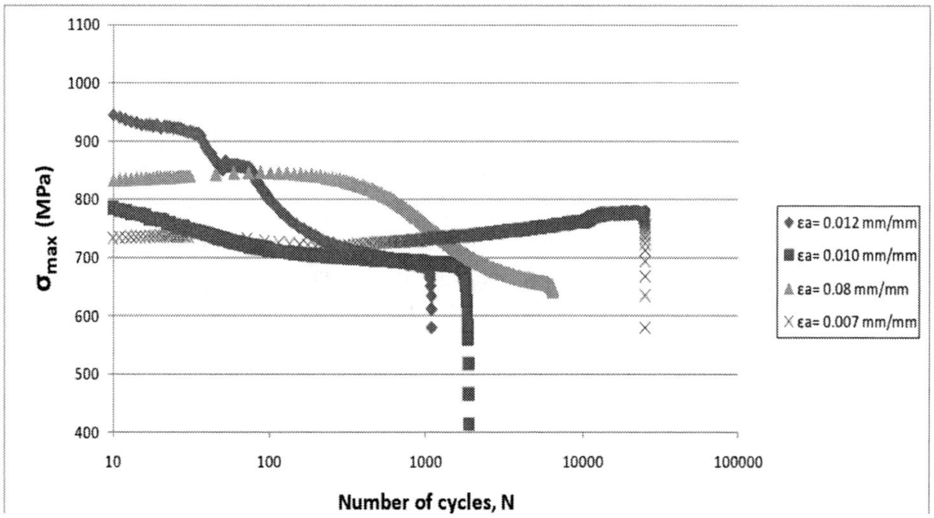

Figure 3. Variation of maximum stress with each number of cycles under fully reversed condition.

With the purpose of quantifying and comparing the degree of the cyclic softening at different levels of ε_a, a simple equation to describe the extent of softening is given below:

$$S = \frac{\sigma_a^{N=1} - \sigma_a^{N=0.5N_f}}{\sigma_a^{N=1}} \quad (2)$$

Where $\sigma_a^{N=1}$ = Stress amplitude of the first fatigue cycle and $\sigma_a^{N=0.5N_f}$ = Half-life stress amplitude.

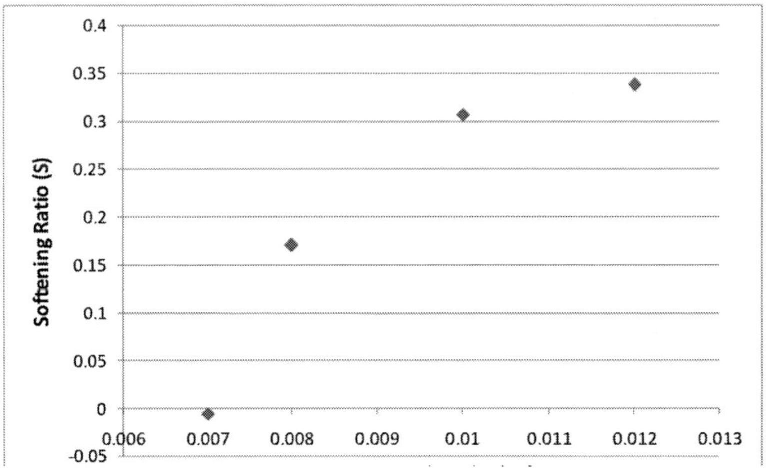

Figure 4. Cyclic softening ratio Vs strain amplitude.

Figure 4 represents the plot of cyclic softening ratio against the strain amplitude. This figure elucidates to the fact that amount of softening is low for the small strain amplitude while high for higher strain amplitude. At ε_a = 0.007 mm/mm, the ratio was negligible, i.e. cyclic softening was nonexistent there which was because the stress response at strain amplitude 0.007 mm/mm is stable and fully elastic [1].

In conclusion, it can be said that the cyclic softening and strain amplitude are correlated to each other. The degree of cyclic softening is directly proportional to strain amplitude. It may be reasonable to suppose that at the microscopic level, the amount of cyclic softening is with the development of persistent slip bands and the mobilization of pinned dislocations which causes severe plastic deformation which further results in increasing the movement of the material and decreasing the stress as stated in [12,18].

3.3. Cyclic stress–strain response and Strain life plot

Several loops of various strain amplitude at half-life is required so as to form the cyclic $\boldsymbol{\sigma} - \boldsymbol{\varepsilon}$ curve. The upper tips of each loop are connected to each other to get the cyclic $\boldsymbol{\sigma} - \boldsymbol{\varepsilon}$ curve and then it is compared with the monotonic tensile curve as illustrated in figure 5.

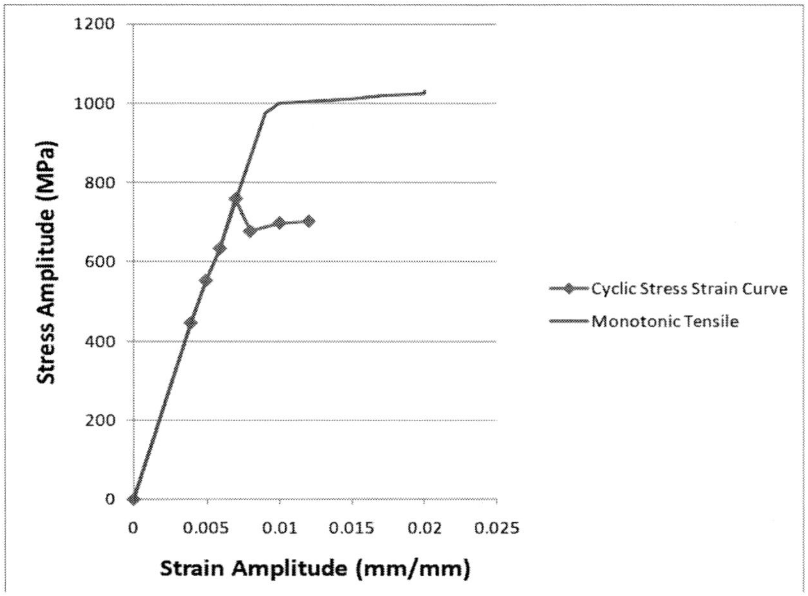

Figure 5. Monotonic and cyclic $\sigma - \varepsilon$ curves.

Material is said to be exhibiting cyclic softening behaviour if the cyclic $\sigma - \varepsilon$ curve lies below the monotonic curve, and cyclic hardening behaviour if the cyclic $\sigma - \varepsilon$ curve lies above the monotonic curve as stated in [23]. As it is very evident from figure 5, cyclic curve lies below the uniaxial tension curve which means material is exhibiting cyclic softening.

Log-log plot of strain amplitude against the number of cycles to failure is plotted in as shown in figure 6.

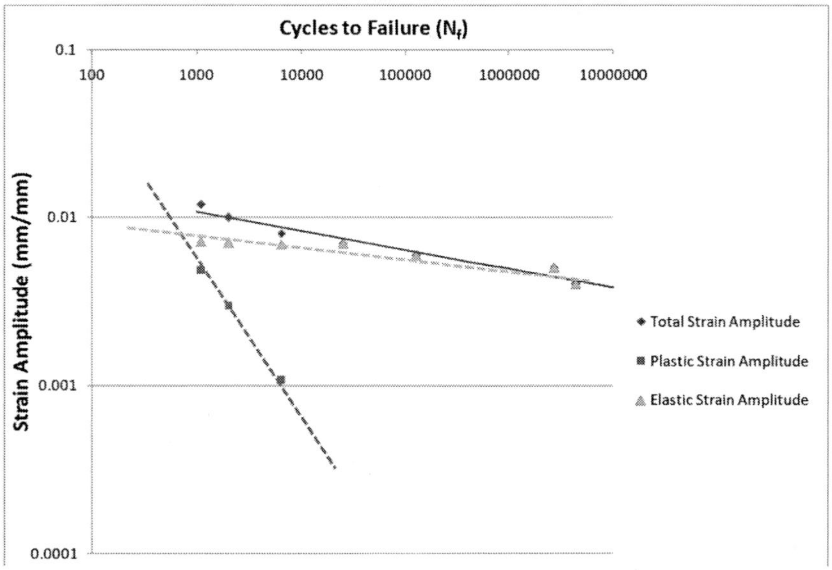

Figure 6. Strain amplitude-life plot.

In order to relate the fatigue life with total strain, Basquin–Manson–Coffin [24-26] proposed the following relationship as given in equation (3). The LCF parameters for symmetrical strain controlled test as enumerated in the following table 3.

$$\frac{\Delta\varepsilon}{2} = \varepsilon_a = \frac{\Delta\varepsilon_e}{2} + \frac{\Delta\varepsilon_p}{2} = \frac{\sigma_f'}{E}(2N_f)^b + \varepsilon_f'(2N_f)^c \quad (3)$$

Where σ_f' = Fatigue strength coefficient, ε_f' = Fatigue ductility coefficient, b = Fatigue strength exponent and c = Fatigue ductility exponent.

Table 3. LCF parameters for symmetrical strain controlled test.

Parameters	Value
σ_f' (MPa)	1317
b	-0.060
ε_f' (MPa)	4.79
c	-0.875

3.4. Massing type behaviour

"Massing cyclic stress-strain describes the skeleton of the stabilized stress-strain curve, which indicates that the loading or unloading branch of the hysteresis loop under different strain amplitudes superimpose with each other" [12]. Massing's hypothesis states that the stabilised hysteresis loop branch may be obtained by doubling the basic material stress-strain curve.

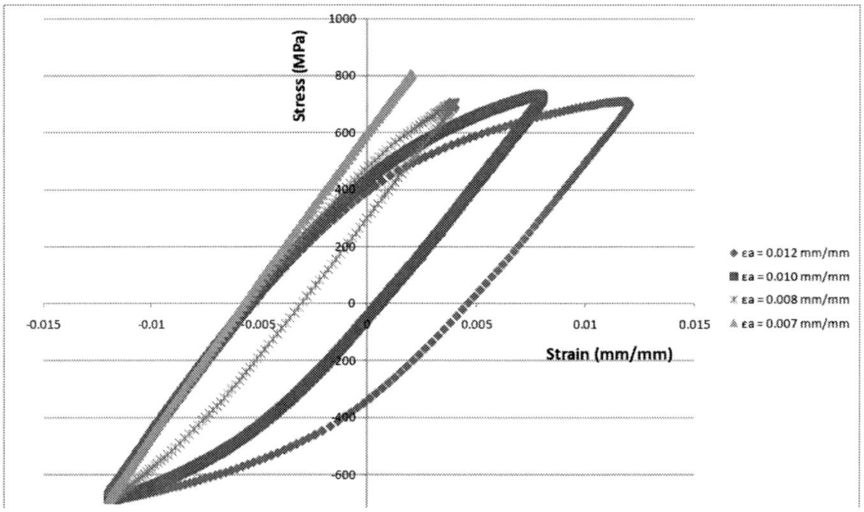

Figure 7. Massing behaviour of the material.

The lower edge of the stabilized hysteresis loop of each strain amplitude is relocated to the same point. It has been noticed that, the loading branches of each loop do absolutely trace the same path, as illustrated in figure 7, which indicates the Massing behaviour of Ti–6Al–4V ELI.

4. Fatigue life estimation based on Modified SWT (Smith, Watson, and Topper) model

In order to include the effect of mean stress in prediction of the number of cycles to failure, Smith-Watson-Topper (SWT) model [27] proposed the following relation:

$$\sigma_{max}\varepsilon_a = \frac{(\sigma_f')^2}{E}(2N_f)^{2b} + \sigma_f'\varepsilon_f'(2N_f)^{b+c} \tag{4}$$

Where σ_{max} = Maximum stress at half-life, N_f = Fatigue life, σ'_f, b, ε'_f and c are the low cycle fatigue parameters.

The values of the properties σ'_f, b, ε'_f and c are determined using Basquin-Coffin-Manson [24-26] relation as given below.

$$\varepsilon_a = \varepsilon_a^{pl} + \varepsilon_a^{e} = \frac{\sigma_f'}{E}(2N_f)^{b} + \varepsilon_f'(2N_f)^{c} \tag{5}$$

Table 3 enumerates LCF parameters for symmetrical strain controlled test. Figure 8 illustrates the S-N curve based on equation (5) for fully reversed strain ratio.

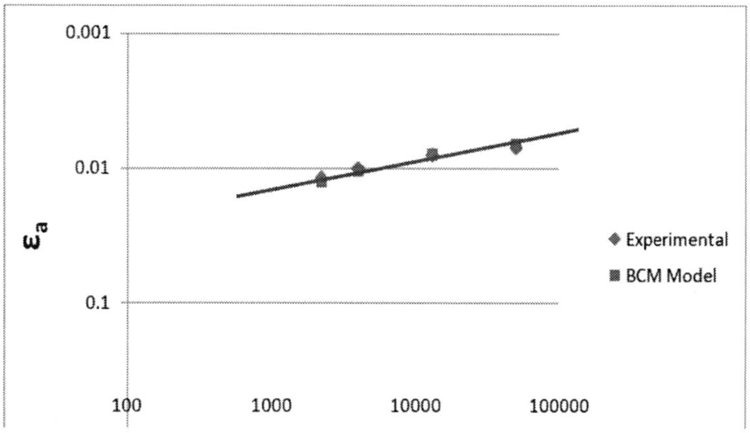

Figure 8. The S-N curve.

The maximum stress can be replaced by the stress amplitude at half-life which is only possible when the material relaxes quickly i.e. mean stress must be zero at the half fatigue life. Thus, equation (4) turns out to be,

$$\sigma_a\varepsilon_a = \frac{(\sigma_f')^2}{E}(2N_f)^{2b} + \sigma_f'\varepsilon_f'(2N_f)^{b+c} \tag{6}$$

Figure 9. displays the log-log plot of left hand side of equation (6) Vs reversals to failure using the low cycle fatigue parameters calculated by using equation (5).

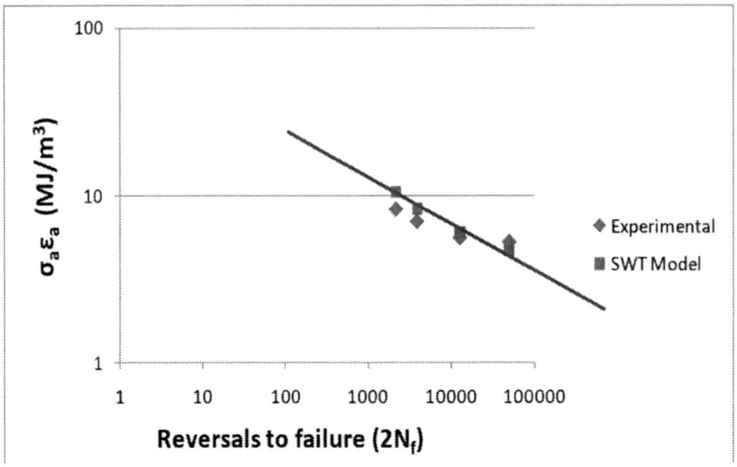

Figure 9. log-log plot of left hand side of equation (6) Vs reversals to failure.

As it is observed in the figure 9, it overestimated the fatigue life at large strain amplitudes hence equation (6) proven to be not a good tool to estimate the fatigue life. Therefore, this study [12] alters the equation (6) to the respective equation below,

$$\sigma_a \varepsilon_a = A(2N_f)^B + G(2N_f)^F \quad (7)$$

Where parameters A, B, G and F are used as a replacement just for the sake of simplicity and can be derived by using linear regression analysis as displayed in table 4. By utilizing the equation (7), Fatigue life has been estimated at each strain amplitude. Figure 10. establishes the very well correspondence concerning equation (7) and experimental fatigue life.

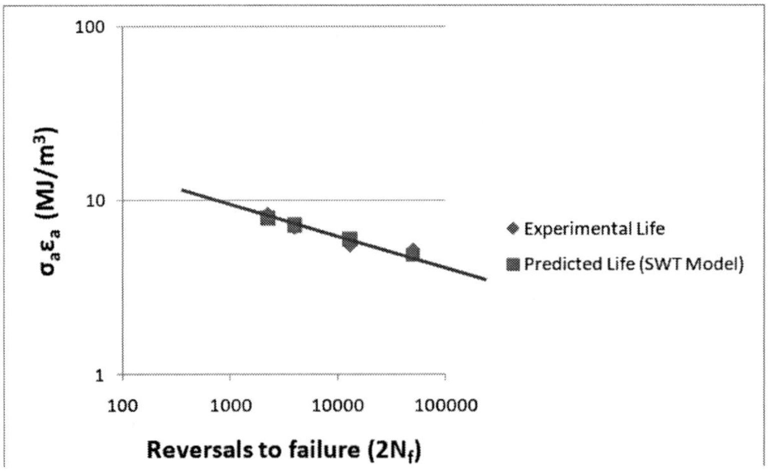

Figure 10. Fatigue test data assessment derived from equation (7).

Table 4. Parameters for equation (7) for the Ti-6Al-4V ELI.

A	B	G	F
-129.5	-2.401	26.57	-0.1575

However, It should be taken into consideration that to determine the stress amplitude at half fatigue life is not feasible or rather impractical in real engineering applications when we compare it to the determination of stress amplitude in the initial cycles. With the purpose to make it easier in engineering applications, the study [12] put forward the equation (8), where the stress amplitude determined at the half-life is replaced with the stress amplitude determined at the first cycle.

$$\sigma_a^{N=1}\varepsilon_a = A_1(2N_f)^{B_1} + G_1(2N_f)^{F_1} \quad (8)$$

Table 5 lists the fitting parameters in equation (8). Figure 11 represents the calculated life derived by using the proposed modified SWT model which is established on the stress amplitude of the first cycle against the experimental fatigue life. The value of R^2 which is an indicative of the accurateness of any model is found to be 0.952 which validates the accurateness of the new model for the given material under study.

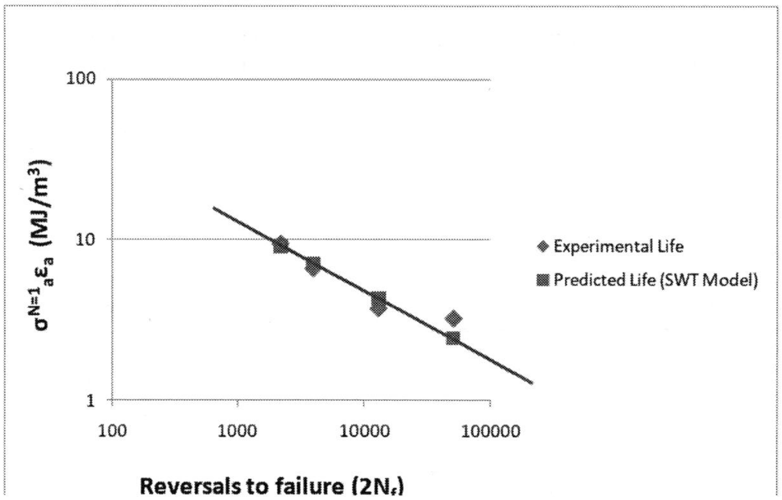

Figure 11. Predicted fatigue life based on modified SWT model Vs Experimental Life.

Table 5. Parameters for equation (8) for the Ti-6Al-4V ELI.

A_1	B_1	G_1	F_1
7.956	-6.061	238.9	-0.4239

5. Fatigue life estimation based on strain energy density

The area within a hysteresis loop is energy dissipated during a cycle (usually in the form of heating). This energy represents the plastic work from the cycle and is termed as Plastic strain energy density (PSED). For the material exhibiting massing type behaviour, it can be calculated using the following equation [33].

$$\Delta W_p = \frac{1-n'}{1+n'}\Delta\sigma\Delta\varepsilon_p \quad (9)$$

Where $\Delta\sigma$ = Stress range, $\Delta\varepsilon_p$ = Plastic strain range and n' = Cyclic hardening exponent. For the material exhibiting non-massing type behaviour, ΔW_p can be calculated using the following equation [34]:

$$\Delta W_p = \frac{1-n^*}{1+n^*}\Delta\sigma\Delta\varepsilon_p + \frac{2n^*}{1+n^*}\delta\Delta\sigma_o\Delta\varepsilon_p \quad (10)$$

Where $n^* =$ hardening exponent of the master curve and $\delta\Delta\sigma_o = \Delta\sigma - \Delta\sigma^*$. Since our material exhibits massing type behaviour, equation (9) was used to compute plastic strain energy density.

Since, the cyclic behaviour at strain amplitude 0.007 mm/mm doesn't show any plastic straining which is very significant for on-setting the damage; hence evaluation of its fatigue life has been excluded for this section. The progression of PSED (Δw) with number of cycles for the strain amplitude of 0.010 mm/mm is shown in figure 12. In figure 12, the Δw curve can be distributed into three phases. The occurrence of these three different stages or phases is an attribution of material behaviour. The first phase is the cyclic softening phase where PSED value keeps on increasing until it reaches to the stabilization phase. The subsequent phase which is the stabilization phase, PSED value remains constant since there no dissipation of the energy until it reaches to the last stage. Once the damage is instigated, there is drastic rise in plastic strain energy density value. Start number of cycles when the stabilization started to occur is defined as stabilized points (Nsta)" [20]. "Concurrently, a 5% rise of the Δw from the maximum value of PSED is pronounced a reliable value at which the damage was initiated (Ni)" [20]. Likewise, a 10% rise of stabilized Δw value from the point at which damage was initiated is considered as the final fatigue fracture (Nf) [20].

Figure 12. The progression of PSED (Δw) with number of cycles.

5.1. Damage Accumulation

The following equation determines the fatigue damage at any cycle using the plastic strain energy density [21]:

$$D_i = 1 - \frac{\Delta w_i}{\Delta w_o} \tag{11}$$

Where $\Delta w_o =$ PSED after cyclic softening and $\Delta w_i =$ PSED of the ith cycle.

The values obtained for Δw, ∑ΔD, N_i, N_f, N_o, ∑ΔN and ∑ΔD/∑ΔN from the LCF experimental tests are summarized in table 6, where N_o and ΔN are total number of stabilized cycles and the total number of cycle material takes to propagate the damage.

Table 6. Fatigue damage values determined from the experimental data.

ε_a (mm/mm)	Δw_o (MJ/m^3)	N_{sta}	N_i	N_o	N_f	$\sum\Delta N$	$\sum\Delta D$	$\sum\Delta D/\sum\Delta N$
0.012	13.9441	556	1021	465	1080	615	0.7198	0.0012
0.01	8.5394	500	1650	1150	1820	670	4.5658	0.0068
0.008	4.2332	3780	5650	1870	6400	4530	42.7102	0.0094

5.2. Estimation of Damage Initiation Parameters

Damage initiation and evolution approaches established on PSED of each cycle is utilized to include the damage analysis into LCF [20]. The following equation defines the damage initiation model:

$$N_o = c_1 \Delta w_o^{c_2} \tag{12}$$

Where, N_o = Total number of stabilized cycles, Δw_o = The accumulated inelastic strain energy of half cycles and c_1, c_2 = Material constants.

Based on equation (12), the logarithmic plot of Δw_0 and fatigue life in the damage initiation stage for the given material is plotted in figure 13. to get the values of c_1 and c_2 as 10524 and -1.13 respectively. The variation is observed to be linear.

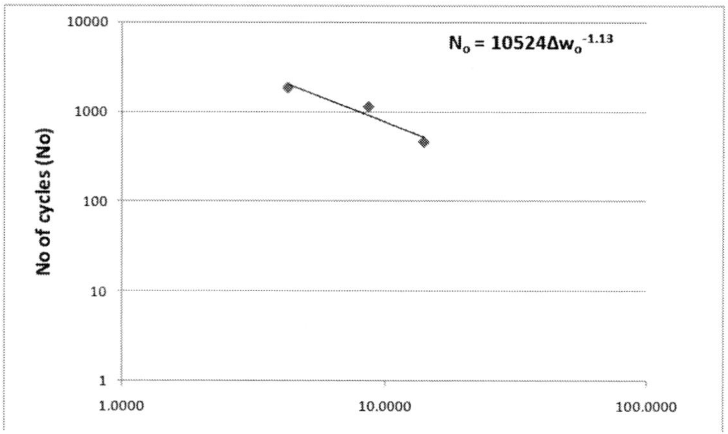

Figure 13. The logarithmic plot of Δw_0 and fatigue life in the damage initiation stage.

5.3. Estimation of Damage Evolution Parameters

Damage evolution to be derived from plastic strain energy density of the half cycle can only be estimated using accumulating material deterioration right after the phase to damage initiation [20]. The following equation defines the rate of the damage (D) of each cycle:

$$\frac{dD}{dN} = \frac{c_3 \Delta w^{c_4}}{L} \tag{13}$$

Where Δw = Plastic strain energy density of the stabilized loop, c_3, c_4 = Material constants and L = 1.587 mm which is the characteristic length.

Based on equation (13), logarithmic relationship of the PSED and Damage values in the degradation stage for the given material is plotted in figure 14 to get the values of c_3 and c_4 as 0.132 and -1.66 respectively.

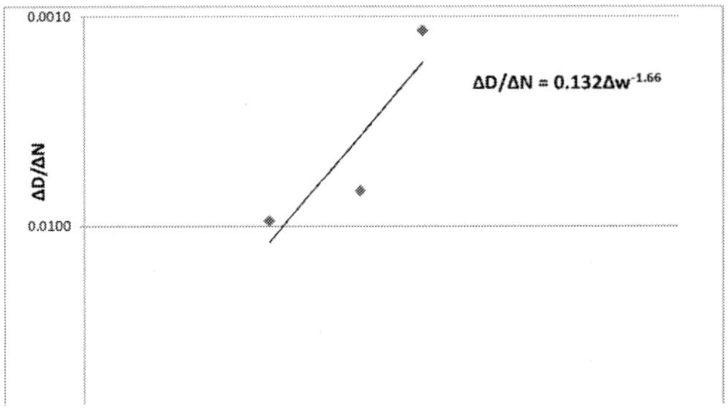

Figure 14. The logarithmic plot of PSED and damage parameter in the damage evolution stage.

5.4. Fatigue life prediction

The final fatigue life is calculated using the subsequent equation:

$$N_f = N_i + N_e = c_1 * \Delta w_o^{c_2} + \frac{L}{c_3 \Delta w^{c_4}} * \frac{1}{1 - c_4} \tag{14}$$

Based on the equation (14), the final fatigue life is determined and is compared with the life found during experimentations as displayed in the table 7. It is noticed that the calculated fatigue life doesn't match greatly with experimental life. Hence, this study proposes the modification in equation for determining the constants c_1 and c_2 by replacing N_o with N_i. Also, it suggests replacing N_o with N_i while calculating ΔN so the new ΔN equals to (N_f - N_i) for determining constants c_3 and c_4. Since, it is the fact that damage develops both by creep and/or fatigue right from the first cycle [20]. Hence all the above modifications are justified.

Table 7. Comparison of the fatigue life determined by experimentally and by equation (16).

Strain Amplitude	Fatigue Life (equation (16))	Fatigue Life (Experimental)
0.012	536	1101
0.010	938	1993
0.008	2064	6457

5.5. Estimation of New Damage Initiation Parameters

Therefore, the equation (14) is corrected to the following equation as proposed,

$$\boldsymbol{N_i = c_1 \Delta w_o^{c_2}} \tag{15}$$

Based on equation (15), logarithmic plot of PSED and fatigue life in the damage initiation stage for the given material is plotted in figure 15 to get the new values of c_1 and c_2 as 43533 and -1.45 respectively. The variation is observed to be linear.

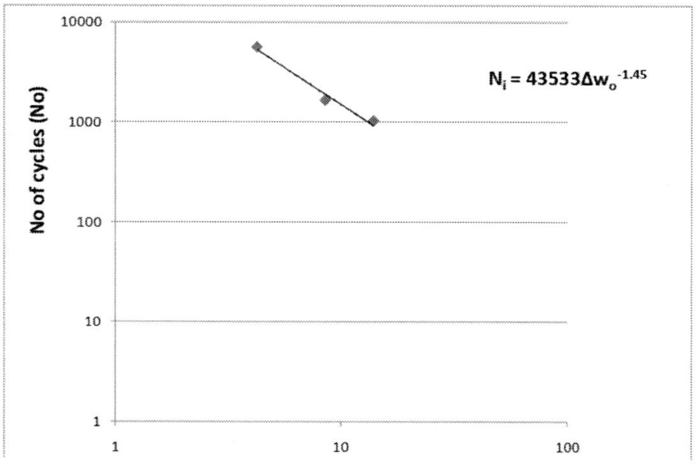

Figure 15. The logarithmic plot of Δw_0 and fatigue life in the damage initiation stage.

5.6. Estimation of New Damage Evolution Parameters

New values of ΔN which are determined by subtracting N_i from N_f along with other fatigue damage values are summarized in table 8 which are required for the estimation of New Damage Evolution Parameters (c_3 and c_4) by using equation (15). Based on equation (13), The logarithmic plot of PSED and damage parameter in the damage evolution stage for the material is plotted in figure 16 to get the new values of c_3 and c_4 as 0.375 and -1.27 respectively.

Table 8. Fatigue damage values determined from the experimental data.

ε_a (mm/mm)	Δw_o (MJ/m^3)	N_{sta}	N_i	N_o	N_f	$\sum\Delta N$	$\sum\Delta D$	$\sum\Delta D/\sum\Delta N$
0.012	13.9441	556	1021	465	1080	59	0.7198	0.0122
0.01	8.5394	500	1650	1150	1820	170	4.5658	0.0269
0.008	4.2332	3780	5650	1870	6400	750	42.7102	0.0569

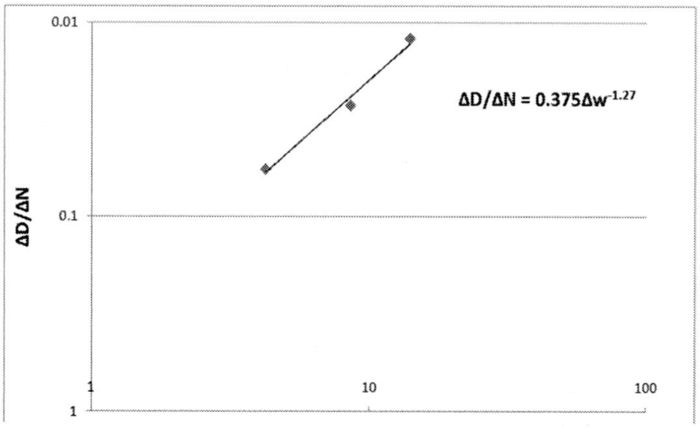

Figure 16. The logarithmic plot of PSED and damage parameter in the damage evolution stage.

5.7. *Final Fatigue life prediction*

Based on the equation (14), the new final fatigue life is determined and compared with the experimental fatigue life and by equation (8) using modified SWT model as displayed in the table 9. It is observed that the new calculated fatigue life matches greatly with experimental life. Hence the good correspondence validates the success and correctness of the new model proposed.

Table 9. Compilation of the fatigue life determined by the experimentation, equation (8) and equation (14).

Strain Amplitude	Fatigue Life (equation (14))	Fatigue Life (Experimental)	Fatigue Life (equation (8))
0.012	954	1101	1183
0.010	1943	1993	2152
0.008	5374	6457	7050
0.007	-	24916	27549

6. Conclusions

The above investigation supports the following conclusions.

- As illustrated in figure 2, It is observed that as strain amplitude increases from 0.0080 to 0.0120 mm/mm, stress is not increasing, but rather a slight decrement. This decrement in stress is mainly attributed to cyclic softening.
- The figure 4 illustrated that the amount of cyclic softening decreased with decreasing strain amplitude. At $\varepsilon_a = 0.007$ mm/mm, the ratio was negligible, i.e. cyclic softening was nonexistent there which was because the stress response at strain amplitude 0.007 mm/mm is stable and fully elastic.
- In conclusion, it can be said that the cyclic softening and strain amplitude are correlated to each other. The degree of cyclic softening is directly proportional to strain amplitude. It has been noticed that material is exhibiting cyclic softening behaviour.
- It has also been observed that the material under study is exhibiting a massing behaviour.
- The study uses a proposed approach to estimate the fatigue life for this particular alloy, where a modified SWT model was used. Here in modified SWT model, the stress amplitude which was previously determined at the half life cycle is replaced with the one which is determined at the first cycle. The result demonstrates a well correspondence with the experimental results.
- This paper proposes the new and modified approach for the estimation of damage initiation and evolution parameters. This approach revises equation (12) to the equation (15) to estimate new damage initiation parameters and also suggests replacing N_o with N_i while calculating ΔN so the new ΔN equals to (N_f - N_i) for determining new damage evolution parameters. By using this model, Fatigue life is predicted which showed well correspondence with the experimental results.

Acknowledgements

To support this study, experimental data was taken from the Harvard Dataverse which was made available by P. E. Carrion and N. Shamsaei with the support of (CAVS), MSU, MS, USA [30]. The conclusions & views set forth indicate the opinions of the authors only.

References

[1] Carrion P 2016 *Theses and Dissertations* (Mississippi: Mississippi State University)
[2] Vandermeulen D, Verlinden B, Rabet L, Houtte P, Humbeeck J, Vantomme J and Habraken A *Ti-6Al-4V: Deformation Map And Modelisation Of Tensile Behaviour.*
[3] Akahori T and Niinomi M 1998 *Materials Science and Engineering A* **243** 237–243
[4] Niinomi M 1998 *Materials Science and Engineering A* **243** 231–236

IOP Conf. Series: Materials Science and Engineering 1272 (2022) 012027 doi:10.1088/1757-899X/1272/1/012027

[5] Wang K 1996 *Materials Science and Engineering A* **213** 134- 137
[6] Carrion P, Shamsaei N, Daniewicz S and Moser R 2017 *International Journal of Fatigue* **99** 87–100
[7] Teoh S 2000 *International Journal of Fatigue* **22** 825–837
[8] Saitova L, Höppel H, Göken M, Semenova I and Valiev R 2009 *International Journal of Fatigue* **31** 322–331
[9] Hein M, Kokalj D, Dias N, Stangier D, Oltmanns H, Pramanik S, Kietzmann M, Hoyer K, Meißner J, Tillmann W and Schaper M 2022 *Metals* **12** 122
[10] Lin C, Ju C and Lin J 2005 *Biomaterials* **26** 2899–907
[11] Zhang S, Li S, Jia M, Prima F, Chen L, Hao Y and Yang R 2011 *Acta Mater* **59** 4690–4699
[12] Feng L and Qian X 2018 *Marine Structures* **61** 343-360
[13] Koh S and Stephen R 1991 *Fatigue Fract. Engng Mater Struct*, **14** 413-428
[14] Manson S 1965 *Exp Mech,* **5** 193–226
[15] Hu Y, Shi J, Cao X and Zhi J 2021 *Materials* **14** 2372
[16] Roy S, Goyal S, Sandhya R and Ray S 2012 *Nuclear Engineering and Design* **253** 219–225
[17] Rong L, Xiang H, Yi G and Yuan W 2012 *Journal Of Iron And Steel Research. International* **19** 47-53
[18] Hao H, Ye D and Chen C 2014 *Materials Science & Engineering A* **605** 151–159
[19] Dowling N and Begley J 1976 *ASTM STP* **590** 82–103.
[20] Hormozi R, Biglari F and Nikbin K 2014 *Fatigue Fract Engng Mater Struct*, **37** 1211–1222
[21] Song W, Liu X, Xu J, Fan Y, Shi D, Yang F, Xia X, Berto F and Wan D *Front. Mater*. **8** 641145
[22] Jie Z, Li Y, Wei X and Zhuge P 2018 *Journal of Constructional Steel Research* **148** 542–550
[23] Stephens R, Fatemi A, Stephens R and Fuchs H 2001 *Metal Fatigue In Engineering*. (New York: Wiley-Interscience publication) 93-119
[24] Basquin O 1910 *In: Proceedings of the American Society for Testing and Materials (West Conshohocken)*
[25] Manson S 1953 *In: Proceedings of the Heat Transfer Symposium (Ann Arbor)* pp. 9–75
[26] Coffin L 1954 *Trans. Am. Soc. Mech. Eng.* **76** 931–950
[27] Smith K, Topper T and Watson P 1970 *J Mater* **5** 767–78
[28] ASTM E606/E606M-12 2012 (West Conshohocken: ASTM International)
[29] ASTM E8/E8M-15a 2015 (West Conshohocken: ASTM International)
[30] Carrion P and Shamsaei N 2016 *Harvard Dataverse* **1**
[31] Golos K and Ellyin F
[32] Hu Y, Shi J, Cao X and Zhi J 2021 *Materials* **14** 2372.
[33] Morrow J 1965 *In Internal Friction, Damping, and Cyclic Plastic* Vol 378 (Philadelphia: ASTM International) pp 45–87.
[34] Moyer E and Sih G 1984 *Eng. Fract. Mech.* **19** 643–652
[35] British Standard BSI: 2006
[36] Hales R, Holdsworth S, O'Donnell M, Perrin I and Skelton R 2002 *Materials at high temperatures* **19** 165-85
[37] Korsunsky A, Dini D, Dunne F and Walsh M 2007 *Int J Fatigue* **9** 1990–5
[38] Halford G 1966 *J. Mater.* **1** 3–18
[39] Feltner C and Morrow J 1961 *J. Basic Eng.* **83** 15–22
[40] Ellyin F and Kujawski D 1984 *Press Vessel Technol* **106** 342–347
[41] Broberg K 1971 *J Mech Phys Solid* **19** 407–18

Evaluation of dipole moments of indole derivative (5-MPIC) by Solvatochromic shift method

Sulochana Devar[1*], Vijayalaxmi Gangadharmath[1], N G Yernale[2], B S Mathada[3], Omnath Patil[1] and S M Hanagodimath[1*]

[*1] Department of Physics, Gulbarga University, Kalaburagi, Karnataka, India
[2] Department of Chemistry, Guru Nanak First Grade College, Bidar, Karnataka, India
[3] PG Department of Chemistry, Organic Chemistry Research Lab, Vijaya College, Bengaluru, Karnataka, India

Corresponding author: sulochana.devar7@gmail.com; smhmath@rediffmail.com

Abstract: The Solvatochromic technique was used to examine the solvent effects of ethyl 5-methyl-3 phenyl-1*H*-indole 2-carboxylate [5-MPIC]. Different solvents were used to record 5-MPIC's absorption and emission spectra at room temperature (300 K). The ground state and excited state dipole moments were estimated experimentally using Lippert's, Bakshiev's and Kawaski Chamma Viallet's equations. The HOMO-LUMO gap and MEP map were also estimated theoretically by using B3LYP/6-31+G (d, p) basis set of Gaussian 16 w program. Dipole moments of excited states reveal a clear polarity difference between the ground and excited states.

Keywords: Dipole moments, Solvatochromic shift, HOMO-LUMO, MEP, Indole derivative.

1. Introduction

Organic compounds' absorption and emission spectra are the topic of a number of studies. The microelectronic and geometric structure of short-lived species may be gleaned from their dipole moments, making them an useful source of information. Electronically excited compounds are important in defining their photo physical and photochemical properties [1].

Indole based compounds acts as efficient dye sensitizer solar cells [2] and has diverse application in the field of biological sciences [3-5], as chemical sensors, fluorescent probes, laser dyes and non-liner optical devices. Indole composites are used as, antimicrobial, anti-cancer, anti inflamentry, anti-anxiety, anti-depressant, rodenticides and fluorescent indicators [6-11]. DFT may be used to investigate the characteristics of a substance and explain some experimental occurrences. So theoretical research may be used to measure experimental results. We also performed DFT in this regard. There has been no research done on this sort of chemical, according to a literature review. This aroused our interest in the indole derivative 5-MPIC, which is the subject of our current study.

2. Materials and methods

The indole derivative namely ethyl 5-methyl-3phenyl-1*H*-indol 2-carboxylate [5-MPIC], was synthesized by using reported method [12]. For the current work, spectroscopic-grade solvents such as Toluene and Benzene, acetonitrile, 1,4-dioxane, dimethyl sulfoxide, dimethyl formamide, and tetrahydrofuran were used. The samples of 5-MPIC molecule have been prepared with concentration 1×10^{-6}M.

Experiments in various solvents at room temperature using a twin beam PG Inst.Ltd., model T-90+, UV-Visible and Hitachi F-2700 Fluorescence analyzer recorded 5-MPIC's absorption and emission frequencies in the range of 250–550 nanometers. 5-MPIC's molecular structure is depicted in figure 1.

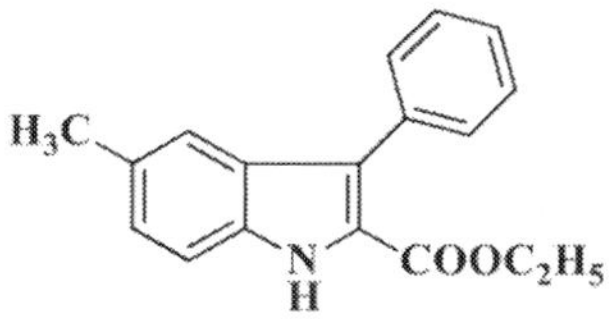

Figure1. Structure of ethyl 5-methyl-3-phenyl-1*H*-indole-2-carboxylate (5-MPIC)

3. Theoretical background

3.1. Evaluation of ground and excited state dipole moments

The three independent equations used for the estimation of ground and excited state dipole moments of the 5-MPICare as follows:

Lippert's equation [13]

$$\bar{\nu}_a - \bar{\nu}_f = m_1 F_1(\varepsilon, n) + \text{Constant} \tag{1}$$

Bakshiev's equation [14]

$$\bar{\nu}_a - \bar{\nu}_f = m_2 F_2(\varepsilon, n) + \text{Constant} \tag{2}$$

Kawaski Chamma Vialette's equation [15,16]

$$\frac{\bar{\nu}_a + \bar{\nu}_f}{2} = m_3 F_3(\varepsilon, n) + \text{Constant} \tag{3}$$

where F_1, F_2 and F_3 are the solvent polarity parameters, which are expressed as

$$F_1(\varepsilon, n) = \left[\left(\frac{(\varepsilon-1)}{2\varepsilon-1}\right) - \frac{(n^2-1)}{(2n^2-1)}\right] \tag{4}$$

$$F_2(\varepsilon, n) = \left[\frac{(2n^2-1)}{(n^2+2)}\right]\left[\left(\frac{\varepsilon-1}{\varepsilon+2}\right) - \left(\frac{n^2-1}{n^2+2}\right)\right] \tag{5}$$

$$F_3(\varepsilon, n) = \left\{\frac{1}{2}\frac{(2n^2+1)}{(n^2+2)}\left[\frac{\varepsilon-1}{\varepsilon+2} - \frac{n^2-1}{n^2+2}\right] + \frac{3}{2}\left[\frac{n^4-1}{(n^2+2)^2}\right]\right\} \tag{6}$$

where,

$\bar{\nu}_a$ is absorption maximum wavelength in cm^{-1}

$\bar{\nu}_f$ is fluorescence maximum wavelength in cm^{-1},

n & ε are refractive index and dielectric constant respectively

From the equations (4) (5) and (6) graphs$(\bar{\nu}_a - \bar{\nu}_f)$versus $F_1(\varepsilon, n)$,$(\bar{\nu}_a - \bar{\nu}_f)$ versus $F_2(\varepsilon, n)$ and $\frac{\bar{\nu}_a+\bar{\nu}_f}{2}$ versus $F_3(\varepsilon, n)$ have been plotted respectively and the slopes m_1, m_2 & are obtained respectively, as shown below.

$$m_1 = \frac{2(\mu_e-\mu_g)^2}{hca^3} \quad (7)$$

$$m_2 = \frac{2(\mu_e-\mu_g)^2}{hca^3} \quad (8)$$

$$m_3 = \frac{2(\mu_e^2-\mu_g^2)}{hca^3} \quad (9)$$

The dipole moments μe and μg indicate the excited-state and ground-state dipole moments of a indole molecule. The letters **'a'** denote the Onsager's cavity radius of the solute molecule, where **'h'** is Planck's constant and **'c'** is the velocity of light. Edward [17] uses the microscopic increment method to estimate **'a'** value. The following Eqs (10) and (11) are obtained by on the basis of Eqs (8) and (9)

$$\mu_g = \frac{m_3-m_2}{2}\left[\frac{hca^3}{2m_2}\right]^{\frac{1}{2}} \text{ For } (m_3>m_2) \quad (10)$$

$$\mu_e = \frac{m_3+m_2}{2}\left[\frac{hca^3}{2m_2}\right]^{\frac{1}{2}} \quad (11)$$

3.2. Molecular-Microscopic Solvent polarity Parameter

H-bonding or polarisation dependency on spectral characteristics may be studied utilizing E_T^N Reichardt [18] proposed that spectral shift and E_T^N are related. Reichardt and Ravi et al. proposed the association between the Stokes shift and, E_T^N in their various theories. Accordingly, equation (12) is obtained.

$$\bar{\nu}_a-\bar{\nu}_f = 11307.6\left(\frac{\Delta\mu^2 a_D^3}{\Delta\mu_D^2 a^3}\right) E_T^N + \text{Constant} \quad (12)$$

where $\Delta\mu = 9D$ and $a_D = 6.2Å$ are the change in dipole moment on excitation. The Stokes shift versus the, E_T^N. plot slope may be used to measure the dipole moment variation (Δμ). The resulting equation is given by Eq (13) [15].

$$\Delta\mu = (\mu_e-\mu_g) = \sqrt{\frac{m\times 81}{\left(\frac{6.2}{a}\right)^3\times 11307.6}} \quad (13)$$

Stokes shift vs $(\bar{\nu}_a-\bar{\nu}_f)$ microscopic solvent polarity E_T^N plot constructed using Eq (12). Solvent polarity parameters (E_T^N) were culled from published study to identify their microscopic equivalent values.

4. Results and Discussion

4.1. Solvent effects on absorption and fluorescence spectra

Figures 2 and 3 show the 5-MPIC absorption and emission spectra of a variety of different solvents. 310–350 nm is the absorption wavelength. Fluorescence spectra peak between 390 and 415 nm. Table 1 shows the molecule's wavenumber in various solvents. The magnitudes of Stokes shifts may differ by around 8639 cm-1 in Stokes shift. It is shown in Tables 1 and 2 that the Stokes shift undergoes a bathochromic change when the solvent polarity increases. Excited state dipole moments are bigger than ground state dipole moments, as shown in Table 4.

.

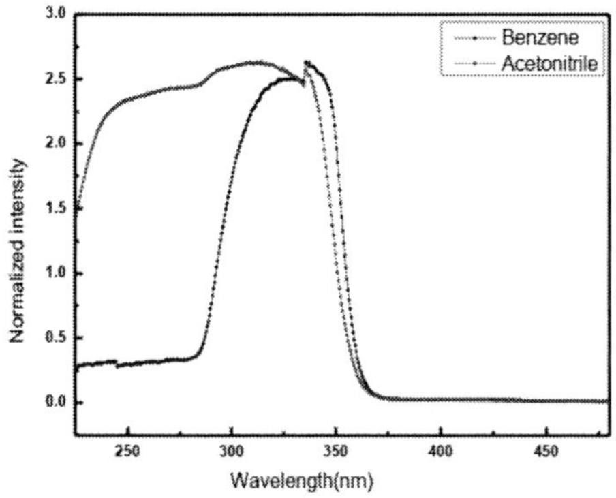

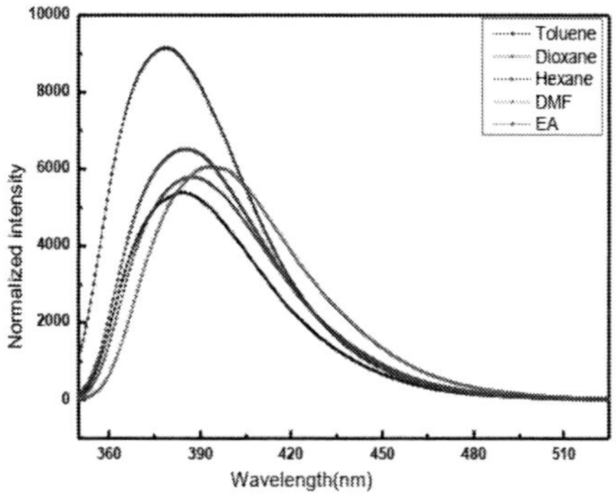

Figure 2. Normalised absorption spectra **Figure 3.** Normalised fluorescence spectra

Lippert, Bakshiev's, Kawaski-Chamma-polarity and variation of stokes shift are all shown in figures 4, 5 and 6. With regard to Stokes shift information, Viallet's functions are linear. Figure 7 shows the change in Stokes shift with the E_T^N parameter as a result of this analysis of the dipole moment. The slopes are computed in order to evaluate μg and μe. Table 3 and 4 lists the slopes, correlation factors, data points, radius, dipole moments, and change in dipole moments for 5-MPIC.

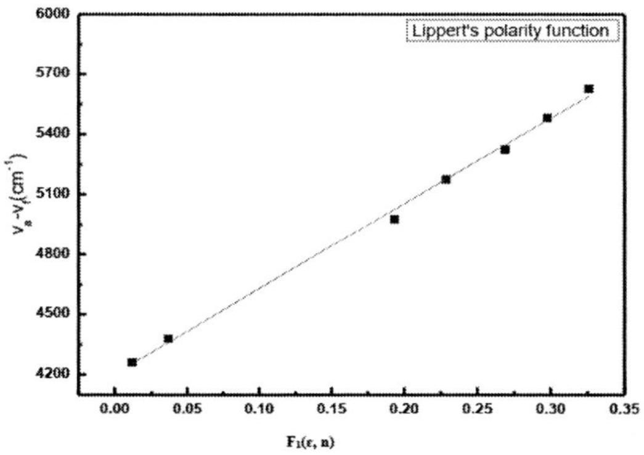

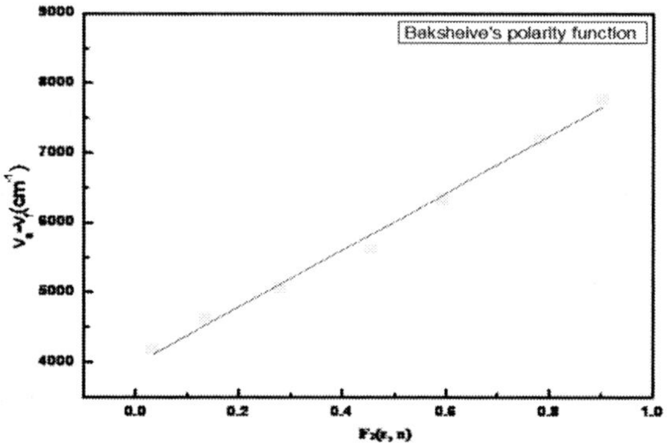

Figure 4. Lippert's polarity function **Figure 5.** Baksheives polarity function

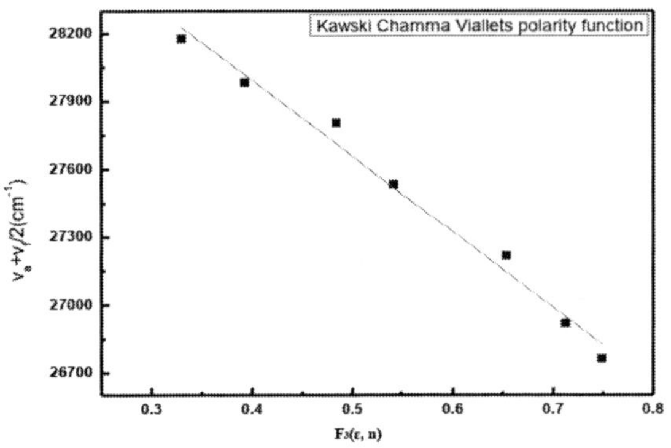

Figure 6. Kawaski ChammaViallets polarity function

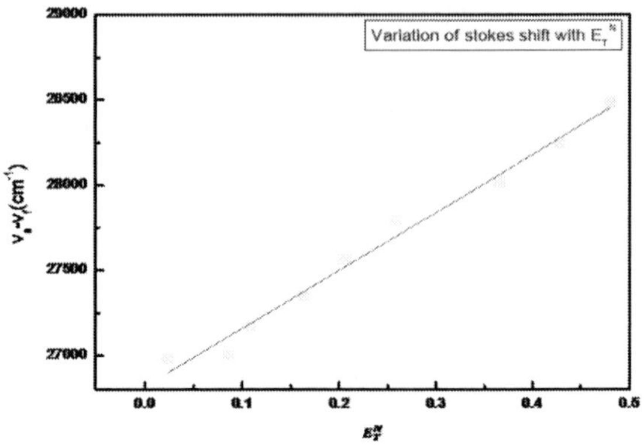

Figure 7. Variation of stoke shift with E_T^N

4.2 Analysis of quantum chemical calculation

4.2.1 Molecular orbital anlysis

As shown by HOMO and LUMO (the highest and lowest unoccupied molecular orbitals, respectively), molecular orbitals can be used to show how reactive chemical, active sites, and kinetic stability are for different molecules. HOMO represents electron giving ability of the molecule and LUMO represents electron acquisition ability. DFT/B3LYP was used to estimate the molecular orbital border based on 6-31+G (d, p). Changes in energy between these molecular orbitals are a key analytical metric in determining the transference characteristics of compounds. 3D graphs of the HOMO-LUMO gap are provided below fig 8.

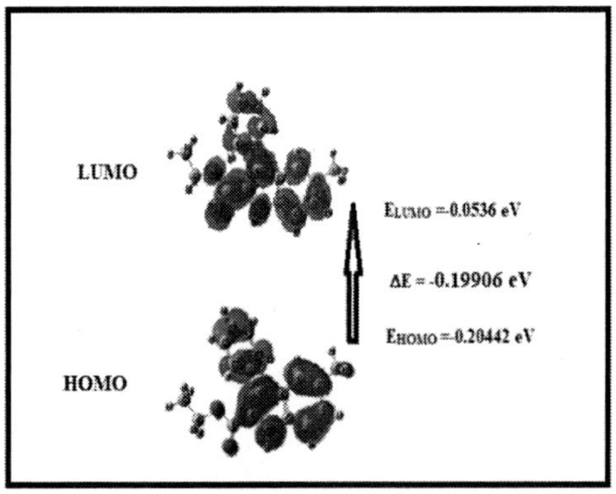

Figure 8. HOMO-LUMO Energy gap

4.2.2. Molecular electrostatic potential

Electrostatic potential maps, also known as electrostatic potential energy maps, represent the charge distributions of molecules in three-dimensional systems. In an electrophilic assault, the lowest electrostatic potential is red, while the maximum is blue in a nucleophilic attack[19-20]. The MEP of compound 5-MPIC is generated and mapped on the geometry optimisation. The colour code of MEP diagram lies between -5320.au. (Dark red) and 5320.a.u (Dark blue) for 5-MPIC.3-D mapped MEP shown in fig 9.

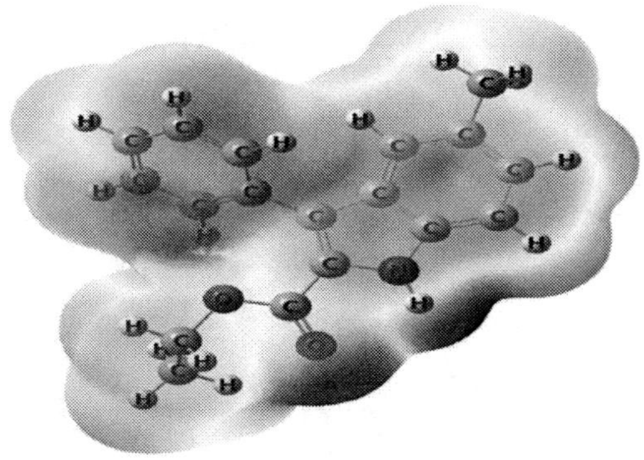

Figure 9. Molecular electrostatics potential map

Table1. Wavenumbers, stokes shift and arithmetic stokes shift values.

Solvents	$\overline{\nu_a}$	$\overline{\nu_f}$	$\overline{\nu_a}-\overline{\nu_f}$	$\frac{\overline{\nu_a}+\overline{\nu_f}}{2}$
Toluene	28571	24390	4181	26480
Benzene	28985	24096	4884	26540
Dioxane	30303	24390	5913	27346
Hexane	31250	25641	5609	28432
Cyclohexane	38461	25641	12820	32051
Acetonitrile	37037	25000	12032	31018
DMF	29411	24390	5021	26900
DMSO	30303	24350	5913	27346
EA	30303	25641	4662	27972
THF	29411	25641	3770	27526

Table 2. Polarity function and microscopic solvent polarity parameter for respective solvents.

Solvents	ε	n	F1(ε, n)	F2(ε, n)	F3(ε, n)	E_T^N
Toluene	2.40	1.497	0.015	0.033	0.744	0.099
Benzene	2.28	1.501	0.002	0.005	0.340	0.111
Dioxane	2.30	1.421	0.217	0.043	0.308	0.164
Hexane	1.80	1.370	0.010	0.242	0.346	0.006
Cyclohexane	2.02	1.426	0.001	0.002	0.289	0.006
Acetonitrile	36.64	1.344	0.305	0.861	0.665	0.46
DMF	38.25	1.430	0.275	0.839	0.711	0.386
DMSO	47	1.479	0.263	0.841	0.699	0.444
EA	6.08	1.372	0.201	0.493	0.499	0.228
THF	7.58	1.407	0.210	0.549	0.551	0.207

Table 3. Statistical treatment of the spectral shift of the compound

Compound	Slope	R square	No of data
	m_1=12914	0.54	7
5-MPIC	m_2 = 4259	0.52	7
	m_3=-5775	0.71	8
	m = 3451	0.40	9

Table 4. Onsager radius, ground and excited state dipole moments of the compound

Compound	Radius(Å)	μ_g^a(D)	μ_g^b(D)	μ_e^c(D)	μ_e^d(D)	μ_e^e(D)	μ_e^f(D)	μ_e^g(D)	μ_e^h(D)
5-MPIC	3.730	0.88	0.77	5.13	8.87	5.43	6.23	4.36	6.62

(1Debye=$3.73X10^{-30}$cm=10^{-18}esucm.)

[a]The ground state dipole moments using Gaussian software

[b]The ground state dipole moment calculated using equation (10)

[c]The excited state dipole moment calculated using equation (11)

[d]The excited state dipole moment calculated from Lippert's equation (7)

[e]The excited state dipole moment calculated from Bakshieves equation (8)

[f]The excited state dipole moment calculated from Kawski Chamma Viallets equation (9)

[g]The Change in dipole moment calculated using equation (10) and (11)

[h]Change in dipole calculated using equation (12)

5. CONCLUSION

For the molecule 5-MPIC, we measured the dipole moments in the ground and excited states from absorption and emission spectra by use of the Solvatochromic shift method. A larger dipole moment is detected in the excited state as compared to that of the ground state. Because the excited state is more polar than the ground state, the dipole moment increases significantly. Solavtochromic shift and microscopic solvent polarity parameter were used to compute and compare the change in dipole moment. Gaussian 16w was used to investigate theoretical ground state dipole moments, HOMO-LUMO, and MEP.

References:

1. J R Lakowicz, Principles of Fluorescence Spectroscopy, Plenum Press, 3rd ed. New York, (2006)
2. Xue-Hua Zhang, Yan Cui, Ryuzi Katoh, Nagatoshi Koumura and Kohjiro Hara *J. Phys. Chem. C* 2010, **114, 42**, 18283–18290
3. B. S. Mathada, N. G. Yernale, J. N. Basha, J. Badiger, An insight into the advanced synthetic recipes to access ubiquitous indole heterocycles, Tetrahedron Letters, Volume **85**, 2021, 153458, https://doi.org/10.1016/j.tetlet.2021.153458.
4. N.G. Yernale, B.S. Mathada, G.B. Vibhutimath, V.D. Biradar, M.R. Karekal, M.D. Udayagiri, M.B. Hire Mathada, Indole core-based Copper (II), Cobalt (II), Nickel (II) and Zinc (II) complexes: Synthesis, spectral and biological study, Journal of Molecular Structure, Volume **1248,** 2022, 131410, https://doi.org/10.1016/j.molstruc.2021.131410.
5. N. G. Yernale, B. H. M. Mruthyunjayaswamy, Metal (II) complexes of ONO donor Schiff base ligand as a new class of bioactive compounds containing indole core: Synthesis and characterization, Int J Pharm Pharm Sci, Vol **8**, Issue 1, 197-204 (2016)
6. T. Hintz, K. K. Matthews, R. Di, Biomedical Research Institute, pp.246, 2015.
7. S. Bhattacharya, J. Banerji, P.C. Ghosh, Indian J. Chem. **15**B (1977) 214.
8. N. Srividya, P. Ramasamy, V. T. Radhakrishnan, Spectrochim.Act. A, Volume **53,** pp.1743, 1997.
9. D. S. Biradar, B. Siddalingeswara, S. M. Hanagodimath, Journal of Molecular Structures, Volume **875**, pp. 108, 2008.
10. Tipperudrappa, D. S. Biradar, S. R. Manohara, S. M. Hanagodimath, S. R. Inamdar, R. J. Manekutla, Spectrochim. Act. A, Volume **69**, pp. 991, 2008.
11. R. Ghazy, S. A. Azim, M. Shaheen, F. E. Mekawey, Spectrochim.Act. A, Volume **60,** pp.187, 2004
12. S. P. Hiremath, B. H. M. Mruthyunjaya swamy, M. G. Purohit, Synthesis of substituted 2-aminoindoles and 2-(2'-phenyl-1',3',4'-oxadiazolyl) amino indoles, Indian Journal of Chemistry B, 1978, 16, 9, 789-792
13. E. Lippert, Z. Naturforsch. 10A, pp.**541**, 1955.
14. N. G. Bakhshiev, Optic Spectroscopy, (USSR) Volume **16**, pp.821, 1964.
15. L. Bilot, A. Kawski, Z. Naturforsch. 17A, pp. 621, 1962;18A, pp. 10 and 256, 1963.
16. Chamma and P. Viallet, Chemical Research Academy of. Science, Paris Ser. C, Volume **270**, pp. 1901, 1970.
17. J. T. Edward, Molecular Volumes and Parachor. Chem. Ind, London, pp. 774 (1956).
18. Reichardt, Chemical, Review, Volume**94**, pp.2319,1994
19. Gangadhar Maddapur, N R Patil, R M Melavanki, Raviraj kusnoor, Journal of Fluorescence Sep 2014
20. S Chandrashekar H R Deepa, R M Melavanki, Chemical data collection 2016

AUTHOR INDEX